STP 1293

Dredging, Remediation, and Containment of Contaminated Sediments

Kenneth R. Demars, Gregory N. Richardson, Raymond N. Yong, and Ronald C. Chaney, editors

ASTM Publication Code Number (PCN): 04-012930-38

ASTM
1916 Race Street
Philadelphia, PA 19103

Printed in the U.S.A.

Library of Congress Cataloging-in-Publication Data

Dredging, remediation, and containment of contaminated sediments/
  Kenneth R. Demars ... [et al.], editors.
      p. cm.—(STP; 1293)
   "ASTM publication code number (PCN): 04-012930-38."
   "Contains papers presented at the symposium of the same name, held
in Montreal, Quebec, Canada on 23–24 June 1994 ... sponsored by ASTM
Committee D-18 on Soil and Rock, in cooperation with Environment-
Canada and the U.S. Environmental Protection Agency"—Foreword.
   Includes bibliographical references and index.
   ISBN 0-8031-2028-1 (alk. paper)
   1. Contaminated sediments—Management—Congresses.   I. Demars, K.
R.   II. ASTM Committee D-18 on Soil and Rock.   III. Canada,
Environment Canada.   IV. United States. Environmental Protection
Agency.   V. Series: ASTM special technical publication: 1293.
TD878.D74 1995
628.1′68—dc20   95-37452
                                                                               CIP

Copyright © 1995 AMERICAN SOCIETY FOR TESTING AND MATERIALS, Philadelphia, PA. All rights reserved. This material may not be reproduced or copied, in whole or in part, in any printed, mechanical, electronic, film, or other distribution and storage media, without the written consent of the publisher.

Photocopy Rights

Authorization to photocopy items for internal or personal use, or the internal or personal use of specific clients, is granted by the AMERICAN SOCIETY FOR TESTING AND MATERIALS for users registered with the Copyright Clearance Center (CCC) Transactional Reporting Service, provided that the base fee of $2.50 per copy, plus $0.50 per page is paid directly to CCC, 222 Rosewood Dr., Danvers, MA 01923; Phone: (508) 750-8400; Fax: (508) 750-4744. For those organizations that have been granted a photocopy license by CCC, a separate system of payment has been arranged. The fee code for users of the Transactional Reporting Service is 0-8031-2028-1/95 $2.50 + .50.

Peer Review Policy

Each paper published in this volume was evaluated by three peer reviewers. The authors addressed all of the reviewers' comments to the satisfaction of both the technical editor(s) and the ASTM Committee on Publications.

The quality of the papers in this publication reflects not only the obvious efforts of the authors and the technical editor(s), but also the work of these peer reviewers. The ASTM Committee on Publications acknowledges with appreciation their dedication and contribution to time and effort on behalf of ASTM.

Printed in Fredericksburg, VA
October 1995

Foreword

This publication, *Dredging, Remediation, and Containment of Contaminated Sediments,* contains papers presented at the symposium of the same name, held in Montreal, Quebec, Canada on 23–24 June 1994. The symposium was sponsored by ASTM Committee D–18 on Soil and Rock, in cooperation with Environment-Canada and the U.S. Environmental Protection Agency. Kenneth R. Demars, University of Connecticut, Storrs, CT; Gregory N. Richardson, G. N. Richardson and Associates, Raleigh, NC; Raymond N. Yong, McGill University, Montreal, Quebec, Canada; and Ronald C. Chaney, Humboldt State University, Arcata, CA served as co-chairmen of the symposium and are editors of the resulting publication.

Contents

Introduction   ix

Overview   1

SEDIMENT CHARACTERIZATION

Keynote Paper: The Fate of Toxic Pollutants in Contaminated Sediments—
R. N. YONG   13
    Discussion   38

Use of Bathymetry for Sediment Characterization at Indiana Harbor—
D. M. PETROVSKI   40

Oxidation of Pyrite in Marine Clays and Zinc Adsorption by Clays—
M. OHTSUBO   50

Heavy Metal Concentration in Bay Sediments of Japan—M. FUKUE, Y. KATO,
T. NAKAMURA, AND S. YAMASAKI   58

DREDGING TRANSPORT AND HANDLING

Removal of Contaminated Sediments: Equipment and Recent Field Studies—
J. B. HERBICH   77

Demonstrations and Commercial Applications of Innovative Sediment
Removal Technologies—J.-P. PELLETIER   112

Data Requirements for Advancing Techniques to Predict Dredge-Induced
Sediment and Contaminant Releases—A Review—D. E. AVERETT   128

Development of Solidification Technique for Dredged Sediments—S. YAMASAKI,
H. YASUI, AND M. FUKUE   136

RESTORATION/REMEDIATION

Field Demonstrations of Sediment Treatment Technologies by the USEPA's Assessment and Remediation of Contaminated Sediments (ARCS) Program—S. GARBACIAK, JR. AND J. A. MILLER   145

Factors Influencing the Development of a Biostimulant for the *In-Situ* Anaerobic Dechlorination of Polychlorinated Biphenyls in Fox River, Wisconsin Sediments—M. B. HOLLIFIELD, J. K. PARK, W. C. BOYLE, AND P. R. FRITSCHEL   155

New *In-Situ* Procedures for Measuring Trace Metals in Pore Waters—H. ZHANG, W. DAVISON, AND G. W. GRIME   170

Mercury-Contaminated Sludge Treatment by Dredging in Minamata Bay—K. YOSHINAGA   182

CONTAINMENT AND ISOLATION

State of the Art: CDF Contaminant Pathway Control—G. N. RICHARDSON, D. M. PETROVSKI, R. C. CHANEY, AND K. R. DEMARS   195

Abating Coal Tar Seepage into Surface Water Bodies Using Sheet Piles with Sealed Interlocks—B. I. COLLINGWOOD, M. D. BOSCARDIN, AND R. F. MURDOCK   220

Modeling Air Emissions from Contaminated Sediment Dredged Materials—K. T. VALSARAJ, L. J. THIBODEAUX, AND D. D. REIBLE   227

Soil-Bentonite Design Mix for Slurry Cutoff Walls Used as Containment Barriers—N. S. RAD, R. C. BACHUS, AND B. D. JACOBSON   239

Testing of a Cement-Bentonite Mix for a Low-Permeability Plastic Barrier—J.-H. DESCHÊNES, M. MASSIÉRA, AND J.-P. TOURNIER   252

Design and Performance Verification of a Soil-Bentonite Slurry Wall for the Hydraulic Isolation of Contaminated Sites—R. HOLLENWEGER AND L. MARTINENGHI   271

MANAGEMENT STRATEGIES

Strategies for Management of Contaminated Sediments—M. R. PALERMO AND J. MILLER   289

Partnering for Environmental Restoration: The Port Hope Harbour Remedial Action Plan (RAP)—S. M. C. WESTON   297

Critical Analysis of Sediment Quality Criteria and Hazard/Risk Assessment Tools—R. GALVEZ-CLOUTIER, R. N. YONG, J. CHAN, AND E. BAHOUT 306
   Discussion 317

Regulatory Strategies for Remediation of Contaminated Sediments—H. ZAR 319

Indexes 329

Introduction

Prior to the 1960s, dredged material disposal alternatives were typically evaluated on the basis of economics or lowest cost rather than environmental impact. This generally meant open water disposal of harbor and channel sediments, and land (or wetland) disposal of river sediments. During the mid to late 1960s, questions were raised about the effects of dredging on water quality. In 1971 the Environmental Protection Agency (EPA) issued the report "Criteria for Determining Acceptability of Dredge Spoil Disposal to the National Waters" as the first effort to identify and classify polluted sediments in the nation's waterways for purposes of controlling dredging and disposal operations. The major concern at that time was to control nutrient loadings such as nitrogen and phosphorous, and sediment quality criteria were published before there was adequate scientific data to define the type and quantity of other contaminants that were associated with dredged material. Sediments that failed these criteria were considered unsuitable for open water disposal and required placement in a confined disposal facility (CDF). Since that time the EPA criteria for classifying polluted sediments has expanded to include numerous heavy metals and PCBs, with the sediment quality criteria undergoing frequent updating. While the development of these criteria was expected to have a positive effect on human health and well-being, they have significantly slowed the process and increased the cost of dredging clean sediments from the nation's waterways.

Thus, there is a need to develop new standards related to dredging or to modify existing standards in an effort to make the dredging of clean sediments and the remediation process for contaminated sediments more efficient. The primary objective of this symposium was to identify tests, methods, procedures, and materials, used in support of dredging, treatment, and containment of contaminated sediments, that are in need of standardization. A secondary objective was to provide a forum for discussion of past dredging practices and future directions including the effects of sediment properties and behavior, equipment requirements, and the impact of regulations.

ASTM sponsored a two-day symposium to fulfill these objectives. The symposium was held on 23–24 June 1994 during the regular committee week in Montreal, Quebec, Canada. Financial assistance for this symposium was provided by the Environment Canada: Contaminated Sediment Removal Program; the U.S. Environmental Protection Agency: Great Lakes National Program Office; the Connecticut Sea Grant Program, and ASTM.

The symposium was organized into five sessions of approximately equal length which consisted of brief individual presentations. Four of the sessions started with state-of-the-art presentations, as summarized below:

1. *Sediment Characterization*—Raymond N. Yong, McGill University, Montreal, Quebec.
2. *Dredging Technologies*—John B. Herbich, Texas A&M University, College Station, TX.
3. *Treatment Technologies*—Stephen Garbaciak, U.S. EPA Great Lakes National Program Office, Chicago, IL.
4. *Containment and Isolation Technologies*—Gregory N. Richardson, G. N. Richardson and Assoc., Raleigh, NC.
5. *Management Strategies*—This session included four brief presentations followed by a panel discussion. The discussants were Rosa Galvez-Cloutier of McGill University; Michael R. Palermo of the U.S. Army Engineer Waterways Experiment Station; Sandra M. C. Weston

of Environment Canada; and Howard Zar of the U.S. Environmental Protection Agency. These individuals were led in the panel discussion by G. N. Richardson. Moderators for the four sessions included R. J. Ebelhar of RUST Environmental and Infrastructure, Cincinnati, OH; J. P. Pelletier of Environment Canada, Toronto, Ontario; D. M. Petrovski of USEPA, Chicago, IL; and P. Selvadurai of McGill University, Montreal, Quebec.

All of the state-of-the-art presentations and the majority of the volume papers presented at the symposium are published in this volume. This volume is divided into five parts of roughly equal length and generally follows the symposium format as outlined above. The book starts with a brief summary paper by the editors. The first four parts follow the symposium format outlined above and each part starts with a state-of-the-art paper followed by the other related papers presented at the symposium. The final part on Management Strategies provides a review of the regulations and environmental factors that drive the sediment remediation process and technology development.

It is obvious from a review of the manuscripts that the process of remediating a site with contaminated sediments is very difficult, complex, and multidisciplinary in nature and that existing regulations, in many respects, often complicate the problem. While regulations specify the criteria for defining a hazardous material, it is not clear how to sample and test a submerged site to define the spatial extent of pollutants, how to select and operate the appropriate dredge, how to treat or contain and isolate the polluted sediments, and finally how to monitor and verify that target levels have been attained. While there has been a small number of contaminated sediment remediations and demonstrations to date, this field is still in its infancy and is likely to increase in importance in the future. It is believed by the editors that the papers in this volume provide a unifying, yet introductory, view of this topic and should remain a valuable reference in the years to come. It is hoped that this book will be a basis for development of standardized tests and equipment for the aquatic environment and will provide a stimulus for future research and innovation in remediation of sites with contaminated sediments.

The editors would like to express their thanks for the contributions of the symposium participants/authors and technical reviewers and would like to acknowledge the support and encouragement provided by ASTM staff and officers of Committee D-18 on Soil and Rock to make this effort a success.

The Editors

Overview

Nature and Extent of Problem

Since the advent of the industrial revolution (starting about 1830), waste products from mines, municipalities, and factories have been discharged and spilled into the nation's waterways as point sources of pollution which supplement the nonpoint discharges of fertilizer, pesticides, and other chemicals that run off from rural areas. There is evidence that many banned chemicals are still entering watercourses at unacceptable concentrations today (Fox 1995). Much of these pollutants become adsorbed to the surface of fine-grained particles that settle and form a polluted sediment at the bottom of the waterway. Polluted sediments may be subsequently buried by clean sediments, remobilized by erosion or disturbance, and resuspended by dredging. These polluted sediments were long considered benign until research in the 1960s and 1970s revealed that trace elements and pollutants in soils and sediments were bioavailable and could bioaccumulate up the foodchain.

The 1971 EPA report "Criteria for Determining Acceptability of Dredge Spoil Disposal to the Nation's Waters" was the first effort to establish criteria for purposes of identifying and classifying polluted sediments, often called the Jensen criteria (Table 1). The primary concern at that time was to control nutrient loadings, such as nitrogen and phosphorous, so the criteria for trace metals, PCBs, pesticides, and other organic pollutants was very limited. These criteria were published before there was adequate scientific data to define the type and quantity of contaminants associated with sediments and that have the potential for detrimental effects on human health and well-being as well as the environment.

While these sediment pollution criteria have undergone significant evolution during the past 25 years, they provided the impetus for routine chemical testing and subsequently biological testing to identify and classify polluted sediments in the nations waterways.

The sediment chemical and biological data that have since been generated in the various EPA regions has become part of a national database. In 1987, the United States and Canada recognized a need to jointly address concerns about persistent toxic contaminants in the Great Lakes. As an annex to a 1972 Great Lakes Water Quality Agreement, the United States and Canada agreed to cooperate with state and provincial governments to "identify the nature and extent of sediment pollution of the Great Lakes System" (EPA 1994). Through these efforts, 43 Areas of Concern (AOCs) were identified where contaminated sediments are considered to significantly impair the beneficial uses of water resources such as drinking, swimming, fishing, and boating. The 43 AOCs are shown in Fig. 1 for the Great Lakes. The primary pollutants in the AOCs are PCBs, PAHs, and metals from a range of industrial sources. This joint U.S.-Canada activity was the first effort to develop a contaminated sediment inventory and was subsequently expanded by the EPA (1994) into a National Sediment Inventory (NSI).

In 1992, Congress passed the Water Resources Development Act which requires EPA, in consultation with the National Oceanic and Atmospheric Administration and the U.S. Army Corps of Engineers, to conduct a national survey of data regarding sediment quality in the United States. The goals of the NSI are to assess the national extent and severity of sediment contamination, to identify areas in need of further assessment, and to identify high-risk areas for appropriate remedial action. The NSI is to be updated every two years with the first report to congress in Summer 1995. The effort to date (Fox 1995) is based on over 3.4

TABLE 1—*Summary of 1971 EPA sediment pollution criteria.*

Chemical Constituent	Concentration Percent (Dry Weight Basis)
Total Volatile Solids (TVS)[a]	6.0
Chemical Oxygen Demand, COD	5.0
Total Kjeldahl nitrogen	0.10
Oil-grease	0.15
Mercury	0.0001
Lead	0.005
Zinc	0.005

[a] TVS % (dry) = 1.32 + 0.98 (COD%).

million records from waterbody segments around the country including sediment chemistry and biological sample measurements of tissue residue, toxicity, benthic abundance, and histopathology. These preliminary analyses have identified over 1700 waterbody segments as Potential Areas of Concern (Fig. 2) with many additional areas likely to be added in the future. The first step in remediating these areas must be to identify and control the source followed by a concerted effort to clean the polluted sediments.

Site/Sediment Characterization

Sediments serve as "sinks" for contaminants discharged into receiving waters. If these contaminants are toxic, they constitute a significant problem since these pollutants have the potential for bioaccumulation and for infecting the overlying waters if they are released from the sediments. Determination of the persistence and fate of these toxic pollutants in contaminated sediments requires a knowledge of the contaminants contained in the sediment, and the manner in which these are "held" within the sediment.

All the papers presented in the session dealing with site and sediment characterization recognized that the nature and concentration of the various contaminants in sediments need

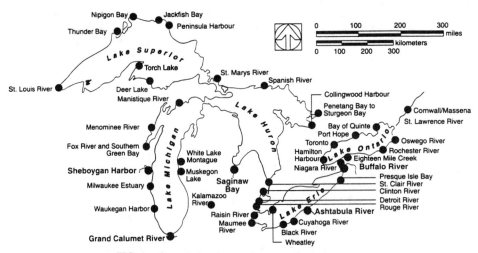

FIG. 1—*Great Lakes Areas of Concern (USEPA 1994).*

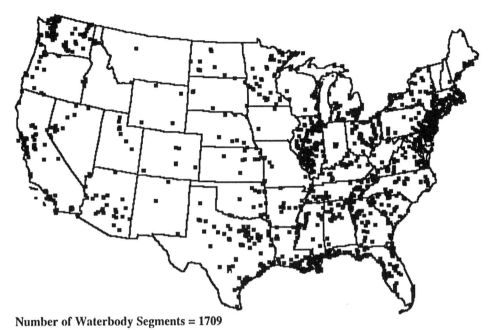

Number of Waterbody Segments = 1709

FIG. 2—*Potential AOCs based on preliminary evaluation of sediment chemistry (Fox 1994).*

to be determined, particularly if remediation procedures are to be designed and implemented. Petrovski has suggested that a bathymetry-based procedure may provide a means for making decisions in respect to sampling sites and numbers, with the aim for reducing the number of samples needed to characterize the large sediment volumes as required to meet RCRA expectations. The rationale is that bathymetric surveys can identify regions of reduced depositional energies, and that since the regions favor the accumulation of finer-grained material, one can deduce that these soil materials will contain higher concentrations of contaminants.

The mechanical properties of contaminated sediments are of interest insofar as they impact on remediation strategies such as capping or natural burial. Decisions in regard to whether the sediment will erode under continued dynamic activity at the sediment interface, and whether the material will remain in a stable configuration under capping loads, require knowledge of the strength and compressibility of the material. There are a number of laboratory and in-place tests that have been developed for soft aquatic sediments such as the vane shear and static or fall cone devices designed to measure the undrained shear strength.

The problems of determination of the fate of pollutants requires knowledge of the relationships established between the pollutants and the reactive surfaces of the sediment fractions, particularly in respect to the bonding forces established and the implications *vis-a-vis* "availability" of the pollutants. Yong noted the importance for determination of distribution and partitioning of the pollutants within the sediment in the assessment of the persistence and fate of the pollutants. Changes in the sediment environment can significantly affect the fate of the pollutants in the sediment. In particular, oxidation-reduction reactions and pH changes will control the chemical forms and mobility and bioavailability of heavy metal pollutants. In addition, it was noted by Yong that biological and geochemical processes within the sediment and especially at the sediment-water interface contribute significantly to the alteration and distribution of the organic pollutants in the sediment, i.e., these processes exert considerable influence on the persistence of the organic pollutants.

The presence and fate of heavy metals in various marine clays and the bay sediments of Japan have been studied and reported by two separate groups, Ohtsubo and Fukue et al. The determination of the level of pollution in the Tokyo and Osaka Bays constituted the primary focus for the study reported by Fukue et al. Of particular importance was the need to develop a scale that would impart information regarding the "degree of pollution." Since remediation priorities and strategies are tied into the "most polluted sites" rationale, the methodology reported by Fukue et al. promises to be a very useful technique for "ranking" of degrees of pollution.

In the Ohtsubo study, the heavy metals attached to marine clays were examined in relation to potential pollution of the ground water should these clays be used in a contemplated bay site reclamation scheme. The primary concern was in relation to the production of sulfuric acid and iron oxide by the oxidation of the pyrite in the clays. As in the concern shown by Yong with respect to pH changes, this ocidation reaction can prove to be very detrimental for pollution control of the underlying ground water.

Dredging Technologies

When dredging is required as part of a remediation project, there are three primary environmental concerns that must be considered for selecting the appropriate dredging equipment. These include the ability to (1) precision-dredge the contaminated layer whether it is thick or thin, (2) minimize resuspension and migration of polluted particles, and (3) minimize the water that is added during dredging so that the contaminants remain associated with the solid fraction. Most traditional maintenance dredging equipment has been designed using economic and reliability criteria for high production and generally do not satisfy the above criteria. However, Herbich has noted that a number of "specialty" dredges have been developed, principally in Japan, Italy, and the Netherlands, for the removal of contaminated sediment.

Precision or surgical dredging involves the removal of the contaminated layer with one or more passes of the dredge and little or no clean sediment added to dilute the pollutants and add to the volume requiring costly treatment or containment. Ideally, the dredgehead should scrape, sweep, or grab the contaminated sediments, thus leaving a clean sediment-water interface. This process involves both the selection of the appropriate dredging equipment and the use of precision navigation and acoustic profiling equipment for horizontal and vertical control as described by Yoshinaga. Herbich reviewed the capabilities of specialty dredges that have been developed and successfully used for removal of contaminated sediments including the Refresher system and Cleanup dredge from Japan, the Dutch-designed Scraper dredge, and the horizontal auger, enclosed clamshell, and Cable-arm bucket dredges available in the United States, to name a few. Each of these dredges has limitations with regard to size, range of excavation depth, water depth, production rate, and sediment resuspension. The efficiency at which they are capable of removing contaminated sediments is highly affected by positioning, operator skill, and cutter geometry in addition to the quantity of bulky debris at the sediment surface which tends to affect the cutter operation. Many cutters may leave a windrowed or dimpled surface of contaminated sediment.

Sediment resuspension has been the focus of many field studies as summarized by Herbich for all the specialty dredges and by Pelletier for demonstrations involving the Cable-arm bucket and Oozer dredges. In general, the closed clamshell bucket dredges and specialty dredges with shields over the cutter exhibit low to moderate resuspension of sediment, compared to open clamshell and conventional cutterhead dredging equipment. In instances where a small amount of resuspension is undesirable, geotextile and sheetpile curtain walls have been used effectively to reduce the migration of suspended particles. According to Averett, particle resuspension and contaminant migration is dependent upon local currents, sediment type and properties, and physical and operational dredging equipment characteristics. He has noted that it is possible to model resuspension and contaminant migration prior to dredging and the results may be used to select appropriate dredging equipment and operational constraints for a specific site. Since current predictive models are new and require calibration, a field monitoring program is recommended to improve resuspension model performance.

During typical dredging activities, water is added to and mixed with the sediment in varying quantities depending upon the dredge type and operational conditions. The added water increases the volume of contaminated material that must be treated and increases the problem of dewatering or consolidating the sediment during disposal in a containment area. Herbich has noted that mechanical dredges tend to add little or no water to the sediment, whereas mechanical-hydraulic dredges, which use the suction side of a centrifugal pump to move the sediment (plus water) to the surface, may have water/solid volume ratios as high as 9 to 10. For practical purposes, hydraulically dredged sediments behave like a fluid that can be easily pumped long distances to a containment area. Several alternative conveyance systems have recently been developed to minimize the slurry water content. These conveyance systems are principally pneumatic although mechanical screw and belt systems have also been developed. The pneumatic systems have shown promise with sediments being conveyed over 400 m with only a slight increase in water content as noted by Yamasaki et al.

Treatment/Remediation Methods

Remedial technologies can be broadly divided into either active remediation or in-place remediation. Active remediation involves dredging of the contaminated sediments and then treating them *ex-situ* under highly controlled conditions. In contrast, in-place remediation involves processes such as isolation by natural burial, capping, and armoring or bioremediation. A treatment/remediation method is selected after a preliminary estimate of the magnitude and extent of sediment contamination and the associated risks to human health and the environment.

The nature of contaminated sediments affects the selection of treatment technologies. The volume of sediments present at contaminated underwater sites tends to be quite high, while the concentrations of contaminants in the sediments tend to be relatively low in comparison to the concentration of contaminants in soils at hazardous waste cleanup sites. In addition, the majority of contamination is typically associated with the fine-grained material. Because most of the contaminants are associated with the finer-grained material, substantial cost savings can be achieved for active remediation by first applying extractive technologies, thus reducing the volume of material requiring further treatment by more expensive destructive methods. Once the volume of material has been reduced, a treatment strategy for the fine-grained material can then be devised. In addition, dredged materials must first undergo a dewatering stage prior to treatment because of the high initial water content as noted by Yamasaki et al.

Treatment technologies can be broken down into two broad categories: (1) those that work on the sediments *in-situ,* and (2) those that process sediment after dredging. In general, *in-situ* treatment technologies for contaminated sediments are less developed than the technologies that can be applied to dredged materials. Any decision to leave sediments in place is highly dependent on an evaluation of the relative risks posed by the sediments left untreated on the bottom, the risks of performing a treatment operation on *in-situ* sediments, and the risks associated with the removal and subsequent disposal or treatment of the contaminated dredged materials.

Treatment technologies for dredged material have been evaluated and demonstrated by a number of groups (Averett et al. 1990). Garbaciak and Miller reported on an extensive body of data from the ARCS program on the relative efficiency of various treatment processes. A review of the proceedings as well as independent literature indicates that all treatment technologies typically leave some degree of contamination in the sediment (Richardson et al. 1995). Based on this work, no single technology has been shown to be effective for all contaminants. Typically, technologies are designed to deal with either the organic contaminants (such as PCBs), as reported by Hollifield et al., or heavy-metal contaminants (Zhang et al.). Some of the technologies (e.g., particle separation and solidification) may effectively treat both types of contaminants, but their applications are limited to sediments with specific characteristics that are not present at all sites. Complications have been shown to arise due to the presence of certain contaminants in application of some technologies, and volatile contaminants may be lost unintentionally during the application of thermal treatment processes. In addition, treatment technologies were effective on only some types of sediment. These limitations suggest that multiple-step treatment process will be necessary to treat sediment.

Containment and Isolation

In the containment and isolation session, Richardson et al. reviewed the regulatory perspective of contaminated sediment disposal using Confined Disposal Facilities (CDFs) in the United States and the basic design considerations to control all potential contaminant migration pathways. The applicable regulations covering the disposal of contaminated sediments is shown to be a function of both the concentration of contaminant and the partitioning coefficient for the contaminant in water. Those sediments that have very low levels of contamination or are contaminated with constituents having a very low solubility in water, e.g., having a low partitioning coefficient, can be controlled by designing the CDF for solids retention. For higher concentrations of contaminants or for contaminants having high partitioning, the CDF must be designed based on hydraulic isolation of the waters released by the sediments.

Contaminants within the dredged material placed within a CDF can leave the system using the following pathways: (1) waterborne pathways, including effluents drained from the waste, effluent seeping through the dikes, leachate seeping into the ground water within the foundation of the CDF, and surface water runoff; (2) via plant or animal uptake; and (3) from airborne emissions. Each contaminant pathway may be controlled by engineered components, i.e., with a liner or by selective placement of the dredged materials.

Waterborne contaminants moving through the dikes or leaching into the subgrade can be controlled either with engineered barriers such as clay or synthetic liners or by placement of clean fine-grained dredged materials initially on the bottom of the CDF to provide hydraulic isolation. Similarly, water runoff, animal and plant uptake, and air emissions can be controlled either through the use of an engineered cover system or the placement of clean dredged material over the contaminated dredged material. The paper by Richardson et al. reviewed alternative methods to obtain pathway control and compares the traditional strategies of the Corps of Engineers and Environmental Protection Agency.

Vertical barriers are the most common engineered barrier systems used in the containment of contaminated dredged materials. Four papers in this session dealt with the design and laboratory evaluation of vertical barriers. Collingwood et al. presented a case study of the use of a watertight sheetpile wall used to control the release of coal tar from past gas plants into the Boston Bay. The sheet metal pilings were fitted with polyurethane rubber seals to act as a vertical barrier. Designed as a tie-back wall, the tie rod penetration proved to be difficult to seal. In general, however, the sheetpile wall was successful in limiting the release of coal tar to the bay waters.

Rad et al. present the laboratory procedures necessary to design and evaluate the performance of a soil-bentonite mix for slurry cutoff walls. Laboratory tests were first conducted to evaluate the workability of possible bentonite slurry mixtures. After selecting a satisfactory slurry mixture based on workability, the laboratory evaluation examines the compatibility of the slurry with the contaminants found in the ground water. Compatibility is evaluated based on the stability of the permeability as permeant passes through samples in the laboratory. A dramatic increase in permeability indicates that the slurry is not compatible with the contaminants. Deschênes et al. review similar laboratory evaluation of a plastic-cement slurry used on a diaphragm wall constructed in Northern Quebec to a depth of 34 m. Guidelines for cement/water ratio and bentonite/water ratios are discussed. Hollenweger and Martinenghi present the laboratory and field history of a 55-m deep slurry wall constructed to encapsulate equipment used to excavate 55 m deep is of particular interest.

Yoshinaga presents the use of a highly watertight revetment (i.e., CDF) to contain dredged materials contaminated with sewage containing methyl mercury. Over 750 000 m^3 of contaminated sediments were removed from Minamata Bay using suction dredges and floating perimeter sediments curtains to limit the contamination resulting from removal of the contaminated sediments. The removal was completed in 1987 and no outbreak of pollution or related disease has occurred since then.

Air emissions coming from contaminated dredged materials was modeled by Valsaraj et al. at four locales: (1) at the dredging site, (2) from dry exposed dredged materials, (3) from the supernatant, and (4) once vegetation is established. Emission rates in mass of volatiles emitted to air per unit time were dependent on chemical concentration in the sediment source, the degree of contact of the concentration in the sediment source, the degree of contact of the dredged material with the air, the air-side mass transfer rate coefficient for the chemical, and the sediment-air partition constant of the chemical. Based on the models, a tentative ranking of the emissions from the four locales is given.

Management Strategies

The strategies for management of contaminated sediments generally pertains to the stewardship of natural and financial resources or the development of cost effective solutions. While many strategies exist, there is no single strategy that pertains to all conditions and projects whether for navigation dredging or for sediment remediation (Palermo and Miller). There are many countries around the world that are faced with contaminated sediment sites and the strategies used to identify and manage these sites differ from country to country, and with organizations or individuals within each country. Yet, there are important questions that are common to the management of all contaminated sediment remediation projects. In this volume, the U.S. and Canadian perspectives are considered as being representative.

The terms "contaminated" and "clean" as they pertain to sediments are somewhat arbitrary. At the national level, agencies such as USEPA, USACE, and Environment Canada continually research and develop guidelines for determining the levels of various pollutants in sediments that cause adverse ecological and human health effects. Sediment quality criteria have been evolving for the last two decades (Galvez-Cloutier et al.) by using a wide range of biological and chemical assessment techniques for evaluation of both water column and benthic impacts. While it is arguable that these criteria are adequately defined, the primary remediation problem is the high cost of field sampling and laboratory testing to define that portion of a site that fails the criteria and requires remediation from the portions that are clean and those that are at a problematic contaminant level, i.e., higher than the remediation criteria. As noted by Petrovski, several hundred core samples may be required for a single site.

Once a site has been identified as having contaminated sediments, a process for evaluating the remediation alternatives must be developed. For navigation dredging projects, Palermo has noted that the USACE and USEPA have developed a technical framework for environmental evaluation of dredged material disposal alternatives. The framework uses a tiered approach that, on a site-specific basis, evaluates the alternatives including: (1) No action, (2) Contain in-place, (3) Treat in-place, (4) Remove and contain, and (5) Remove and treat.

For contaminated sediment remediation projects, management strategies are not well established, but the USACE and USEPA are presently cooperating on a strategy. Zar further notes that a primary role of USEPA is remediation and enforcement of environmental regulations. The legal authorities for remediation and enforcement are provided in legislation of the Clean Water Act, CERCLA/Superfund, RCRA, TSCA, the Oil Pollution Act of 1990, and many state regulations.

Technical and scientific evaluations would be expected to dominate any remediation strategy; however, Weston has noted that other factors are also important, such as public support for the project. In Canada each remediation project must develop a Remedial Action Plan (RAP) which is a concensus of all affected groups such as business, industry, commerce, and the general public. The common goal is to develop an environmentally sound plan that reflects the views of the community and satisfies the overlapping jurisdictions of the many national, provincial, and local agencies involved.

Conclusions

This symposium has shown that there is a need to modify some existing ASTM standards and to develop some new standards to aid the process of dredging, remediation, and containment of soft contaminated sediments. The following conclusions are justified from these symposium papers:

1. Contaminated sediment remediation projects are multidisciplinary in nature and are driven by federal and state regulations. ASTM standards may provide a basis for improved communication and understanding between the involved parties.

2. Site investigations to characterize the nature and extent of contamination are difficult and costly. A tiered-testing approach that has been proposed to identify suspect sites should be standardized.

3. There remains an urgent need to develop *in-situ* chemical sensors such as the polyacrylamide gel probe for discrete data or possibly fiber optic sensors with telemetry for continuous data monitoring before, during, and after disposal.

4. There is a need to develop a number of analytical and stochastic models to support scientific and engineering evaluations of a remediation option such as the level of turbidity associated with a given dredge/production rate or mass-balance analysis of pollutants during dredging, placement in a CDF, and consolidation.

5. Natural sediments, by nature, are soft and difficult to sample and handle. Dredged sediments have entrained water and exhibit fluid-like behavior. Standardized tests and methods are needed to sample, handle, store, and prepare these materials prior to physical and chemical testing to obtain representative test results.

6. Contaminated sediments usually contain oil and grease or other organic components with the potential to generate gas and odor following dredging. Test methods are needed to define problematic conditions.

References

Averett, D. E., Perry, B. D., Torrey, E. J., and Miller, J., 1990, "Review of Removal, Containment and Treatment Technologies for Remediation of Contaminated Sediments in the Great Lakes," USEPA and USACOE Miscellaneous Paper EL-90-25, Chicago, IL.

Fox, C., 1995, EPA Office of Science and Technology, personal communication.

Richardson, G. N., Chaney, R. C., and Demars, K. R., 1995, "Design, Performance and Monitoring of Dredged Material Confined Disposal Facilities-Guidance Document," USEPA Region 5, Chicago, IL (in press).

USEPA, 1994, "Assessment and Remediation of Contaminated Sediments (ARCS) Program-Final Summary Report," Great Lakes National Program Office, EPA 905-S-94-001, Chicago, IL.

USEPA, 1994, "Framework for the Development of the National Sediment Inventory," Office of Water, EPA 823-R-94-003, August.

Sediment Characterization

Raymond N. Yong[1]

The Fate of Toxic Pollutants in Contaminated Sediments

REFERENCE: Yong, R. N., "**The Fate of Toxic Pollutants in Contaminated Sediments,**" *Dredging, Remediation, and Containment of Contaminated Sediments, ASTM STP 1293,* K. R. Demars, G. N. Richardson, R. N. Yong, and R. C. Chaney, Eds., American Society for Testing and Materials, Philadelphia, 1995, pp. 13–39.

ABSTRACT: Sediments function as "sinks" for various kinds of contaminants (pollutants and nonpollutants) discharged into the receiving waters. Toxic pollutants in the sediments constitute a significant concern inasmuch as they can infect the waters above the sediment if they are released from the sediments. Hence the persistence and fate of these toxic pollutants need to be determined. At least two sets of interests can be identified in the contamination of sediments as a whole: (1) assessment of the "storage" capacity (for contaminants) of the sediments, and the potential for "mobilization" or release of contaminants into the aqueous environment, particularly into the overlying water, and (2) development of a strategy for removal of the contaminants from the sediments that would be most appropriate (i.e., compatible with the manner in which the contaminants are retained in the sediment) and cost-effective. Both sets of interests require a knowledge of the distribution of the contaminants, i.e., characterization of the contaminants contained in the sediment, and the manner in which these are "held" within the sediment, i.e., "bonded" to the various sediment solid fractions (constituents).

KEYWORDS: persistence, fate, toxic pollutants, heavy metals, organic chemicals, selectivity, partitioning, functional groups, adsorption

Sediments function as "sinks" and reservoirs for various kinds of contaminants (pollutants and non pollutants) discharged into the receiving waters, ranging from runoffs from agricultural and urban activities to industrial waste stream discharges, leachates, or as "dumped" solid waste materials. The major sinks are oxides of iron, manganese, sulfides, carbonates, and natural organic matter. High concentrations of hydrocarbons and other man-made organics, heavy metals, and pesticides, have been reported to be in the waters and sediments in lakes and rivers in areas associated with anthropogenic and agricultural activities (Environment Canada 1991; Natale et al. 1988; Nair and Aziz 1988; Wakeham and Carpenter 1976; Galvez-Cloutier et al. 1993). As reservoirs, the contaminated sediments pose a serious problem because of their potential for recontamination of the overlying water through release of pollutants from the sediments. An understanding of the various factors and conditions necessary to prevent the contaminated sediments from releasing their contaminants, and/or the need for sediment cleanup requires knowledge of the many interactions occurring at the sediment-water interface and within the sediment itself.

Whereas the terms "contaminants" and "pollutants" have often been used interchangeably in the literature, strictly speaking, the term "contaminants" refers to substances that are foreign to the sediment environment (system) and that may or may not pose any health or

---

[1] William Scott Professor of Civil Engineering and Applied Mechanics, and scientific director, Geotechnical Research Centre, McGill University, Montreal, Canada.

environment threat to the system. A "pollutant" on the other hand, represents a substance that has been judged to pose a health and/or environmental threat to the system. "Toxic pollutants" is a term that is used to indicate inorganic and/or organic pollutants that have been assessed by means of toxicity tests or established criteria *vis-a-vis* known distress to biological receptors. When such pollutants are found in sediments, these pose a significant problem inasmuch as the pollutants can infect the waters above the sediment through pollutant release or through resuspension of the polluted sediments. In this paper, the terms "toxic pollutant" and "pollutant" refer to those substances contained in the "priority list" or have been previously assessed as toxic, and are used when specific issues are being discussed, whereas the term "contaminants" will be used generally in the broader discussions.

A knowledge of the persistence and fate of contaminants and, in particular, toxic pollutants, is critical if control of the quality of the receiving waters is to be obtained. One needs to recognize at least two sets of concerns in respect to the problem of sediment contamination: (1) assessment of the "storage" capacity (for contaminants) of the sediments, and the potential for "mobilization" or release of contaminants into the aqueous environment, particularly into the overlying water, and (2) development of a strategy for removal of the contaminants from the sediments that would be most appropriate (i.e., compatible with the manner in which the contaminants are retained in the sediment) and cost-effective.

In both of the preceding sets of interests, evaluation of the fate of the pollutants involves determination or assessment of the distribution of the contaminants, i.e. characterization of the contaminants contained in the sediment, and the manner in which these are "held" within the sediment, i.e., "bonded" to the various sediment solid fractions (constituents). By and large, considering that sediments are derived from erosion and transport processes, it is not unusual for these sediment solids to be comprised principally of soil fractions such as soil minerals, amorphous material, and soil organic matter including organic debris. Thus, whereas the adsorption sites of the sediment solids can be expected to attract and hold the contaminants, determination of the adsorption and retention capacity of the sediment fractions must account for the presence of inorganic and organics ligands which will compete for adsorption of the contaminants. Although the factors that influence the adsorption, retention, and immobilization of the contaminants are to a very large extent similar to those encountered in the accumulation and transport of contaminants in fine-grained soils [e.g., distribution of the various soils fractions (types and amounts), redox potential, buffering capacity, temperature, pH, salinity], the aqueous environment within which the sediment resides requires one to pay attention to the equilibrium composition of the chemical species and ion-associations. Chemical reactions, e.g., acid-base, precipitation, complexation, and oxidation-reduction, need to be considered in the determination of the distribution, mobility, and potential release of the contaminants that constitute the toxic pollutants.

When a sediment is formed, the passage of time will render the sediment pH to near neutral and the sediment environment to anaerobic. Thus, deposition and "hosting" of contaminants in the sediments will be subject to this same sediment environment. Changes in the sediment environment can alter the fate of the contaminants in the sediment. Taking the case of heavy metal pollutants, it is well established that the resultant oxidation-reduction and pH conditions of the sediment-water system will control the chemical forms and mobility of the pollutants. When pH changes occur, the mobility and bioavailability of the "trapped" toxic pollutants will be affected. In the case of organic pollutants, biogeochemical processes are considered to be very important in the distribution of the organics in the sediments. Biological and geochemical processes within the sediment and especially at the sediment-water interface contribute significantly to the alteration and distribution of the organic pollutants in the sediment, i.e., these processes exert considerable influence on the persistence of the organic pollutants. Thus, for example, in petroleum hydrocarbon (PHC) contaminated sediments, microbial degradation can slowly remove n-alkanes from the PHC contaminated sediments, leaving behind the more resistant cycloalkanes and cycloalkenes.

In this keynote paper, the problems of determination of the fate of pollutants are examined not from the viewpoint of ecotoxicology, but from the geoenvironmental engineering perspective. The relationships established between the pollutants and the reactive surfaces of the sediment fractions are considered, particularly in respect to the bonding forces established and the implications *vis-s-vis* "availability" of the pollutants. Whereas the availability of pollutants in benthic ecosystems constitutes one of the major sets of concerns, it is recognized that problems and questions in regard to bioaccumulation potential, adaptation of benthic invertebrate to sediment pollution, and/or degree of toxicant tolerance with respect to pollutant-specific mechanisms are beyond the scope of this study. The problem of particular interest in this paper is the manner in which the pollutants are held to the sediment fractions, and more specifically, the distribution and partitioning of the pollutants within the sediment with respect to the manifestation of the persistence and fate of the pollutants.

Contaminants and Interactions in the Sediment

Various kinds and forms of interactions occurring between pollutants and the individual sediment fractions will determine the fate of the pollutants. In sediments contaminated by organic chemicals, since these are generally susceptible to degradation by biotic processes, determination of the fate of the pollutant chemicals is most often considered in terms of the resistance to degradation of the pollutants and/or their products. When evidence shows that a particular organic pollutant resists biodegradation, the pollutant is identified as a recalcitrant (organic chemical) pollutant, and the study of the fate of the pollutant includes determination of the persistence of the pollutant. In the case of sediments contaminated by inorganic pollutants, the study of their fate is generally oriented towards their bioavailability. By and large, this means the study of the retention of heavy metals by the sediment fractions, and the manner in which these metals are "held" by the various sediment fractions, inasmuch as these are most the most common inorganic pollutants contaminants arising from anthropogenic activities. The influence of the dynamics of the sediment-water interface and the various processes contributing to the establishment of the bioavailability of the pollutants depends not only the bonds established between the pollutants and the sediment fractions, but also the "extractability" of the pollutants, i.e., degree of ease or difficulty in "detachment" (amount of energy required?) of the metal pollutants.

The difficulties in seeking to establish the various abiotic and biotic processes responsible for pollutant presence, fate, and persistence lies not only with the means and methods for analyses, but also with the various dynamics of the problem. Whereas numerous studies have been focused on the presence and distribution of inorganic or organic pollutants in sediments, analyses of contaminated sediments in rivers and lakes generally show the presence of both organic or inorganic pollutants. For example, from the data obtained (International Joint Commission 1989; Upper Great Lakes 1988; Hodge and Bubelis 1991), it is shown that the sediments from the St. Clair River region contain toxic organic compounds and heavy metals. Similar kinds of pollutants have been found in many other sites along the Great Lakes. In the waterway bypassing the Lachine rapids in the St. Lawrence River flowing by the island of Montreal, for example, generally referred to as the "Bay and Canal Lachine," the presence of arsenic, cadmium, chromium, copper, cyanide, mercury, nickel, lead, zinc, polychlorinated biphenyls (PCBs), monocyclic aromatic hydrocarbons (MAHs), and polycyclic aromatic hydrocarbons (PAHs) has been reported (Galvez-Cloutier et al. 1993; Techsult and Roche 1993). The information shown in Fig. 1, obtained from Galvez-Cloutier (1994), shows the

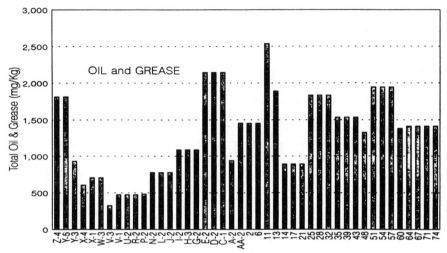
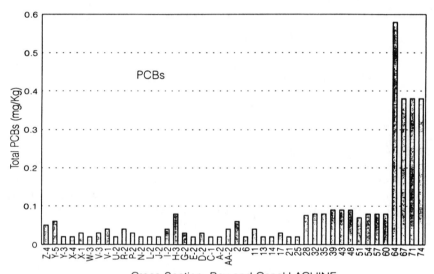

FIG. 1—*Concentration of total oil/grease* (top) *and PCBs* (bottom) *in Bay and Canal Lachine.*

concentrations of "oil and grease" (Fig. 1a) and PCBs Fig. 1b. Two of the heavy metals (lead and zinc) which have been reported previously (Galvez-Cloutier et al. 1993) are shown in Fig. 2.

The determination of the persistence and fate of the toxic pollutants requires one to consider not only the dynamics at the sediment interface, and the other microenvironmental factors such as pH, ligands present, redox potential, and nature of the sediment particulates and their reactive surfaces, but also the synergistic-antagonistic relationships established by

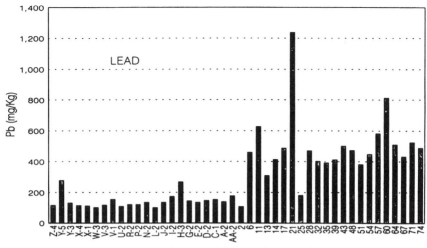
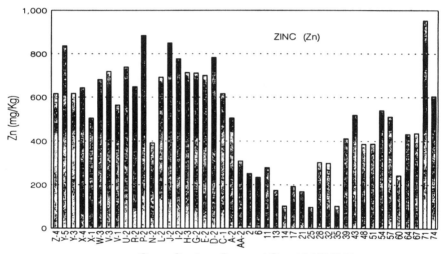
FIG. 2—*Concentration of lead* (top) *and zinc* (bottom) *in Bay and Canal Lachine.*

the presence of the myriad of contaminants. The results of the various activities that will develop sediment-bound pollutants include: (1) sorption, occurring principally as a result of ion-exchange reactions and van der Waals forces, and chemical adsorption (chemisorption) which involves short-range chemical valence bonds; (2) complexation with ligands, and (3) precipitation, i.e., accumulation of material (solutes, substances,) on the interface of the soil solids to form new (insoluble) bulk solid phases. The microenvironmental conditions will dictate which of the processes may be more dominant than the others. It will suffice to say that distinguishing between physical and chemical adsorption, and the results of complexation or precipitation insofar as processes responsible for the "binding" the sediment-bound pollutants is not easily accomplished.

Inorganic Contaminants and Sediment Adsorption Mechanisms

By and large, adsorption reactions or processes are governed by the surface properties of the sediment solids (inorganic and organic), and the chemistry and physical-chemistry of the contaminants, e.g., cations, anions, and nonionic molecules. The net energy of interaction due to adsorption of a solute ion or molecule onto the surfaces of the sediment fractions is the result of both short-range chemical forces such as covalent bonding, and long-range forces such as electrostatic forces. Contaminant cations and anions which are specifically or nonspecifically adsorbed by the sediment fractions can interact in both the diffuse double layer (DDL) and in the Stern layer. Interaction of the indifferent ions in the DDL results in reduction of the zeta potential (ζ), and are nonspecific in that the interactions do not reverse the sign of ζ. Specific adsorption of the ions in the inner Helmholtz plane will make the ions considerably less available for re-release into the overlying waters. This becomes very important in considerations of bioavailability. Adsorption of inorganic contaminant cations is related to their valencies, crystallinities, and hydrated radii. From Coulomb's law, one would expect that cations with a smaller hydrated size or large crystalline size would be preferentially adsorbed—everything else being equal.

From the results of studies with soil suspensions consisting of clay soil fractions and heavy metals (Farrah and Pickering 1976a, 1976b, 1976c, 1977a, 1977b, 1979; Maguire et al. 1981; Yong et al. 1990; Yong and Sheremata 1991; Yong and Galvez-Cloutier 1993; Yong et al. 1993; Yong and Phadungchewit 1993), it is seen that accumulation of the heavy metal pollutants increases with increasing pH. Precipitation of the heavy metals at around neutral pH and above results in the formation of compounds such as hydroxides, sulfates, and chlorates species. At acidic pH values, heavy metals become mobile and adsorption onto clay soil particles becomes less effective due to competition at the exchange sites from H^+ ions. The amount of heavy metals retained and selectivity of retention depend on soil and its composition.

Table 1 shows the results of selectivity tests reported in the literature and those obtained by Yong and Phadungchewit (1993). As noted in the table, selectivity depends on soil composition and interaction mechanisms established with the contaminants. In respect to contamination of sediments by heavy metals, the ease of exchange or the strength with which the metals of equal charge are held by the sediment fractions is by and large inversely proportional to the hydrated radii, or proportional to the unhydrated radii of the metals (Bohn et al. 1971). Prediction of the order of selectivity on the basis on unhydrated radii for some of the heavy metals found for example in the Bay and Canal Lachine (lead, zinc, copper, and cadmium) would indicate the following order: Pb^{2+} (0.120 nm) > Cd^{2+} (0.097 nm) > Zn^{2+} (0.0074 nm) > Cu^{2+} (0.072 nm). However, the studies by Yong and Phadungchewit (1993) show that the position of cadmium has been exchanged with that of copper, and that for the four heavy metals chosen, the selectivity order appears to be Pb>Cu>Zn>Cd. The hydrolysis of aqueous metal cations at high pH values results in a suite of soluble metal complexes. The precipitation of the metal hydroxides onto the sediment fractions results in removal of the metal ions from the interstitial water, and is experimentally indistinguishable from the results obtained from removal of metals from solution by adsorption. It is possible to relate the adsorption affinity of heavy metals in sediment fractions to the pK of the first hydrolysis product of the metals, as has been suggested (Forbes et al. 1974) for studies with soils. K is the equilibrium constant for the reaction in Eq 1 when $n = 1$

$$M^{2+} (aq) + nH_2O \Leftrightarrow M(OH)_n^{2-n} + nH^+ \tag{1}$$

Selectivity order based on ranking of the pK values will show Pb(6.2) > Cu(8.0) >

TABLE 1—*Adsorption selectivity of heavy metals in different soils.*

Material	Selectivity Order	References
Illite Soil	Pb>Cu>Zn≈Cd	Yong and Phadungchewit (1993)
Montmorillonite Clay		
pH > 3	Pb>Cu>Zn>Cd	
Champlain Sea Clay		
pH < 4.5	Pb>Cu>Zn>Cd	
pH > 4.5	Pb>Cu>Zn=Cd	
Kaolinite Clay (pH 3.5–6)	Pb>Ca>Cu>Mg>Zn>Cd	Farrah and Pickering (1977)
Kaolinite Clay (pH 5.5–7.5)	Cd>Zn>Ni	Puls and Bohn (1988)
Illite Clay (pH 3.5–6)	Pb>Cu>Zn>Ca>Cd>Mg	Farrah and Pickering (1977)
Montmorillonite Clay (pH 3.5–6)	Ca>Pb>Cu>Mg>Cd>Zn	Farrah and Pickering (1977)
Montmorillonite Clay (pH 5.5–7.5)	Cd=Zn>Ni	Puls and Bohn (1988)
Al Oxides (amorphous)	Cu>Pb>Zn>Cd	Kinniburgh et al. (1976)
Mn Oxides	Cu>Zn	Murray (1975)
Fe Oxides (amorphous)	Pb>Cu>Zn>Cd	Benjamin and Leckie (1981)
Goethite	Cu>Pb>Zn>Cd	Forbes et al. (1974)
Fulvic Acid (pH 5.0)	Cu>Pb>Zn	Schnitzer and Skinner (1967)
Humic Acid (pH 4–6)	Cu>Pb>Cd>Zn	Stevenson (1977)
Japanese Dominated by Volcanic Parent Material	Pb>Cu>Zn>Cd>Ni	Biddappa et al. (1981)
Mineral Soils (pH 5.0), (with no organics)	Pb>Cu>Zn>Cd	Elliot et al. (1986)
Mineral Soils (containing 20 to 40 g/kg organics)	Pb>Cu>Cd>Zn	Elliot et al. (1986)

Zn(9.0) > Cd(10.1), where the numbers in the brackets refer to the pK values, corresponding very closely to the selectivity order of heavy metal retention found previously (Yong and Phadungchewit 1993), particularly for the illite and montmorillonite soils at pH values above 5 and 3, respectively.

Sediment Fractions and Heavy Metal Accumulation

The mechanisms responsible for accumulation of heavy metal contaminants in the sediment are sensitive to pH of the immediate sediment-interstitial water environment, to a very large extent because of the solubility of the hydroxide species of the heavy metals as shown in the schematic diagram in Fig. 3. In the diagram, M indicates metal ions, and MOH refers to hydroxide compounds. To assess the influence of pH variation in relation to the various soil fractions that constitute the sediment on retention or adsorption of heavy metal contaminants, the results obtained from soil suspension studies with heavy metal contaminants and variable pH conditions can be used. Figure 4 (*left*) shows the influence of soil type on accumulation of lead from a soil suspension with 200 ppm (1.93 meq/100 g soil) lead introduced as $PbCl_2$. The results for the kaolinite soil (Phadungchewit 1990) have been combined with those obtained by Galvez-Cloutier (1989). Using the results reported previously (Yong and Galvez-Cloutier 1993) for lead solubility studies, the results for lead concentration at 200 ppm are superposed onto the soil suspension accumulation curves shown in Fig. 4 (*left*). As noted from the Pb solubility curve, total lead in solution form remains up to about pH 4.3 to 4.5. From this point onward, some precipitation begins to appear, and

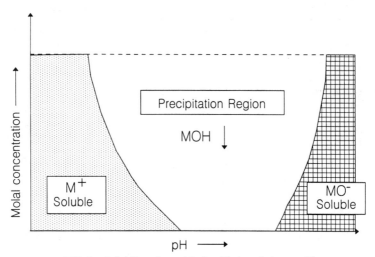

FIG. 3—*Solubility of metal hydroxide in relation to pH.*

if one assumes that there is no soil particle influence on lead precipitation in the soil solution, the precipitation effect deduced from the solubility results can be "subtracted" from the accumulation results portrayed by the solid curves shown in Fig. 4 (*left*). The net result of this accountability for precipitation of lead is shown in Fig. 4 (*right*) for the kaolinite and illite soil suspensions. From selective sequential extraction studies (Yong et al. 1993; Tessier

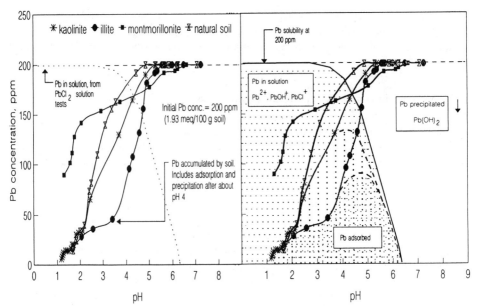

FIG. 4—*Lead (Pb) accumulated by four different soil suspensions* (left) *and lead adsorbed by kaolinite and illite soils in the soil suspensions* (right).

et al. 1979), the distribution or partitioning of the lead amongst the soil fractions in the soil suspension can be deduced as shown in the right-side diagram in Fig. 5 for lead retained by the solid fractions of the illite soil material. The diagram illustrates the proportion of lead retained by the respective soil fractions (minerals, natural organic matter, oxides/hydroxides, carbonates) in the illite soil-suspension in relation to the pH of the system, through mechanisms that range from ion-exchangeable to precipitation and/or coprecipitation. The influence of pH is pronounced when the precipitation pH of lead is reached (and beyond). Thus, the exchangeable ions which can be considered to be nonspecifically adsorbed and ion exchangeable, i.e., they can be replaced by competing cations, become reduced in quantity when the precipitation pH of lead is exceeded because other competing mechanisms begin to dominate, as shown in the left-side diagram in Fig. 5. At pH values of 1 to 4.3, lead is present in the solution as a free cation (Pb^{2+}). The dominant mechanism of lead retention is by cation exchange and the amounts retained increases as pH increases. However, when the soil solution pH increases to a certain level, lead begins to form hydroxy species, resulting thereby in the beginning of lead retention by the hydroxide fractions. The lead precipitated or coprecipitated as natural carbonates can be released if the immediate environment is acidified. For the situation where the lead is accumulated by amorphous or poorly crystallized iron, aluminum and manganese oxides, the bonding is expected to be relatively strong. This observation is of considerable importance since it is not unusual for sediments to contain ferromanganese nodules, ranging from completely crystalline to completely amorphous occurring as coatings on detrital particles and as pure concretions in measurable quantities. Their varying degree of crystallization provides the opportunity for several kinds of bonding mechanisms with heavy metals contaminants, ranging from exchangeable forms via surface complexation with functional groups (e.g., hydroxyls, carbonyls, carboxyls, amines, etc.) and interface solutes (electrolytes), moderately fixed via precipitation and coprecipitation (amor-

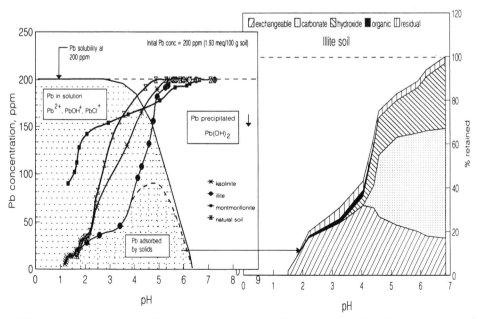

FIG. 5—*Lead adsorbed by illite soils in soil suspension* (left) *and distribution of lead accumulated* (*includes adsorbed and precipitated*) *by the different fractions in the illite* (right).

phous) and relatively strongly bound. Precipitation of heavy metals in contaminated sediments is a significant mechanism for removal of the contaminants from the interstitial water in the sediment. The mechanism includes the transfer of solutes from the aqueous phase to the sediment fraction interface, resulting thereby in accumulation of a new substance in the form a new soluble solid phase. The process which occurs in two stages, nucleation and particle growth, is governed by Gibbs phase rule which restricts the number of solid phases that can be formed. The net accumulation at the sediment fraction interface is considered as a separate mechanism from the mechanism of the formation of a new solid (precipitated) phase. It is difficult to distinguish between the two mechanisms because the chemical bonds formed in both processes can be closely similar. High concentrations of solutes are needed for precipitation to occur in the interstitial water since the process requires the ionic activity of the solutes to exceed the solubility product.

The heavy metals retained in the "residual fraction," shown in the right-side diagram in Fig. 5, are generally considered to be retained within the lattice of silicate minerals, and can only become available after digestion with strong acids at elevated temperatures. The residual materials consist of silicates and other resistant materials, and determination of the metal associated with this fraction, which is not considered to be significantly large, is important in completing mass balance calculations.

It is interesting to compare the development of the different heavy metal species for the situation where more than one heavy metal species is present in the contaminant mixture in the sediment. The results from calculations using the chemical speciation model MINTEQ, developed by the U.S. EPA (U.S. Environmental Protection Agency 1988) are shown in Table 2 for the study with mixed contaminants involving the addition of equal proportions of heavy metals lead, copper, zinc, and cadmium at concentrations of 1.0×10^{-3} mol/L and inorganic ligands reported by Yong et al. (1993). Because the contaminants contain more than the single lead species, the precipitation pH for lead is seen to be different from that reported in the left-side graph shown in Fig. 4. The calculations in Table 2 suggest that if the sediment environment pH values is increased, the form of heavy metal species in the interstitial water will change from a simple to a more complex form. In the case of lead for example, the free cation Pb^{2+} percentage decreases between the region of pH 5 and pH 6. A significant proportion of the lead precipitates at pH 6 and more species of lead are formed in the interstitial water. This observation accords well with the information shown in Figs. 4 and 5. As the sediment environment pH continues to increase, one would expect that a higher percentage of metals will precipitate, particularly for the lead and copper species of heavy metals. A somewhat different picture is observed with the cadmium, from the calculations shown in Table 2. The amounts of cadmium retained in the hydroxide phase is seen to be smaller than those of the other metals, because of the formation of hydroxy species at high pH values, with little evidence of precipitation even at pH 8. As shown previously (Yong and Sheremata 1991), the formation of Cd-Cl complexes will hinder precipitation. One can conclude that retention by sediment fractions consisting of soil materials will not only be sensitive to the surface characteristics of the sediment fractions, but also to the many interactions occurring between the various contaminants.

Fate of Heavy Metals in Sediment

The presence of contaminants other than heavy metals in sediments, and also inorganic and organic ligands, offers opportunities for heavy metal interactions with other contaminant species and formation of various complexes, resulting in competition for adsorption sites and metals. Tests on sediment samples recovered from the Bay and Canal Lachine (Galvez-Cloutier 1994; Chan 1993, Yong and Galvez-Cloutier 1995) show the major contaminants

TABLE 2—*Calculations from MINTEQ for heavy metal species distribution.*

Metal Species	pH							
	1	2	3	4	5	6	7	8
Pb								
Percent precipitated	0	0	0	0	0	70.4	99.7	100
Percent dissolved	100	100	100	100	100	29.6	0.3	0
Percent Distribution of Components Among Dissolved Species								
Pb^{2+}	88.3	88.1	88.1	88.1	88.0	88.6	78.8	40.5
$PbCl^+$	9.6	9.8	9.8	9.8	9.8	10.1	9.3	4.9
$PbNO_3^+$	2.1	2.0	2.0	2.0	2.0	2.1	1.9	1.0
$PbOH^+$						1.1	9.9	51.8
$Pb(OH)_2^0$								1.7
Cu								
Percent precipitated	0	0	0	0	0	92.5	99.8	100
Percent dissolved	100	100	100	100	100	7.5	0.2	0
Percent Distribution of Components Among Dissolved Species								
Cu^{2+}	99.2	99.2	99.2	99.2	99.1	97.0	43.8	
$Cu(OH)_2^0$						1.2	52.7	98.6
$CuOH^+$							2.9	
Zn								
Percent precipitated	0	0	0	0	0	0	0	97.3
Percent dissolved	100	100	100	100	100	100	100	2.7
Percent Distribution of Components Among Dissolved Species								
Zn^{2+}	99.2	99.2	99.2	99.2	99.2	99.2	98.3	85.7
$ZnOH^+$								6.4
$Zn(OH)_2^0$								6.4
Cd								
Percent precipitated	0	0	0	0	0	0	0	0
Percent dissolved	100	100	100	100	100	100	100	100
Percent Distribution of Components Among Dissolved Species								
Cd^{2+}	78.9	78.5	78.5	78.5	78.5	77.6	77.3	76.0
$CdCl^+$	20.5	20.9	20.9	20.9	20.9	21.7	21.9	22.1
$CdOH^+$								1.0

identified in Table 3. The contaminants have been listed with respect to the sites in the Bay and the Canal themselves. Since the Bay is upstream from the Canal, it was useful to study the distribution of the contaminants as one progressed downstream. The major industries were located on both sides of the Canal, one side of the Bay, and upstream from the Bay. At least eight of the contaminants exceed established limits published by the Ministries of Environments of several Provinces in Canada.

To assess the distribution of heavy metals in the sediment and also between sediment fractions, the selective sequential extraction procedure can be used to study the problem, as in the right-side graph in Fig. 5. Using lead as an example, the information obtained (Galvez-Cloutier 1994) shows the partitioning between sediment fractions for lead (Fig. 6) in relation to the pH of the system. In examining the results portrayed in the figure, it is useful to note that the pH range of the sediment samples recovered ranged from 5.7 to 6.8. From the results shown, one can deduce that as the system approaches pH 4, precipitation processes are apparently set in motion and that hydroxy species begin to form, as is witnessed by the increasing amount represented by the "oxide" in the figure. This appears to "stabilize" at about pH 4.5. Precipitation and/or coprecipitation as natural carbonates also appear to begin as pH 4 is approached, also as shown in the increasing "carbonate" shown in the figure. The results shown in the figure suggest that at the pH range of 5.7 to 6.8 for the sediment, the major amount of lead "retained" appears to be "residual," i.e., tied up within the lattice of the silicate minerals. This may not be inconsistent since the composition of the sediment appears to contain a range of from 4 to 27.5% w/w natural organic matter, with the remaining proportion shown as solids (minerals). The minerals in the sediments, shown in descending order of abundance are illites, quartz, kaolinites, and feldspars, with significantly lesser proportions of chlorites and amphobiles.

To further assess the potential release of the heavy metal contaminants, i.e., to determine reflux capability of heavy metal pollutants in the sediment as a requirement for development of sediment remediation strategies, one requires information in regard to the "ease" by which

TABLE 3—*Bay and Canal Lachine contaminant concentrations.*

	Contaminants (given as range of values in mg/L)	
	(Lachine) Bay	(Lachine) Canal
Chromium	28.3 to 70.8	23.4 to 73.4
Arsenic	0.8 to 4.8	1.2 to 10.1
Cadmium	1.2 to 6.4	2.0 to 7.6
Copper	31.0 to 82.0	71.0 to 141.0
Mercury	0.26 to 1.05	0.47 to 1.8
Nickel	24.0 to 50.0	57.0 to 75.0
Lead	101.2 to 268	107.6 to 1237.2
Zinc	309.2 to 836.8	97.5 to 952.4
MAH	ND (not detect. <0.01)	ND to 8.3
PAH	ND to 0.3	3.4 to 43.4
Oil and Grease	328 to 2148	892 to 2573
PCB	0.02 to 0.08	0.02 to 0.58
TKN	1243 to 5562	1731 to 4773
TOC	4.23 to 14.79	2.37 to 11.33
Phosphorous	595 to 2557	636 to 2125

Na^+ = 49.2 to 1171.8; Ca^{2+} = 40.9 to 4305.1; Cl^- = 2.0 to 61.1
SO_4^{2-} = 0.4 to 76.0; Mg^{2+} = 1484.9 to 21 239.4;
K^+ = 150.9 to 9429.3; HCO_3^+ = 12.2 to 183.0

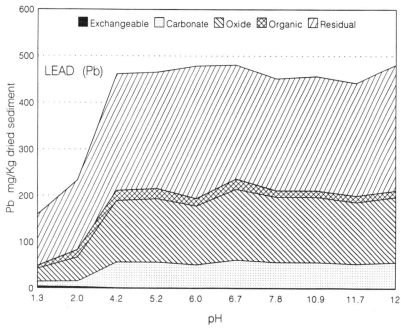

FIG. 6—*Selective sequential extraction study of lead distribution in sediment sample recovered from Bay and Canal Lachine.*

the pollutants can be released from the sediment environment. The results shown in Figs. 5 and 6 need to be examined in relation to the extraction procedures implemented to determine the distribution of lead shown in the diagrams. Figure 7 shows the extractant reagents used on the illite soil previously discussed. A quick comparison of the results shown in Figs. 6 and 7 suggests that the complex contaminant and sediment fraction situation has affected the precipitation processes, i.e., the process appears earlier in the sediment sample. Although this is not unusual, much remains to be determined before proper assessments can be made inasmuch as more information and study is needed to provide for the complete sets of chemical interactions occurring in the sediment. Using Fig. 7 as a control situation, one observes that with the selective sequential extraction technique, step 1 used potassium nitrate as the competing replacement ion, i.e., the lead released will be the exchangeable ions accumulated within the system. This procedure suggests that these ions will be "easily" released from the sediment environment, and will constitute the potential reflux ions. The illite soil sample shown in Fig. 7 indicates that reflux of lead will be substantial in contrast to the sediment sample shown in Fig. 6 where the exchangeable ions are almost nonexistent at pH 4 and above.

As one proceeds with the extraction process, steps 2, 3, and beyond require more aggressive extraction treatment to remove the lead. While it is speculated that these kinds of aggressive removal opportunities may not exist in contaminated sediments under normal circumstances, there are insufficient studies that have addressed "origin" of refluxed contaminants to determine actual source. Until such information becomes available, one can only rely on such extraction techniques to deduce "extractability" of the contaminants. Let it suffice to say that when one implements digestion (step 5), total destruction of the sample occurs—a situation not likely to be encountered in most circumstances. Thus, if the results

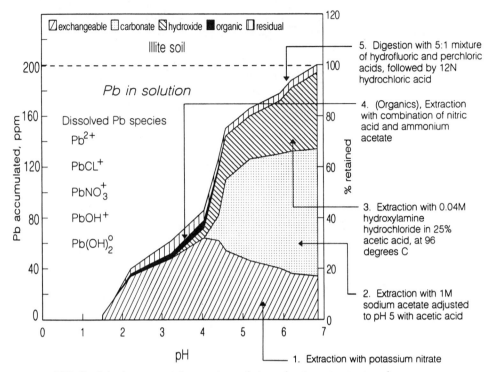

FIG. 7—*Selective sequential extraction technique showing extractants and sequence.*

in Fig. 6 (sediment sample) are examined, it would be apparent that the lead contained in the sediment system and identified as residual would be "locked" within the system.

The lead in solution shown in Fig. 7 contains the dissolved species previously determined by the geochemical model MINTEQ (Table 2). Similar procedures can be implemented for the contaminated sediment situation to determine the dissolved species for the heavy metals contained in the sediment (Yong and Galvez-Cloutier 1995). These species will be resident in the interstitial water of the sediment, and are subject to communication with the overlying water, particularly when physical, density, and thermal gradients exist in the near-sediment environment. The phenomenon of "overturning" is cited as a good example of the results of such gradients (particularly temperature gradients). One needs to evaluate both the species and the concentrations or proportions of those species that are toxic.

Functional Groups and Bonding for Organic Chemical Pollutants

A simple initial characterization of organic chemical pollutants distinguishes between organic acids/bases and nonaqueous phase liquids. The latter, (i.e., NAPLs), which are liquids that exist as a separate fluid phase in an aqueous environment, are not readily miscible with water, and are in general distinguished in terms of NAPL density greater than or less than water. Because DNAPLS are heavier than water, the pollutants in the DNAPLs family in sediments include those associated with anthropogenic sources, e.g., chlorinated hydrocarbons such as PCBs, carbon tetrachloride, 1,1,1-trichloroethane, chlorophenols, chlorobenzenes, and tetrachloroethylene. The interstitial water in a sediment which contains dissolved substances such as free salts, solutes, colloidal material, and/or organic solutes, in combination with sediment-interface dynamics, will control the transfer and removal of solutes from the aqueous phase to the solid surfaces. The "bonding" relationships between pollutants and the sediment fractions needed to establish "sediment-bound" pollutants need to be examined not only in terms of the constituents in the interstitial water (inorganic and organic ligands), but also in terms of the relationships developed between the chemically reactive groups of the pollutants and those present in the sediment fractions.

The binding mechanisms developed between organic compounds and soil fractions, which have been extensively studied in the terms of clay-organic complexes, can be used as a guide in the assessment of the basic processes controlling organic chemical pollutant adsorption fate in sediments. Some of the basic mechanisms of interaction and bonding include: (1) ion exchange involving organic cations (anions) and "charged" surfaces (reactive surfaces) of sediment fractions, (2) hydrogen bonding, (3) π bonding, (4) covalent bonding, (5) ion-dipole and coordination, and (6) van der Waals forces.

The functional groups for sediment fractions and contaminants, which are chemically reactive atoms or groups of atoms bound into the structure of a compound, are either acidic or basic. In the case of organic chemicals, the nature of the functional groups in the (organic) molecule, shape, size, configuration, polarity, polarizability, and water solubility are important in the adsorption of the organic chemicals by the sediment fractions (soil colloids). Since many organic molecules (amine, alcohol, and carbonyl groups) are positively charged by protonation (adding a proton or hydrogen), surface acidity of the sediment fractions becomes very important in the adsorption of these ionizable organic molecules. The adsorption of the organic cations is related to the molecular weight of the organic cations. Large organic cations are adsorbed more strongly than inorganic cations by clays because they are longer and have higher molecular weights. Depending on how they are placed, and depending on the pH of the sediment-fluid system, the functional groups will influence the characteristics of organic compounds, and will thus contribute greatly in the determination of the mechanisms of accumulation, persistence, and fate of these compounds in soil.

Whereas the hydroxyl functional group is the dominant reactive surface functional group for most of the sediment fractions (clay minerals, amorphous silicate minerals, metal oxides, oxyhydroxides, and hydroxides), the natural organic matter (NOM) will contain many of the same functional groups identified with organic chemicals, e.g., hydroxyls, carboxyls, amines, and phenols, as seen in Fig. 8. In contaminants, the hydroxyl group is present in two broad classes of compounds: (1) alcohols, e.g., methyl, ethyl, isopropyl, and n-butyl, and (2) phenols, e.g., monohydric (aerosols) and polyhydric (obtained by oxidation of acclimated activated sludge (pyrocatechol, trihydroxybenzene). Alcohols can be considered as hydroxyl alkylcompounds (R-OH), and are neutral in reaction since the OH group does not ionize, whereas phenols are compounds that possess a hydroxyl group attached directly to an aromatic ring (Sawyer and McCarty 1978). Adsorption of hydroxyl groups of alcohol can also be obtained through hydrogen bonding and cation-dipole interactions. One notes that most primary aliphatic alcohols form single-layer complexes on the negatively charged surfaces of the sediment fractions, with their alkyl chain lying parallel to the surfaces of the fractions. Double-layer complexes are also possible with some short-chain alcohols such as ethanol.

Organic chemical pollutants with (1) functional groups having a C—O bond, e.g., carboxyl, carbonyl, methoxyl, and ester groups, and (2) nitrogen-bonding functional groups, e.g., amine and nitrile groups, are fixed or variable-charged organic chemical compounds since they can acquire a positive or negative charge through dissociation of H^+ from or onto the functional groups, dependent on the dissociation constant of each functional group and the pH of the sediment system. The situation *vis-a-vis* fate of the organic chemical pollutant becomes of considerable concern inasmuch as when a high pH regime replaces an original

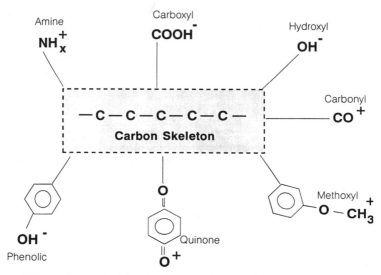

FIG. 8—*Some typical functional groups for natural organic matter (NOM).*

lower pH regime in the sediment, H^+ dissociation could occur, resulting thereby in the development of negative charges for the organic compounds. If cation bonding between the organic chemical and the sediment fraction was initially responsible for the attachment of the organic chemical pollutant to the sediment, the reversal in charge of the pollutant will result in possible release of the pollutant to the overlying waters.

Carbonyl compounds (aldehydes, ketones, and carboxylic acids) most often possess dipole moments because the electrons in the double bond are unsymmetrically shared. Whereas they can accept protons, the stability of complexes between carbonyl groups and protons is considered to be very weak. The carboxyl group of organic acids (benzoic and acetic acids) have the ability to interact either directly with the interlayer cation or by forming a hydrogen bond with the water molecules coordinated to the exchangeable cation associated with the sediment fractions. Adsorption of organic acids depends on the polarizing power of the cation. Because of their ability to donate hydrogen ions to form basic substances, most carboxyl compounds are acidic—weak acids as compared to inorganic acids.

The amino functional group NH_2, which is found in primary amines and which may be aliphatic, aromatic, or mixed, can be adsorbed with the hydrocarbon chain perpendicular or parallel to the reactive surfaces of the sediment fractions, depending on their concentration. The phenolic functional group, which consists of a hydroxyl attached directly to a carbon atom of an aromatic ring, can combine with other components such as pesticides, alcohol, and hydrocarbons to form new compounds; e.g., anthranilic acid, cinnamic acid, ferulic acids, gallic acid, and *p*-hydroxy benzoic acid.

The various petroleum fractions in petroleum hydrocarbons (PHCs) are primarily constituted by nonpolar organics with low dipole moments (generally less than one) and dielectric constants less than three. It has been shown (Hoffman and Brindley 1960) that adsorption of nonionic organic compounds by soil material typical of sediment fractions is governed by the CH activity of the molecule; the CH activity arises from electrostatic activation of the methylene groups by neighboring electron-withdrawing structures, such as $C=O$ and $C=N$. Molecules possessing many $C=O$ or $C=N$ groups adjacent to methylene groups would be more polar and hence more strongly adsorbed than those compounds in which such groups are few or absent (Tehng 1974). An examination of the chemical structure of the representative group petroleum hydrocarbons (MAHs and PAHs) shown in Fig. 7, indicate that there are no electron-withdrawing units such as C=O and C=N associated with the molecules. Accordingly, one would expect that the PHC molecules would be weakly adsorbed (mainly by van der Waals adsorption) by the soil functional groups and do not involve any strong ionic interaction with the soil.

Partitioning and "Sediment-bound" Organic Chemical Pollutants

The distribution of pollutants within the sediments (i.e., between the sediment solids and sediment interstitial water) is generally given in terms of distribution coefficients K_d, i.e., ratio of the concentration of pollutants "held" by the sediment fractions (i.e., sediment-bound) to the concentration of pollutants "remaining" in the interstitial water. The distribution coefficient is generally obtained from sediment-suspension tests with target pollutants and specified (or actual) sediment fractions, and gives information in relation to the adsorption of the test organic chemical pollutant by the sediment fractions. Whereas the test data obtain relate to the distribution of the set of test pollutants and sediment fractions, the mechanisms responsible for bonding, and the ease by which the sediment-bound pollutants can be "released" from the sediment fractions are not addressed by the test technique or manner of data interpretation. Quantitative determination of the "bonding strength" is not easily accomplished. Instead, the bonding mechanisms and intensities established between the pollutants are generally determined indirectly via theoretical considerations of adsorption mechanisms and interactions between the various functional groups, or via qualitative assessments using selective sequential extraction techniques.

The water solubility of an organic chemical pollutant is of significant importance in the control of the fate of the pollutant, since it determines the partitioning of the pollutant to the sediment fractions and the biota. Within the sediment fractions itself, the water solubility of the pollutant affects bonding and distribution between the various sediment fractions, and the transformations occurring as a result of various processes associated with oxidation/reduction, hydrolysis and biodegradation. The equilibrium partition coefficients, i.e., coefficients pertaining to the ratio of the concentration of the specific pollutant in other media to that in water, are considered to be well correlated to water solubilities of most organic chemicals. With respect to biota, the equilibrium partition coefficient, which is identified as the bioconcentration factor, is reported to be well correlated to the fat or lipid content for fish species (Schnoor 1982). Chiou et al. (1977), for example, have reported good correlations between the solubilities of organic compounds and their octanol-water partition coefficient K_{ow}. Equation 2 shows the relationship given in terms of the solubility S, where S is expressed either as ppm or ppb.

$$\log K_{ow} = 4.5 - 0.75 \log S \quad \text{(ppm)}$$
$$\log K_{ow} = 7.5 - 0.75 \log S \quad \text{(ppb)} \tag{2}$$

The K_{ow} octanol-water partition coefficient has been widely adopted as a significant parameter in studies of the environmental fate of organic chemicals, and found to be sufficiently correlated not only to water solubility, but also to soil/sediment adsorption coefficients, and bioconcentration factors for aquatic life (Neely et al. 1974; Metcalf et al. 1973). The use of an octanol-water partition coefficient is predicated on the assumption that octanol is a good surrogate for lipid material. In the experimental measurements reported, it appears that the assumption is valid, and that for biota and soils/sediments, the octanol is considered to be the surrogate for the fat of the organisms (biota) and for the organic matter (soil/sediment). Chemicals with low K_{ow} (e.g., less than 10) may be considered relatively hydrophilic; they tend to have high water solubilities, small sediment (or soil) adsorption coefficients, and small bioconcentration factors for aquatic life. Conversely, chemicals with a high K_{ow} value (e.g., greater than 10^4) are very hydrophobic, and thus demonstrate low water solubilities (Lyman et al. 1982). Solvent systems that are almost completely immiscible (e.g., alkanes-water) are fairly well behaved and if the departures from ideal behavior exhibited by the more polar solvent systems are not too large, a thermodynamic treatment of partitioning can be applied to determine the distribution of the organic chemical without serious loss of accuracy. A brief discussion of some of the aspects of the thermodynamics of the partitioning process can be found in Yong et al. (1992).

Aqueous concentrations of hydrophobic organics, such as PAHs and compounds such as nitrogen and sulphur heterocyclic PAHs, and some substituted aromatic compounds indicate that the accumulation of the hydrophobic chemical compounds is directly correlated to the organic content (natural organic matter NOM) of the sediment fractions. The dominant mechanism of organic adsorption is the hydrophobic bond established between the organic chemical and the NOM, and it is often argued that the term "partitioning" should strictly be applied to such bonding mechanisms. Studies have shown that whereas the variability in sorption coefficients between different sediments (and soils) may be due to characteristics of sediment fractions (surface area, cation exchange capacity, pH, etc.), and the amount and nature of the organic matter present, a good correlation of sorption (expressed via a distribution coefficient) with all of these variables with the organic carbon content can be obtained. There is the need however to pay attention to such controlling factors as temperature, pH of the system, sediment particle size and distribution, salinity, and NOM presence. As stated previously, considering the octanol to be a surrogate for organic matter, one should expect that the partition coefficient K_{ow} can be related to the organic content coefficient K_{oc}. This has been adopted, and a Freundlich type of linear distribution coefficient K_d has been found to be relatively useful, as previously indicated by other researchers for subsurface and surface soils (Means et al. 1982). The water solubility at room temperature (25°C), together with the log K_{oc} and log K_{ow} values reported in the literature are given in Fig. 9 for a representative group of MAHs and PAHs, typical of those present in petroleum hydrocarbon (PHCs). Adsorption of the hydrocarbons by the active sediment fractions' surfaces occurs only when the water solubility of the PHCs is exceeded and the hydrocarbons are accommodated in the micellar form. Instead of using the K_{ow} and K_{oc} partition coefficients, the accommodation concentration of hydrocarbons in water is sometimes used to reflect the partitioning tendency of organic substances between the aqueous and sediment solids. Hydrocarbon molecules with lower accommodation concentrations in water (i.e., higher K_{oc} values) would be partitioned to a greater extent onto the sediment fractions than in the aqueous phase. From the results of Meyers and his coworkers (1978, 1973), one can expect to obtain a general inverse relationship between the accommodation concentration of the hydrocarbons and the proportion (percent) adsorbed; i.e., the lower the accommodation concentration of the hydrocarbon in water, the greater the tendency of the organic compound to be associated with the reactive surfaces of the sediment fractions. The important consequence of such a relationship is that the aromatic fraction of petroleum products that are the most toxic would have the least affinity for the reactive surfaces associated with the sediment fractions. As might be expected, a study of adsorption data of hydrocarbons show that anthracene is substantially adsorbed, as can be confirmed by the high K_{oc} value and the very low solubility of the organic compound in water, as shown in Fig. 9; presumably the higher accommodation concentrations of the aromatic hydrocarbons inhibits their association with the clay particles (Mey-

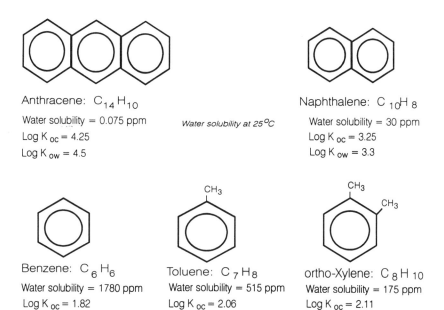

FIG. 9—*Some typical monocyclic and polycyclic aromatic hydrocarbons (MAHs and PAHs) associated with petroleum hydrocarbons (PCHs).*

ers and Oas 1978). The information shown in Fig. 9 would indicate that the MAH/PAH association with the reactive surfaces in sediments could be: Anthracene > Naphthalene > ortho-Xylene > Toluene > Benzene.

Persistence of Organic Chemical Pollutants

The activities of benthic organisms have the distinct potential for alteration of the physical and chemical nature of sediments and the interactions established between the sediment fractions and contaminants. The biogenic processes that are of importance are biodeposition, particle transport, fluid transport, stabilizing mechanisms, and macrofaunal-microbial interactions. In situations where concern is directed towards possible redistribution of pollutants, one needs to pay attention to biodeposition and bioturbation. These processes impact directly not only on the rate of flux of pollutants between water and sediment, but also on the distribution of pollutants within the sediments.

In respect to organic chemical pollutants in sediments, biogeochemical processes are important in determining the distribution of hydrocarbons in sediments through selective removal and/or selective production. Microbial degradation can slowly but preferentially remove n-alkanes from a petroleum-contaminated sediment, leaving behind the more resistant isoprenoids, cycloalkanes and cycloalkenes, and aromatics. Relative rates of microbial degradation proceed as: n-alkanes > branched alkanes > cyclic alkanes. A combination of diffusion, water solubilization and transport from sediments, evaporation, and microbial degradation is responsible for observed changes in aromatic hydrocarbon concentrations and composition.

The low water solubilities of organochlorine compounds, such as PCBs and DDT, combined with their very slow rate of microbial degradation, make these compounds recalcitrant. Because of their low solubilities, they tend to concentrate in the sediment. Since the lower chlorinated isomers of PCBs are more readily degraded, the higher chlorinated compounds will dominate as the persistent compounds of PCBs found in the sediments.

In addition to the MAHs representative of the PHCs shown in Fig. 9, the chlorinated hydrocarbons which also are considered as MAHs, e.g., chloro-, dichloro-, trichloro-, pentachloro-, and hexachlorobenzene (Fig. 10), have been found to be quite persistent, i.e., their presence in sediments has been well established even though their use has been severely restricted in recent times. Analysis of sediment cores (Oliver and Nicol 1982, 1984) indicate that these MAHs have been accumulating in the sediment in Lake Ontario since the early 1900s. There appears to be little evidence of either microbial oxidation or anaerobic dehalogenation of chlorobenzenes (C_6H_5Cl) in sediments. Bosma et al. (1988) suggest that trichlorobenzenes ($C_6H_3Cl_3$) can be transformed to dichlorobenzenes ($C_6H_4Cl_2$) in some sediments under anaerobic conditions with half-lives ranging from a few days to over 200 days. Dichlorobenzene, also known as ortho-dichlorobenzene, is used primarily as solvent for carbon removal and degreasing of engines. With the K_{oc} value as shown in Fig. 10, the dichlorobenzene partitions well to sediments, and particularly the organic fractions [natural organic matter (NOM)], and because of its low anaerobic degradation, it is very persistent. Although there exist three isomers of trichlorobenzene ($C_6H_3Cl_3$), 1,2,4-, 1,2,3- and 1,3,5-, the isomer 1,2,4- is most common. The low water solubilities, high log K_{ow} and log K_{oc} values, indicate that it partitions well to the sediment fractions, and as in the case of the dichlorobenzene, the trichlorobenzene is well adsorbed by the NOM in the sediments and will persist and accumulate under anaerobic conditions in sediments. The similarly high

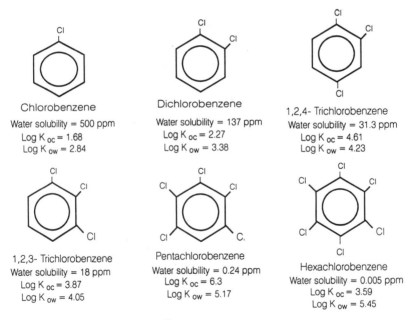

Water solubility at 25°C

FIG. 10—*Chlorinated benzenes.*

values of K_{oc} for pentachlorobenzene (C_6HCl_5) and hexachlorobenzene (C_6Cl_6) are also indicative of the ability to partition to sediments, in common with the trichlorobenzenes. The pentachlorobenzene which has been identified in waste streams from pulp and paper mills, iron and steel mills, inorganic and organic chemical plants, petroleum refineries, and activated sludge waste water treatment plants (Meyers and Quinn 1973; Laflamme and Hites 1978) appears to have the highest K_{oc} value of the various chlorobenzenes. The low water solubilities of the dichloro-, trichloro-, pentachloro-, and hexachlorobenzenes combined with their respective high K_{oc} values indicate that they can adsorb onto suspended particulates in the receiving waters. When the particulates sediment, it would appear that the adsorbed chlorobenzene pollutants persist and become entrapped within the sediment and also by overlying sediments. From the studies conducted in the Great Lakes, it would appear that some desorption from resuspended bottom sediments does occur and may be an important continuous source of the chlorobenzenes to water (Oliver 1984; Oliver et al. 1989). From available studies (e.g., Callahan et al. 1979; Mansour et al. 1986; Mill and Haag 1986; Oliver and Pugsley 1986), chemical and biological degradation are not considered as important removal processes of the chlorobenzenes from the overlying water or the contaminated sediments.

The effects of biodegradation, or the resistance to biodegradation as an indication of the persistence of the organic chemicals in polluted sediment, have been recorded in many instances. As an example, one can cite the case of sediment contamination by pentachlorophenol (PCP) that has been reported by Crosby (1972) to be relatively soluble in water at pH 6 and above, and which can be degraded by photochemical and microbial action. The relative solubility of the chemical suggests a low potential for accumulation of PCP in the sediment. However, there is evidence (Munakata and Kuwahara 1969) that shows the presence of substantial amounts of PCP associated with the particles in the sediments studied, suggesting thereby that PCP may not be readily degradable in the presence of particle bonding. On the other hand, the study reported by Pierce et at. (1980), in regard to the degradation of PCP over a two-year period following a major oil spill into a creek leading into a 60 acre (24.3 ha) lake, shows a reduced presence of PCP from an original maximum concentration of about 1.35 mg/kg air dry sediment to about 0.2 mg/kg in the shallow contaminated creek, and also the degradation products such as pentachloroanisole (PCA) and 2,3,4,5-, 2,3,4,6,- and 2,3,5,6-tetrachlorophenol (TCP). However, when the contaminated creek results are considered in conjunction with those obtained from sediments in a deeper portion of the lake, the conclusion reached is that because little production of TCP occurred in the creek sediment, and because the TCP component was highly variable in the anaerobic sediment in the deeper portion of the lake, it was possible that reductive dechlorination of PCP to TCP could have occurred and that the TCP could subsequently be converted or removed.

Anaerobic dehalogenation of some organic chemicals can have detrimental effects *vis-a-vis* increase levels of threat to the benthic microsystem because of a greater facility in communication of the transformed chemical to the overlying water. A good case in point is the degradation of tetrachloroethylene (PCE, C_2Cl_4) to trichloroethylene (TCE, C_2HCl_3), to dichloroethylene (DCE, $C_2H_2Cl_2$) and to vinyl chloride (VC, C_2H_3Cl), as shown in Fig. 11. Also shown in the diagram are the permissible exposure level (PEL) in air for the various compounds, the IDLH (immediately dangerous to life or health) values, and the log K_{oc} and log K_{ow} values. Beginning with PCE, where the log K_{oc} value indicates good partitioning to the sediment, it is observed that as PCE degrades to TCE and onward to VC, the log K_{oc} values diminish considerably to a very low value for the vinyl chloride. This indicates that as the PCE continues to degrade, more of the chemical substance is released to the overlying water, particularly for VC, where both the log K_{oc} and water solubility values are very high. Considering that VC is a carcinogenic substance, and that the PEL in air is 1 ppm, with a

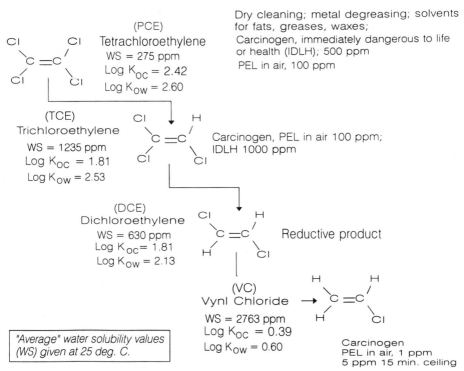

FIG. 11—*Anaerobic dehalogenation of tetrachloroethylene resulting in production of intermediate products trichloroethylene, dichloroethylene, and vinyl chloride.*

5 ppm 15 min ceiling, one can appreciate the greatly increased health threat posed by the degradation of PCEs entrapped and held within sediments.

Concluding Remarks

The evidence collected in regard to contaminated sediments shows the presence of heavy metals and various organic chemicals. The potential for release of these pollutants from a contaminated sediment needs to be assessed if proper measures are to be undertaken to implement secure "storage" or remediation of the contaminants. In the case of heavy metals, the metals that have received the most attention with regard to retention by soil materials typical of those found in sediments include lead (Pb), cadmium (Cd), copper(Cu), zinc (Zn), nickel (Ni), chromium(Cr), and mercury (Hg). Using information gained from soil suspension experiments conducted with these contaminants, the detailed studies on kaolinite, illite, and montmorillonite clay suspensions (Farrah and Pickering 1976a, 1976b, 1976c, 1977a, 1977b, 1979; Maguire et al. 1981; Yong et al. 1990; Yong and Sheremata 1991; Yong and Galvez-Cloutier 1993; Yong et al. 1993; Yong and Phadungchewit 1993) indicate that the accumulation of heavy metal ions by the various clay suspensions increased with pH. From the studies, one would expect that when the precipitation pH of the metal ions in a contaminated sediment is exceeded (by the sediment pH), metal hydroxy species tend to be formed, and a significant increase in the amount of the metals can be accumulated within the sediment. Since heavy metal accumulation by the sediment fractions include both cation exchange and precipitation, and since precipitation is dependent on the pH of the sediment system, one would expect that a direct relationship exists between heavy metal uptake and both the cation exchange capacity (CEC) of the sediment and precipitation. It is useful to note at this stage that should the pH regime in the sediment reverse and decrease to pH levels below the precipitation pH of the heavy metals, solubilization of the precipitates could occur, i.e., a reversal of the processes would be expected. The retention capacity observed in the sediment fractions consisting of clay minerals shows the following order: kaolinite < illite < montmorillonite, corresponding to the CEC of the clay minerals. At acidic pH values, heavy metals become mobile and adsorption onto clay soil particles becomes less effective due to competition at the exchange sites from the H^+. The amount of heavy metals retained and selectivity of retention depend on soil and its composition.

The various accumulation (sorption) processes that contribute to a sediment-bound organic chemical pollutant include partitioning (hydrophobic bonding) and accumulation, through adsorption mechanisms involving the clay minerals and other soil particulates such as carbonates, and amorphous materials. In summarizing the more prominent properties affecting the sorption of organic molecules on sediment fractions (other than NOM), the following can be listed:

Sediment Fractions: surface area, nature of surfaces, surface charge (density, distribution and origin), surface acidity, exchangeable ions on the reactive surfaces, configuration of the reactive surfaces.
Organic Molecules: charge, size, shape, flexibility, polarity, water solubility, polarizability.
Sediment Environment: temperature, inorganic/organic ligands available, pH, salinity, physical gradients (fluxes).

The similarity between many of the NOMs and organic chemicals is seen as a strong basis for the hydrophobic bonding that allows one to use the relationships discussed between organic content and partition coefficients. Some examples of these similarities between the NOMs and organic chemical pollutants (OCPs) can be seen as follows (Hopper 1989):

Aromatic NOMs
  Phenylalanine
  Vanillin
  Lignin
  Tannins
NOMs (Sugar)
  Glucose
  Cellulose
  Sucrose
  Pectin
  Starch
Aliphatic NOMs
  Fatty acids
  Ethanol
  Acetate
  Glycine
  Cyanides

Aromatic OCPs
  Benzenes, toulenes
  Xylenes
  Chlorophenols
  PAHs, phenols, napthalenes, phthalates
OCPs (Sugar)
  Cyclohexane
  Cyclohexanol
  Chlorocyclohexanes
  Heptachlor
  Toxaphene
Aliphatic OCPs
  Alkanes
  Alkenes
  Chloroalkenes
  Chloroalkanes
  Cyanides, nitriles, paraffins

The anaerobic dehalogenation of chlorinated organics can sometimes lead to problems of greater health threats not only because of the increased toxicity of the substance itself, but also because of the greater potential for release from the sediment. The particular case in point demonstrated in Fig. 11 indicates that should one choose to "bury" a sediment contaminated by PCE, via isolation techniques within the sediment environment using geomembranes as has been proposed for canal Lachine, for example, subsequent anaerobic degradation would have the capability for releasing the more toxic degradation products—to the detriment of the overlying water and public health.

Acknowledgment

This study was supported by a grant from the Natural Sciences and Engineering Research Council (NSERC) of Canada, Grant A-882.

References

Bohn, M. D. A., Posner, D. M., and Quirk, J. P., 1971, *Soil Chemistry*, John Wiley and Sons, New York.
Bosma, T. N. P., van der Meer, J. R., Schraa, G., et al., 1988, "Reductive Dechlorination of All trichloro- and dichlorobenzene Isomers," *FEMS Microbiol. Ecol.*, Vol. 53, pp. 223–229.
Callahan, M., Slimak, M., Gabel, N., et al., 1979, "Water-Related Environmental Fate of 129 Priority Pollutants," Vol. II, Office of Water Planning and Standards, Office of Water and Waste Management, Washington, DC, U.S. EPA (EPA 440/4-79-029b).
Chan, J., 1993, "A Comparative Study of Three Computerized Geochemical Models," Master of Engineering thesis, McGill University, Montreal, Canada.
Chiou, G. T., Freed, V. H., Schmedding, D. W., and Kohnert, R. L., 1977, "Partition Coefficient and Bioaccumulation of Selected Organic Chemicals," *Environmental Science and Technology*, Vol. 11, No. 5.
Crosby, D. G., 1972, "Photodegradation of Pesticides in Water," *Advances in Chemistry Series*, Vol. 111, pp. 173–188.
Environment Canada, 1991, *The State of Canada's Environment*, Environment Canada.
Farrah, H. and Pickering, W. F., 1976a, "The Sorption of Copper Species by Clays: I. Kaolinite," *Australian Journal of Chemistry*, Vol. 29, pp. 1167–1176.
Farrah, H. and Pickering, W. F., 1976b, "The Sorption of Copper Species by Clays: II. Illite and Montmorillonite," *Australian Journal of Chemistry*, Vol. 29, pp. 1177–1184.
Farrah, H. and Pickering, W. F., 1976c, "The Sorption of Zinc Species by Clay Minerals," *Australian Journal of Chemistry*, Vol. 29, pp. 1649–1656.
Farrah, H. and Pickering, W. F., 1977a, "Influence of Clay-Solute Interactions on Aqueous Heavy Metal Ion Levels," *Journal of Water, Air, and Soil Pollution*, Vol. 8, pp. 189–197.
Farrah, H. and Pickering, W. F., 1977b, "The Sorption of Lead and Cadmium Species by Clay Minerals," *Australian Journal of Chemistry*, Vol. 30, pp. 1417–1422.
Farrah, H. and Pickering, W. F., 1979, "pH Effects in the Adsorption of Heavy Metals Ion by Clays," *Chemical Geology*, Vol. 25, pp. 317–326.
Forbes, E. A., Posner, A. M., and Quirk, J. P., 1974, "The Specific Adsorption of Inorganic Hg(II) Species and Co(III) Complex Ions on Goethite," *Journal of Colloid and Interface Science*, Vol. 49, pp. 403–409.
Galvez-Cloutier, R., 1989, "Clay Suspension as a Buffering System for Accumulation of Lead as a Soil Pollutant," Master of Engineering thesis, McGill University, Montreal, Canada.
Galvez-Cloutier, R., 1994, "A Study of Heavy Metals Accumulation Mechanisms in Lachine Canal Sediments," Ph.D. thesis, Department of Civil Engineering, McGill University, Montreal, Canada.
Galvez-Cloutier, R., Yong, R. N., and Potter, H. A. B., 1993, "The Geochemical Distribution of Heavy Metals and Inorganic Ligands as a Critical Factor in the Clean-up of Contaminated Sediments," *Proceedings, Joint CSCE-ASCE National Conference on Environmental Engineering*, Montreal, pp. I: 711–718.
Hodge, T. and Bubelis, P., 1991, "Great Lakes Basin: Pulling Back from the Brink," *The State of Canada's Environment*, Chap. 18, Environment Canada.

Hoffman, R. F. and Brindley, G. W., 1960, "Adsorption of Nonionic Aliphatic Molecules from Aqueous Solutions on Montmorillonite. Clay Organic Studies II," *Geochimica et Cosmochimica Acta*, Vol. 20, pp. 15–29.

Hopper, D. R., 1989, "Remediation's Goals: Protect Human Health and the Environment," *Chemical Engineering*, August, pp. 94–110.

International Joint Commission, 1989, "1989 Report on Great Lakes Water Quality, Windsor," International Joint Commission, Great Lakes Water Quality Board.

Laflamme, R. E. and Hites, R. A., 1978, "The Global Distribution of Polycyclic Aromatic Hydrocarbons in Recent Sediments," *Geochim. Cosmochim. Acta*, Vol. 42, pp. 289–303.

Lyman, J. L., Reehl, W. F., and Rosenblatt, D. H., 1982, *Handbook of Chemical Property Estimation Methods*, McGraw-Hill Book Co., New York.

Maguire, M., Slavek, J., Vimpany, I., et al., 1981, "Influence of pH on Copper and Zinc Uptake by Soil Clays," *Australian Journal of Soil Research*, Vol. 19, pp. 217–229.

Mansour, M., Scheunert, R., Viswanathan, R., and Korte, F., 1986, "Assessment of the Persistence of Hexachlorobenzene in the Ecosphere," *Proceedings, International Symposium of the International Agency for Research on Cancer, IARC Scientific Publication No. 77*, Lyon, France, pp. 53–59.

Means, J. C., Wood, S. G., Hassett, J. J., and Banwart, W. L., 1982, "Sorption of Amino and Carboxyl-Substituted Polynuclear Aromatic Hydrocarbons by Sediments and Soils," *Environmental Science and Technology*, Vol. 15, No. 2, p. 93.

Metcalf, R. L., Booth, G. M., Schuth, C. K., et al., 1973, "Uptake and Fate of di2-ethylhexyphthalate in Aquatic Organisms and in a Model Ecosystem," *Environmental Health Perspectives*.

Meyers, P. A. and Oas, T. G., 1978, "Comparison of Associations of Different Hydrocarbons with Clay Particles in Simulated Seawater," *Environmental Science and Technology*, Vol. 132, pp. 934–937.

Meyers, P. A. and Quinn, J. G., 1973, "Association of Hydrocarbons and Mineral Particles in Saline Solution," *Nature*, Vol. 244, pp. 23–24.

Mill, T. and Haag, W., 1986, "The Environmental Fate of Hexachlorobenzenes," *Proceedings, International Symposium of the International Agency for Research on Cancer, IARC Scientific Publication No. 77*, Lyon, France, pp. 61–66.

Munakata, K. and Kuwahara, M., 1969, "Photochemical Degradation Products of Pentachlorophenol," *Residue Reviews*, Vol. 25, pp. 13–23.

Nair, N. B. and Aziz, P. K. A., 1988, "Heavy Metal Distribution in the Kadinamkulam Backwater, Kerala (India), *Proceedings, Hazardous Waste Detection, Control, Treatment*, Elsevier Science Publishers, pp. A:909–918.

Natale, O. E., Gomez, C. E., de D'Angelo, A. M. P., and Soria, C. A., 1988, "Waterborne Pesticides in the Negro River Basin (Argentina)," *Proceedings, Hazardous Waste Detection, Control, Treatment*, Elsevier Science Publishers, pp. A:879–908.

Neely, W. B., Branson, D. R., and Blau, G. E., 1974, "Partition Coefficient to Measure Bioconcentration Potential of Organic Chemicals in Fish," *Environmental Science and Technology*, Vol.8, No. 13.

Oliver, B. G., 1984, "Distribution and Pathways of Some Chlorinated Benzenes in the Niagara River and Lake Ontario," *Water Pollution Research Journal, Canada*, Vol. 19, pp. 47–59.

Oliver, B. G. and Nicol, K. D., 1982, "Chlorobenzenes in Sediments, Water, and Selected Fish from Lakes Superior, Huron, Erie, and Ontario," *Environmental Science and Technology*, Vol. 16, No. 8, pp. 532–536.

Oliver, B. G. and Nicol, K. D., 1984, "Chlorinated Contaminants in the Niagara River, 1981–1983," *Science, Total Environment*, Vol. 39, pp. 57–70.

Oliver, B. G., Charlton, M. N., and Durham, R. W., 1989, "Distribution, Redistribution, and Geochronology of Polychlorinated Biphenyl Congeners and Other Chlorinated Hydrocarbons in Lake Ontario Sediments," *Environmental Science and Technology*, Vol. 23, pp. 200–208.

Oliver, B. G. and Pugsley, C. W., 1986, "Chlorinated Contaminants in St. Clair River Sediments," *Water Pollution Research Journal, Canada*, Vol. 21, pp. 368–379.

Phadungchewit, Y., 1990, "The Role of pH and Soil Buffer Capacity in Heavy Metal Retention in Clay Soils," Ph.D. thesis, McGill University, Montreal, Canada.

Pierce, R. H., Gower, S. A., and Victor, D. M., 1980, "Pentachlorophenol and Degradation Products in Lake Sediment," *Contaminants and Sediments*, R. A. Baker, Ed., Vol. 2, pp. 43–56.

Sawyer, C. and McCarty, P., 1978, *Chemistry for Environmental Engineering*, McGraw-Hill Company, New York.

Schnoor, J. L., 1982, "Field Validation of Water Quality Criteria for Hydrophobic Pollutants," *Aquatic Toxicology and Hazard Assessment (5th Symposium), ASTM STP 766*, American Society for Testing and Materials, Philadelphia, pp. 302–315.

Techsult and Roche, 1993, *Environmental Assessment: Lachine Canal Decontamination Project*, Vol. 2.

Tessier, A., Campbell, P. G. C., and Bisson, M., 1979, "Sequential Extraction Procedure for the Speciation of Particulate Trace Metals," *Analytical Chemistry*, Vol. 51, No. 7, pp. 844–851.
Theng, B. K. G., 1974, "The Chemistry of Clay-Organic Reactions," John Wiley and Sons, New York.
U.S. Environmental Protection Agency, 1988, "MINTEQ—Metal Speciation Equilibrium Model for Surface and Groundwater," Center for Exposure Assessment and Modeling, Office of Research and Development, Environmental Research Laboratory, Georgia.
Upper Great Lakes Connecting Channels Study, Management Committee, 1988, "Upper Great Lakes Connecting Channels Study," Vol. 1, Executive Summary.
Wakeham, S. G. and Carpenter, R., 1976, "A Comparative Survey of Petroleum Hydrocarbons in Lake Sediments," *Mar. Poll. Bull.*, Vol. 7, No. 11, pp. 206–211.
Yong, R. N. and Galvez-Cloutier, R., 1993, "Le Contrôle du pH Dans les Mécanismes d'accumulation Lors de l'intéraction de Kaolinite-plomb," *Proceedings, Environnement et Géotechnique*, Paris, pp. 309–316.
Yong, R. N. and Galvez-Cloutier, R., 1995, "Partitioning of Heavy Metals in Contaminated Sediments: A Case Study," paper prepared for ASCE Specialty Conference GEOENVIRONMENT 2000, New Orleans.
Yong, R. N., Galvez-Cloutier, R., and Phadungchewit, Y., 1993, "Selective Sequential Extraction Analysis of Heavy Metal Retention in Soil," *Canadian Geotechnical Journal*, Vol. 30, pp. 834–847.
Yong, R. N., Mohamed, A. M. O., and Warkentin, B. P., 1992, "Principles of Contaminant Transport in Soils," Elsevier Publishing Co., Amsterdam.
Yong, R. N. and Phadungchewit, Y., 1993, "pH Influence on Selectivity and Retention of Heavy Metals in Some Clay Soils," *Canadian Geotechnical Journal*, Vol. 30, pp. 821–833.
Yong, R. N. and Sheremata, T. W., 1991, "Effect of Chloride Ions on Adsorption of Cadmium from a Landfill Leachate," *Canadian Geotechnical Journal*, Vol. 28, pp. 378–387.
Yong, R. N., Warkentin, B. P., Phadungchewit, Y., and Galvez, R., 1990, "Buffer Capacity and Lead Retention in Some Clay Materials," *Journal of Water, Air, and Soil Pollution*, Vol. 53, pp. 53–67.

DISCUSSION

Dennis W. Lawson[1] *(written discussion)*—This outstanding symposium represents a landmark in the proper handling of contaminated sediments and it is important that the nature and intent of associated government processes be fully appreciated. At the formal closing of the symposium, some disparaging comments of admittedly perhaps a harmless nature were inadvertently cast on governments in general, and Canadian governments in particular, which were probably inappropriate to an international symposium. Some commentary is warranted as ASTM is held in high regard throughout society.

The closing remarks were presumably of a personal nature and would likely not be taken as reflecting the positions or opinions of ASTM, sponsoring agencies, or the speaker's own affiliation. The comments of concern were undoubtedly well intended and could probably be taken in jest, or as an example of over-exuberant academic flair. Nevertheless, the remarks were unnecessary and could be construed as detracting from the overall goodwill and better understanding which professional meetings are intended to foster.

With the possible exception of one otherwise stimulating presentation that has already been discussed, the presentations in symposium Session III on Management Strategies served to illustrate the necessity of appropriately dealing with public concerns in a prudent manner. This is done within well conceived government efforts to engage technical professionals in addressing scientifically complex, expensive, emotional, and often polarizing public issues. The quest is the common good as reflected in the consensus or compromise that arises from

---

[1] Environment Canada, Prairie and Northern Region, Regina, Saskatchewan, Canada S4P 4K1.

the expression of individual perspectives. One such collection of perspectives will be the proceedings of this symposium. ASTM embodies voluntary professional efforts of an analogous nature to government processes for conducting public hearings.

Canadian and American governments have established world renowned procedures for effectively dealing with disputes along their common border. Federal and state/provincial governments have also been working hard to harmonize their processes for environmental impact assessments and ensuing regulatory actions. Closing symposium remarks that acknowledged the difficulties that confront the streamlining of these cumbersome democratic mechanisms could have served a useful societal purpose. It is hoped that attention can be directed towards concluding all future such symposia with inspiring summary remarks that encourage the better resolution of environmental issues.

Government processes for conducting environmental impact assessments are mere extensions of time-honored and time-consuming diplomacy and civility that technocrats must learn to observe, analyze, follow, tolerate, reformulate, improve, trust, respect, protect, and pass on to future generations of technical graduates. Our symposia should set a good example of professional dialogue leading to effective dispute resolution. In choosing between sincerity and politeness we need to exercise the discretion that is appropriate to each unique situation, and there is usually no clear choice, other than, perhaps, to take some middle ground. It is really not so much what we say, but how we approach and compose the saying of it. By maintaining our professional decorum we can help to preserve and improve important social institutions.

David M. Petrovski[1]

Use of Bathymetry for Sediment Characterization at Indiana Harbor

REFERENCE: Petrovski, D. M., "**Use of Bathymetry for Sediment Characterization at Indiana Harbor,**" *Dredging, Remediation, and Containment of Contaminated Sediments, ASTM STP 1293,* K. R. Demars, G. N. Richardson, R. N. Yong, and R. C. Chaney, Eds., American Society for Testing and Materials, Philadelphia, 1995, pp. 40–49.

ABSTRACT: In 1992, U.S. EPA, Region 5, sampled sediments within the Federal Navigation Project at Indiana Harbor, IN. Lack of a disposal site has precluded dredging since 1972, resulting in the accumulation of over 750 000 m^3 of highly contaminated sediment. The Federal Project covers approximately 1.08 km^2 of both enhanced and secondary sediment accumulation. The purpose of the sampling effort was to characterize these sediments under Subtitle C of the Resource Conservation and Recovery Act (RCRA). Several approaches common to regulatory characterization were considered and rejected in favor of a bathymetry based procedure. Bathymetric surveys were used to identify 14 areas of thick sediment accumulation. Such areas are indicative of reduced water velocities which favor the accumulation of finer-grained sediment having a strong tendency to be associated with higher contaminant concentrations. Samples obtained from these locations should contain contaminant concentrations that exceed the mean concentrations for the project sediments. Consequently, a regulatory decision based upon these samples should be conservative. Bathymetry may provide a mechanism to reduce the number of samples necessary to characterize large sediment volumes, while maintaining an acceptable level of confidence in any derived regulatory decision.

KEYWORDS: Indiana Harbor, sediment characterization, bathymetry, regulatory determinations, environmental media, navigation projects, dredging

Introduction and Purpose of the Characterization Effort

The Federal Navigation Project at Indiana Harbor is an artificial deep-draft navigation channel and sheltered harbor located in Lake County, IN on the southern shore of Lake Michigan (Fig. 1). Initially a drainage ditch, the Federal Project was originally authorized by the River and Harbor Act of 1910. Subsequently, the Federal Project has been repeatedly widened, deepened, and dredged to maintain authorized navigational depths. Sediments that enter the Grand Calumet River and the Indiana Harbor Canal tend to accumulate in the artificially deepened federal navigation channel, reducing depths and ultimately restricting navigation traffic. In order to maintain adequate navigational depths, the U.S. Army Corps of Engineers (U.S. ACOE) is authorized to dredge these sediments when necessary.

From 1955 to 1972, approximately 75 000 m^3 of sediments were dredged annually from the Federal Project at Indiana Harbor (U.S. ACOE 1986). Until 1966, dredged materials were dumped directly into Lake Michigan at approved open-lake disposal areas. During the next several years, maintenance dredgings were placed at several lake-fill disposal sites in the vicinity of the project. Since 1972, problems associated with the identification of an acceptable disposal site have precluded dredging. This has resulted in the accumulation of over

---

[1] Environmental scientist, U.S. Environmental Protection Agency, Region 5, Chicago, IL 60604.

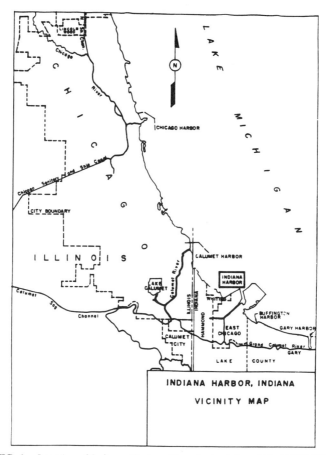

FIG. 1—*Location of Indiana Harbor. Map obtained from U.S. ACOE 1986.*

750 000 m³ (one million cubic yards) of highly contaminated sediments within the limits of the Federal Project.

As the Federal Project at Indiana Harbor has not been dredged since 1972, it is believed that the federal channel is no longer functioning as a trap for the sediments that enter the project. In essence, sediment input into the Federal Project ultimately results in sediment output to Lake Michigan. This is supported by a chronologic review of the bathymetric surveys for the project. These surveys reveal that the rate of sediment accumulation in the project was greatest between 1972 and 1980. Sediment accumulation rates after 1980 have decreased notably (U.S. ACOE 1986). Currently, sediments that would settle within the limits of the project if dredged to authorized depths are discharging to the lake. The highly contaminated nature of the sediments make their release to Lake Michigan undesirable. Contaminants that enter Lake Michigan are quickly dispersed by wave action and near-shore currents, rendering subsequent capture and remediation unlikely. The U.S. ACOE has estimated that 75 000 to 150 000 m³ of contaminated sediments are currently being discharged from the mouth of Indiana Harbor annually (U.S. ACOE, in press). According to the U.S. ACOE, restoring and maintaining the navigation channel at authorized depths would create a sediment trap capable of reducing this release rate to Lake Michigan by 50 to 70% (U.S. ACOE, in press).

The project sediments impact the overlying water column and the lake through the release of contaminants into solution from *in-situ* and resuspended sediments, and the suspension and the migration of the contaminated sediment particles themselves (U.S. ACOE, in press). These processes are enhanced through the turbulence caused by ship traffic, point source discharges along the Project, and storm events. It should be noted that the surface hydraulics of the project are predominantly artificial. The U.S. Geological Survey has estimated that over 90% of the dry weather flow in the Grand Calumet River and the Indiana Harbor Canal is from municipal and industrial discharges (USGS 1993).

Recognizing the environmental significance of dredging and maintaining the authorized depths of the Federal Project at Indiana Harbor, the United States Environmental Protection Agency (U.S. EPA), Region 5 signed a Memorandum of Understanding (MOU) with the Chicago District of the U.S. ACOE in October 1991. The MOU designated the U.S. EPA as a cooperating agency for the dredging project. In conformance with its role as a cooperating agency, Region 5 set out to determine the regulatory status of the project sediments under Subtitle C of the Resource Conservation and Recovery Act (RCRA) via the Toxicity Characteristic Leaching Procedure (TCLP).

The purpose of this paper is to describe the methodology used for this characterization effort. The methodology used reduced the number of samples collected, while maintaining an acceptable level of confidence in the regulatory determination. This approach may have generic application for the characterization of large sediment volumes blanketing extensive previously dredged areas.

The Federal Navigation Project at Indiana Harbor

The Federal Project at Indiana Harbor covers approximately 1.08 km^2 (266 acres). As shown on Fig. 2, the project is divided by the U.S. ACOE into two sections: the harbor portion and the canal portion. The authorized depth of the outer harbor is 8.5 m (28 ft), while the authorized depth of the channel segment of the harbor is 8.2 m (27 ft). The authorized depth of the entrance channel of the project is 8.8 m (29 ft). The surface area of the outer harbor and the entrance channel is 0.74 km^2 (183 acres). The channel portion of the harbor is approximately 80 m in width and 1100 m in length. The second section of the project, the canal portion is approximately 50 m in width and is 4300 m in length, with an authorized depth of 6.7 m (22 ft). All depths are referenced to the International Great Lakes Low Water Datum, which is 576.8 ft (175.8 m) above the mean water level at Father Point, Quebec (U.S. ACOE, in press).

Nature of the Project Sediments

The area surrounding the Federal Project at Indiana Harbor is home to one of the most significant concentrations of heavy industry in the world. The Grand Calumet River and the Indiana Harbor Canal drain approximately 174 km^2 (67 miles^2) of highly urbanized and industrialized areas in northwest Indiana (U.S. ACOE 1986). Due to previous sediment sampling efforts and the highly urbanized and industrial nature of the watershed, the bottom sediment in all portions of the project are known to be associated with a variety of contaminants. These contaminants include free phase oil, polychlorinated biphenyls (PCBs), polycyclic aromatic compounds (PAHs), volatile organic compounds (VOCs), and heavy metals. Dry weight oil and grease concentrations for 0.91 m (3 ft) core samples collected by the

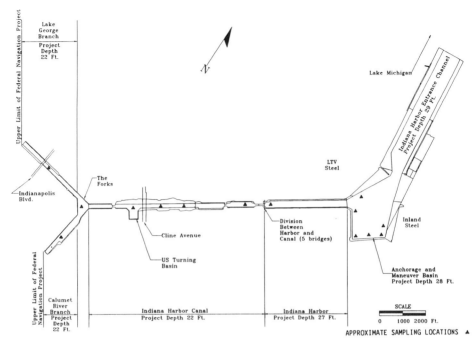

FIG. 2—*Federal Navigation Project at Indiana Harbor with the approximate sediment sampling locations indicated. Duplicate field samples were obtained for the sampling points in the Lake George Branch and the southern corner of the outer harbor. Map obtained from U.S. ACOE 1994.*

U.S. ACOE in 1979 from 13 locations within the project averaged 56 146 mg/kg (U.S. ACOE 1986).

Most of the project sediments consist of silt and clay, which are supplemented with increasing amounts of sand and gravel in the outer harbor as the lake entrance is approached. Generally, the U.S. ACOE has noted that the sediments along the central sections of the canal contain coarser-grained materials than the canal edges (U.S. ACOE 1986).

Current contaminant sources for the Federal Project include: municipal and industrial discharges, combined sewer overflows, runoff from urban and industrial areas, contaminated sediment migration from the upstream river reaches into the limits of the project, potential erosion of contaminated soil and fill along the unrestrained portions of the channel banks, and the discharge of contaminated groundwater. Municipal and industrial discharges, combined sewer overflows, and urban runoff have been estimated to contribute over 110 000 m^3 of sediment to the Indiana Harbor/Canal and Grand Calumet River annually (U.S. ACOE, in press). Many of the sources of contamination lie upstream of the Federal Project. Significant portions of the existing contamination are considered to be the result of spills and point-source releases which predate most of the current federal and state environmental legislation including the Clean Water Act. During the early years of this century, the Gary, Whiting, and Hammond areas in Indiana were apparently exempt from the then-existent state restrictions on industrial discharges to surface water (Colton 1985).

Regulatory Status of the Project Sediments Prior to This Characterization Effort

The sediments currently within the limits of the Federal Project have been sampled and analyzed repeatedly by the U.S. ACOE, U.S. EPA, and other organizations (U.S. ACOE

1986, in press). These sampling efforts have resulted in a determination that classified the project sediments as "heavily polluted," under the U.S. EPA 1977 Guidelines for the Pollutional Classification of Great Lakes Harbor Sediments (U.S. EPA, Region 5, 1977). In addition, two reaches of the channel have been found to contain PCB concentrations exceeding 50 ppm, and would be regulated under the Toxic Substance Control Act (TSCA) if dredged. In 1973, the Grand Calumet River and the Indiana Harbor and Canal were designated a Problem Area by the International Joint Commission (IJC) for the Great Lakes. In 1981, the IJC designated Indiana Harbor and the Grand Calumet River as one of the Areas of Concern (AOC) around the Great Lakes (U.S. ACOE, in press).

In 1983, at the request of the Indiana Board of Health, five composite samples of the project sediments were analyzed under RCRA via the Extraction Procedure (EP) Toxicity Test (U.S. ACOE 1986). Superseded by the Toxicity Characteristic Leaching Procedure (TCLP), the EP toxicity test was used until September 1990 under RCRA to define the hazardous waste characteristic for toxicity. All of the EP-toxicity test results for all parameters were found to be below the regulatory thresholds. Consequently, the project sediments were determined to be unregulated under RCRA Subtitle C if dredged. The replacement of the EP-Toxicity Test by the TCLP created the need for the characterization effort described in this paper.

Environmental Media Characterization under RCRA

The RCRA characterization of the project sediments at Indiana Harbor involved several elements that are common to most characterization efforts for contaminated environmental media. These elements include a determination of the sample parameter to be evaluated, a definition of what will constitute a "sample/member" of the population to be characterized, specification of the number of samples required, and identification of the locations where these samples will be collected. Guidance under RCRA regarding the last three issues is provided in Chapter 9 of SW-846 (U.S. EPA 1986). Further recommendations regarding sample spacing, and therefore sample number, are also provided in a State of Michigan guidance document for the "clean closure" of RCRA hazardous waste management units (Michigan Department of Natural Resources 1994). Clean closure requires the removal of all hazardous wastes and hazardous constituents from the unit in question. This generally is documented by sampling after the removal of all material considered to be contaminated has been completed. A discussion of each of these elements as it related to the RCRA characterization effort at Indiana Harbor is provided below.

The RCRA TCLP

As noted above, in 1990 the U.S. EPA finalized the TCLP as the replacement for the EP-toxicity test. The TCLP is used to define the hazardous waste characteristic of toxicity, one of four hazardous waste characteristics provided for under RCRA Subtitle C. An aqueous extraction test, the TCLP is a measure of the potential of a waste to leach contaminants into ground water. In addition to altering the extraction procedure, the TCLP test added 25 organic compounds to the list of regulated parameters. The TCLP methodology is given in Appendix II of 40 CFR Part 261. Dredging sediments that exhibit a hazardous waste characteristic such as toxicity is defined as "generation" and the sediments would require handling in accordance with RCRA Subtitle C.

Due to the documented significance of the sediment contamination (U.S. ACOE 1986) and unknowns regarding the efficiency of the TCLP extraction procedure to mobilize the project contaminants, Region 5 determined that TCLP testing was warranted. In conformance with the region's role as a Cooperating Agency, the Office of RCRA in Region 5 in cooperation with the Indiana Department of Environmental Management (IDEM) conducted the sediment sampling and the data review. The TCLP analysis was done by the Lockheed Analytical Laboratory located in Las Vegas, NV. The regulatory interpretation of the data was provided by Region 5 and IDEM.

Sample Definition and Collection

While stating that regulatory samples should be representative of the medium being sampled and the physical size of the sample should be "as large as practical," SW-846 provides little specific guidance regarding these issues. These factors can be especially critical for contaminated sediments due to the pronounced degree of vertical and lateral heterogeneity exhibited by contaminant concentrations. The RCRA Program in Region 5 had previously decided that sediment samples collected for regulatory purposes should ideally consist of a sediment core sample composited from a recovered sediment interval 1.52 m (5 ft) in length. It should be noted that sites with 1.52 m of sediment sufficiently contaminated to warrant TCLP testing are not common in the Great Lakes. However, since the project at Indiana Harbor had not been dredged for 20 years, adequate intervals of sediment accumulation were known to be available.

Each of the sampling locations selected were associated with at least 1.52 m of sediment accumulation between the sediment-water interface and the authorized project depth. At each sampling location, a 10.2 cm (4 in.) sediment core was collected. The sediment sampling device was extended from the sediment-water interface to 0.61 m (2 ft) below project depth. None of the lengths of the recovered sediment cores were less than 1.52 m or exceeded 2.8 m (109 in.). The interval of recovered sediment resampled for analysis alternated between adjacent sampling sites. For example, if the upper 1.52 m of recovered sediments were composited for analysis at a given location, at the next sampling point, sediments from the lower 1.52 m interval were composited. To document the exact location of each sampling point, a Global Positioning System (GPS) reading was obtained.

The sediment cores were collected using a vibricore sampling unit. A 10.2 cm aluminum coring tube with a dedicated plastic liner was attached to the vibricoring unit. The dedicated plastic liner prevented cross-contamination between the sampling locations. Upon penetration to the desired depth, the coring unit was withdrawn and the plastic-encased sediment sample was removed from the coring tube. After the recovered core length was measured, the plastic liner was cut with minimal disturbance of the adjacent sample. Sediments from all sections of the 1.52 m interval selected for compositing were collected with a decontaminated and dedicated stainless steel spoon and placed in sampling vessels for shipment to the laboratory.

Number of Samples

There are two approaches that are common under RCRA to address this issue. Chapter 9 of SW-846 outlines one of these methods, using the following equation to determine the appropriate sample number (U.S. EPA 1986):

$$n = t^2 s^2/(RT - X)^2$$

where

n = appropriate number of samples,
t = Student's t value for a one-tailed confidence interval, a probability of 0.1 and the appropriate number of degrees of freedom,

s = standard deviation of the sample values,
RT = regulatory threshold, and
X = mean concentration of the sample values.

Use of this equation requires the population mean and the standard deviation to be estimated. This is obviously problematic as these are the parameters that the characterization effort is attempting to define. In addition, if the estimated population mean approaches the regulatory threshold and the estimate of the standard deviation is increased, the number of samples required rises rapidly. Conservative estimates could result in a large number of unnecessary samples and coupled expenditures, while less conservative values would entail additional sampling trips. Obtaining additional sediment cores is costly, as all of the logistical support required for the initial effort must be duplicated for each subsequent event.

Once the number of samples to be collected has been determined, sampling locations would be selected in a manner that would attempt to ensure that the samples were representative of the sediment within the project. This would generally involve a randomization procedure to select specific sampling sites.

The second method that can be used to determine an adequate number of samples is provided under the clean closure guidance for RCRA hazardous waste units (Michigan Department of Natural Resources 1994). This document gives a procedure that results in recommendations for sample spacing/sample density and therefore the total number of samples. These sampling recommendations are generally used to quantify the necessary number of samples needed to document that all of the hazardous wastes and hazardous constituents have been removed from the hazardous waste unit under consideration. The document classifies units on the basis of their surface area. Units greater than 1.21×10^{-2} km^2 (3 acres) in surface area are considered large. For "large" units, the interval between adjacent sampling sites is determined from the following equation:

$$\text{Grid Interval} = \{A\pi/SF\}^{1/2}$$

where

A = area to be characterized, ft^2, and
SF = surface factor (length of the area to be characterized), ft.

NOTE: this equation is dimensionally inconsistent. Input into this equation should be in ft^2 and ft for the area and the surface factor, respectively.

Use of this equation for the Federal Project at Indiana Harbor is complicated by the odd shape of the project. However, a minimal estimate of the required sample number can be obtained by assuming the project is perfectly square in shape. Inserting a length of 3400 ft (1036 m) for the surface factor (length of a side of the square) for a square 1.16×10^7 ft^2 (1.08 km^2) in surface area into this equation results in a grid interval of 103 ft (31.5 m). Placing evenly placed sampling points every 31.5 m on a 1.08 km^2 site requires over 1000 samples. Attempts to account for the actual shape of the project significantly reduce the grid interval and can increase the required number of samples by approximately a factor of 7. Once the number of required samples is known, a rectangular sampling grid is generally superimposed upon the area to be characterized to determine the specific sampling locations.

The analytical costs, field time, and logistical challenges associated with collecting thousands of samples precluded using this approach at Indiana Harbor. The current cost of one TCLP extraction and analysis of the extract for all regulatory parameters commonly exceeds $1000.

Use of Bathymetry Data in Sediment Sampling

As briefly reviewed above, the lack of knowledge regarding the sediment population parameters and the large size of the area to be characterized hindered the use of either of these approaches at Indiana Harbor. Therefore, another sampling methodology was sought. This need was met through the use of project bathymetry (water depth) maps that were obtained from the Chicago District of the U.S. ACOE. The U.S. ACOE periodically conducts bathymetric surveys at Indiana Harbor to determine when maintenance dredging will be required and to assess the volume of material in the project that would need to be removed to recover authorized depths.

Immediately after dredging, a navigation channel will have an artificial and relatively well-defined geometry. However, with time and sedimentation, the geometry of the channel will change, and the water depths initially present after dredging will decrease. Dredging will increase the cross-sectional area of the channel and reduce the water velocities. Consequently, the dredged channel section will function as a preferential collection area for materials entering the project. This has been observed at Indiana Harbor, where the U.S. ACOE noted that the rate of sediment accumulation in the project was greatest immediately after the last dredging episode in 1972 (U.S. ACOE 1986).

Sediment deposition in the channel will not occur uniformly, as some channel portions will be more quiescent and will therefore be associated with higher rates of sedimentation than the more turbulent areas. In addition to precipitation events, significant channel turbulence at Indiana Harbor is caused by ship traffic and outfall releases. The USGS has determined that over 90% of the dry weather flow in the project is from municipal and industrial discharges. The turbulence associated with outfall release events from point sources along the project were observed during previous sampling visits by the author. Review of the bathymetry data indicated that the greatest accumulations of sediment within the project were along the boundaries of the navigation channel and the near-shore portions of the outer harbor. Water depths were greatest and sediment accumulation least along the central sections of the channel and the outer harbor.

Beyond accumulating greater total sediment, the more quiescent areas in the project also accumulate finer-grained material. The presence of coarser-grained material along the more turbulent central portions of the canal and finer-grained material along the more quiescent canal margins has been observed by the U.S. ACOE (U.S. ACOE 1986). For a variety of reasons, the finer-grained materials in most industrial channel or harbor settings will generally be associated with higher concentrations of contaminants than will their coarser-grained counterparts. These reasons include: a greater percentage of clay and silt sized material, increased quantities of carbonaceous material, and higher specific surface values (ratio of sample particle surface area to sample mass). A methodology that can preferentially select locations associated with higher levels of sediment accumulation would also select the more quiescent channel reaches associated with the deposition of finer-grained more contaminated deposits. Contaminant data obtained from sediments collected from these locations should approximate a local contaminant maximum for the adjacent reaches of the project. Laboratory results derived from sediment samples selected in this manner found to be below the applicable regulatory threshold(s) would mitigate the necessity to sample and analyze the nearby less-contaminated, coarser-grained sediments. In addition, a regulatory conclusion based upon results derived from the finer-grained more contaminated sediments should be conservative to some extent.

In essence, this methodology will allow areas associated with enhanced contaminant concentrations to be identified, without knowing the hydraulic details of the channel in question. Upon identification, the sampling effort can focus on these more contaminated sediment accumulations, potentially reducing the required number of samples and associated expenditures. Should sediment samples identified in this manner meet the applicable regulatory criteria, failure of the lesser contaminated adjacent sediments accumulations would be appear unlikely.

The deviation from the precept requiring representative samples outlined in Chapter 9 of SW-846 should be noted. A sample selection methodology based upon bathymetry will yield samples that are nonrandom and biased toward the selection of sediments with elevated contaminant concentrations. The analytical results obtained from these samples will not be representative of the mean contaminant concentration of the project sediments, and the statistical manipulation of the sample results should be approached with caution.

Results

Bathymetry maps for the Federal Project at Indiana Harbor dated 10 January 1992 were obtained from the Chicago District of the U.S. ACOE. Review of the maps resulted in the selection of 14 sediment sampling locations. Each sampling location was identified on the basis of a local water depth minimum and therefore a local maximum for sediment accumulation thickness. During the first week of June 1992, 16 sediment samples were collected from these 14 locations. These consisted of 14 investigative samples and 2 field duplicates. The locations where the sediment samples were obtained are shown on Fig. 2. Field duplicate samples were collected at the sampling point located on the Lake George Branch and the sampling point located in the southern corner of the outer harbor. Each sampling location was associated with at least 1.52 m of sediment accumulation between the sediment-water interface and the authorized project depth. As previously described, if the length of the recovered sediment core exceeded 1.52 m, a 1.52 m interval was selected for TCLP analyses.

With the exception of the duplicate sample obtained from the outer harbor, all of the TCLP results were found acceptable for regulatory use and to be below the applicable regulatory thresholds. As discussed further below, the exception was associated with both a regulatory excursion and a laboratory error. Upon review of the TCLP data, U.S. EPA and IDEM determined that, excluding the sediments represented by this problematic result, all of the project sediments would not be regulated under RCRA Subtitle C if dredged.

The results for the duplicate sample obtained from the southern corner of the outer harbor included an analytical result for benzene of 575 $\mu g/L$, which exceeded the TCLP regulatory threshold of 500 $\mu g/L$. In contrast to the rest of the TCLP results, these data were also coupled with a laboratory calibration error which made their interpretation difficult. The U.S. EPA/IDEM laboratory review emphasized the potential of regulated levels of contamination in this area of the outer harbor and the likelihood that these sediments would fail the TCLP for benzene if properly tested. Therefore, in the absence of further testing, the regulatory interpretation concluded that the sediments represented by these data would require handling as a characteristic hazardous waste under Subtitle C of RCRA if dredged.

One of the sampling locations was later found to be outside the limits of the Federal Project. This sampling location is indicated on Fig. 2, and is the central sampling location between Cline Avenue and the "Division Between the Harbor and Canal." None of the TCLP results for this sample, or the adjacent locations immediately up and down stream, exceeded any of the TCLP regulatory thresholds. Due to the tendency for quiescent areas to be associated with higher contaminant concentrations, the results obtained were viewed as a conservative representation for the materials in the adjacent more turbulent federal channel. For these reasons, further TCLP sampling and analysis in this portion of the project was not considered necessary.

Conclusions

U.S. EPA, Region 5, sampled the sediments within the Federal Navigation Project at Indiana Harbor Indiana in 1992 to determine their regulatory status under RCRA Subtitle C. Dredging has been precluded for over 20 years due to the lack of an acceptable disposal site. This has resulted in the accumulation of approximately 750 000 m^3 of highly contaminated sediments within the 1.08 km^2 Federal Project. Patterns of sediment deposition within the project are extensively determined by ship traffic and point source discharges, resulting in areas of enhanced and secondary accumulation. Commonly used approaches to media characterization were considered and rejected in favor of a procedure based on bathymetry.

Bathymetry data from areas previously dredged can be useful for the identification of locations of enhanced sediment accumulation. Such areas are generally associated with finer-grained sediment and elevated contaminant concentrations. Use of this method is straightforward, and it allows the characterization effort to focus on the more contaminated sediment accumulations. Despite a limited number of samples, a regulatory decision based upon the analysis of sediments identified in this manner should be conservative to some extent. Lesser numbers of samples will accelerate remedial projects and reduce expenditures.

References

Colton, C., 1985, *Industrial Wastes in the Calumet Area, 1868-1970, An Historical Geography*, Illinois State Water Survey, Illinois Department of Energy and Natural Resources.
Michigan Department of Natural Resources, April 1994, *Guidance Document for Verification of Soil Remediation.*
U.S. ACOE, February 1986, *Draft Environmental Impact Statement, Indiana Harbor Confined Disposal Facility and Maintenance Dredging*, Lake County, IN, pp. 8–22.
U.S. ACOE, March 1994, *Draft Indiana Harbor and Canal Sediment Trap Investigation/Sediment Sampling and Probings.*
U.S. ACOE, in press, *Letter Report and Draft Environmental Impact Statement, Maintenance Dredging and Disposal Activities, Indiana Harbor and Canal*, Appendix discussing the "No Action" Alternative, Lake County, IN.
U.S. EPA, Region 5, 1977, *Guidelines for the Pollutional Classification of Great Lakes Harbor Sediments.*
U.S. EPA, November 1986, SW-846, *Test Methods for Evaluating Solid Waste*, third ed., chap. 9.
USGS, 1993, *Geohydrology and Water Quality of the Calumet Aquifer in the Vicinity of the Grand Calumet River/Indiana Harbor Canal*, Northwestern Indiana, Water-Resources Investigations Report 92-4115.

Masami Ohtsubo[1]

Oxidation of Pyrite in Marine Clays and Zinc Adsorption by Clays

REFERENCE: Ohtsubo, M., "**Oxidation of Pyrite in Marine Clays and Zinc Adsorption by Clays,**" *Dredging, Remediation, and Containment of Contaminated Sediments, ASTM STP 1293,* K. R. Demars, G. N. Richardson, R. N. Yong, and R. C. Chaney, Eds., American Society for Testing and Materials, Philadelphia, 1995, pp. 50–57.

ABSTRACT: The oxidation of pyrite in marine clays exposed to a subaerial environment was confirmed and was determined to be well correlated with decreased adsorption of zinc by the clays. The production of sulfuric acid and iron oxide by this oxidation and the accompanying decrease in pH was demonstrated based on an investigation of the chemistry of the marine clay profile and laboratory incubation tests for remoulded clay samples. Both pH decrease and the production of iron oxides reduced the zinc adsorption capability of the clays. This suggests that the zinc adsorbed by the marine clays would be released into the pore water due to exposure of the sediment surface to the atmosphere.

KEYWORDS: marine clays, pyrite oxidation, iron oxide, heavy metal adsorption, zinc

Heavy metals accumulated in marine clay sediments could be a source of ground water contamination when the bay area is subjected to reclamation or the dredged marine clays are spread on the ground. Movements of heavy metals with water in soils require that the metal be in a soluble or suspended phase or associated with mobile particulates. The concentrations of metals in soil solution are affected by the chemical parameters such as pH, ionic strength, ion adsorption sites, and iron oxide content through changes in the amount of the metals retained on clay surfaces (Dowdy and Volk 1983; Harter 1983; Stahl and James 1991b). When marine clays are exposed to an aerobic environment due to dredging or land draining by reclamation, acidification of the clays occurs by pyrite oxidation in the clays (Ritsema and Groenenberg 1993), which would affect the heavy metal adsorption capability of the clays. The objective of this study was to confirm the production of sulfuric acid and iron oxide by the oxidation of pyrite in the marine clays, and to discuss the adsorption of zinc as a function of the pH and iron oxide content of the clays.

Materials and Methods

Materials

The production of sulfuric acid and iron oxide due to pyrite oxidation was demonstrated by an investigation into the chemistry of the marine clay profile at Ariake and Nakaumi and by laboratory tests of the clays. The air-dried clay samples were subjected to zinc adsorption tests with respect to the effects of pH and iron oxide content. The main clay minerals were

---

[1]Associate professor, Department of Agricultural Engineering, Kyushu University, Fukuoka-shi, Japan.

identified to be smectite, kaolinite, and illite for clays at Ariake, and illite, kaolinite, and vermiculite for clays at Nakaumi.

Methods

Pyrite (FeS_2)—10 mL of 30% H_2O_2 solution containing 3 g of potassium chlorate ($KClO_3$) were added to 1 g of the air-dried samples and then heated to dissolve sulfate in the soil. The sulfur in the sulfate was determined by titrating against 0.1M-NaOH using bromocresol purple as the indicator, and was then converted to the percentage of pyrite by weight.

Iron Oxides (mainly Fe_2O_3)—Because ferrihydrite (poorly-ordered iron oxide) was predominant in Ariake clays (Torrance et al. 1986), catecholdisulphonic acid di-sodium salt (Tiron) which selectively dissolves amorphous iron oxides in alkaline medium (Kodama and Jaakkimatien 1981) was employed for the extraction of the iron oxides from the samples.

Sulfuric Acid (SO_4^{2-})—The solutions for the determination of SO_4^{2-} concentrations were obtained by centrifugation and filtration of 10% clay-water suspensions by weight prepared by adding distilled water to the fresh samples. The SO_4^{2-} concentration was measured by a colorimetric method based on the reaction of solid barium chloranilate (commercial CLB tablets) with sulfate ions (Bertolacini and Barney 1957).

pH—The soil pH was measured in a 1:4 (by weight) mixture of soil and distilled water.

Zinc Adsorption—0.1 g of air-dried soil placed in a container was saturated with sodium by three washings with 20 mL of 0.5M $NaNO_3$ solution adjusted to appropriate pH in the range of 4 to 7, and additional three washings with 20 mL of 0.04M $NaNO_3$ solution of the appropriate pH values. After removing the supernatant by centrifugation, 20 mL of the $Zn(NO_3)_2 \cdot 6H_2O$ solution contained 20 μg Zn was added to the soil sample, shaken for 24 h, and the pH of the soil suspensions was finally determined. A clear extract was obtained from the soil suspensions by centrifugation or filtering, and the zinc concentration in the extract was measured by atomic absorption spectroscopy. The same procedure was employed, using the $NaNO_3$ and zinc solutions of pH 6.3, to examine the effect of iron oxide on zinc adsorption capability of Ariake clay samples. The amount of zinc adsorbed by the soil was calculated by extracting zinc present in the extract from the zinc added initially to the soil. The percentage of zinc adsorbed by the soil was expressed as the ratio of the amount of zinc adsorbed against that initially added to the soil.

Results and Discussions

Chemistry of the Clay Profile and Pyrite Oxidation

Figure 1 shows the chemistry of the clay profile at Ariake. This area was reclaimed in 1963. The ground water table was at the depth of 40 cm, and the pyrite and iron oxide contents were extremely different between the zone above the ground water table and the lower zone below this level. The pyrite content was less than 0.1% above 40 cm, while greater than 0.5% below this depth. The iron oxide content was greater than 1% above 40 cm and decreased with depth below 40 cm. The iron oxide content thus appeared to vary inversely with the pyrite content, and was plotted against the pyrite content (Fig. 2). A significant negative correlation was observed between the two parameters. The reduction in the pyrite due to an increase in the iron oxide can be explained in terms of the pyrite oxidation incurred after the reclamation of the clay sediment, which is described by the following equation:

$$4FeS_2 + 15O_2 + 14H_2O \longrightarrow 4Fe(OH)_3 + 16H^+ + 8SO_4^{2-} \qquad (1)$$

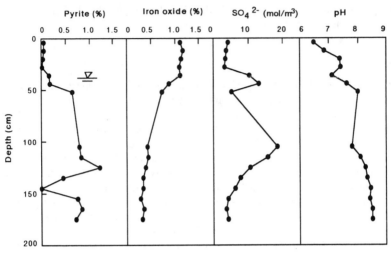

FIG. 1—*The chemistry of Ariake clay sediment.*

According to the equation, the SO_4^{2-} was produced as well as the iron hydroxide due to pyrite oxidation. Since the SO_4^{2-} concentration produced is proportional to the amount of iron hydroxide, the SO_4^{2-} concentrations of 20 to 110 mol/m³, which were estimated based on the iron oxide content in Fig. 1, would have been produced by reclamation. The measured SO_4^{2-} concentrations in Fig. 1, however, gave smaller values than the SO_4^{2-} concentrations estimated. The reasons for this could be a leaching out of SO_4^{2-} and/or a reduction in SO_4^{2-} due to the production of gypsum ($CaSO_4 \cdot 2H_2O$) from the reaction of SO_4^{2-} and calcite ($CaCO_3$) of 1 to 2% present in Ariake marine clays (Kawasaki 1988), which is described by the following equation:

$$CaCO_3 + 2H^+ + SO_4^{2-} + H_2O \longrightarrow CaSO_4 \cdot 2H_2O + CO_2 \qquad (2)$$

The presence of gypsum in Ariake marine clays has been indicated by the differential thermal

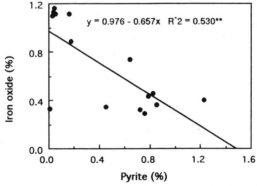

FIG. 2—*The correlation of iron oxide and pyrite contents for Ariake clay.*

analysis of the clays (Kawasaki 1988). The reduced SO_4^{2-} concentrations led to the greater pH values than 6.5 (Fig. 1).

Thus the observed low SO_4^{2-} concentrations and the high pH values in the clay profile are suggested to have occurred in the period after reclamation. To demonstrate the initial high production of SO_4^{2-} and the accompanying pH decrease by pyrite oxidation for marine clays, the changes in pH and SO_4^{2-} concentration of the clays from Nakaumi, when they are exposed to subaerial conditions, are shown in Figs. 3 and 4.

Figure 3 indicates the changes in the pH with time after reclamation at the two different depths of the clay sediment at Nakaumi. The ground water table varied between 40 and 80 cm during the year. The pH of the clay at the 15 cm depth, which was above the ground water table, decreased to values under 4.0 within two years, while the pH of the clays at 70 cm depth remained above 7.0.

Figure 4 indicates the changes in SO_4^{2-} concentration and pH with time obtained from a remolded Nakaumi clay sample that was incubated in containers without changing the water content. The SO_4^{2-} concentration of the sample increased from 3.2 to 13 mol/m^3 while the pH decreased from 7.6 to 4.6 for 12 months. The two parameters thus obtained were plotted in Fig. 5, where a negative correlation was observed between the pH and SO_4^{2-} concentration. These results demonstrated that both iron oxide and SO_4^{2-} were produced by pyrite oxidation in the clay according to Eq 1, leading to the decrease in the pH of Ariake and Nakaumi clays.

Zinc Adsorption

The pH decrease and the production of iron oxides due to pyrite oxidation in marine clays would both affect the adsorption of zinc by the clays; the effects of pH and iron oxide content on the zinc adsorption for Ariake and Nakaumi marine clays are discussed below. Figure 6 indicates the amount and percentage of zinc adsorbed by the clays as a function of pH for Ariake and Nakaumi marine clays, montmorillonite, and kaolinite. For pure clay minerals, the amount of zinc adsorbed was more pH-dependent for kaolinite than for montmorillonite. This can be explained in terms of the difference in pH-dependency of charge where only 5 to 10% of the negative charge on montmorillonite is pH-dependent, whereas

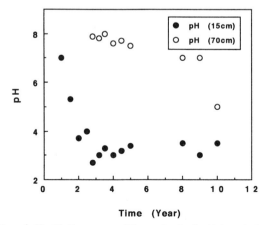

FIG. 3—*The variation of pH with time at two different depths for Nakaumi clay sediment (reproduced from data in Adachi 1992).*

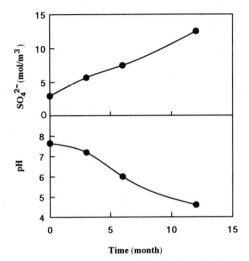

FIG. 4—*The variation of SO_4^{2-} concentration and pH with time for incubated Nakaumi clay samples (reproduced from data in Adachi 1992).*

50% or more of the charge developed on kaolinite can be pH-dependent. For Ariake clay, the zinc adsorbed decreased from 60 to 37% with a decrease in the pH from 6.3 to 4.7. Nakaumi clay also exhibited a marked decrease in the amount of zinc adsorbed from 53 to 11% against a decrease in the pH from 5.8 to 4.4. Thus, the adsorption curves for the two marine clays were largely pH-dependent, similar to the adsorption curve for kaolinite, which can be attributed to the pH-dependency of charge of Ariake and Nakaumi clays.

Although the variation in zinc adsorption with pH can be explained in terms of the change in charge characteristics, in a similar manner as for the alkaline metals calcium, magnesium, potassium, and sodium, the nature of zinc adsorption on pH-dependent charge allows the formation of a bond that is chemical rather than electrostatic in nature. This has been demonstrated by laboratory studies to separate exchangeable and non exchangeable zinc adsorbed by soils where cation exchange capacity (CEC) increased with increasing pH and the zinc was retained predominantly in non exchangeable forms over a pH range of 3 to 8 (Stahl and James 1991a).

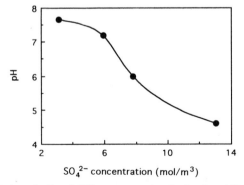

FIG. 5—*The correlation of pH and SO_4^{2-} concentration for incubated Nakaumi clay sample.*

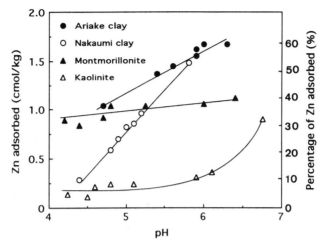

FIG. 6—*The amount and percentage of zinc adsorbed for marine clays and pure clay minerals as a function of pH (data for pure clay minerals from Yamamoto 1983).*

The pH of submerged marine clays is above 7 under sea water (Ohtsubo et al. 1983), but Figs. 3 and 4 indicated that the pH of the clays decreased to below 4.0 within two years when the clay sediment surface was exposed to the atmosphere, and also when remolded clay samples were incubated under aerobic conditions. These observations, along with marked decreases in the amount of zinc adsorbed due to decreases in pH (Fig. 6), suggested that the zinc adsorbed by clays under sea water would be released into the soluble phase due to oxidation of pyrite in the clays when clay sediments were reclaimed or dredged clays were spread on the ground. The increased zinc in the soluble phase could be a potential source of ground water contamination through soil water movement and metal diffusion.

Figure 7 shows the correlation of the percentage of zinc adsorbed and the iron oxide content for Ariake marine clay. A negative correlation exists between the two, where the zinc adsorbed decreased about 20% due to an increase in the iron oxide of only 1%. This can be explained in terms of the electrostatic interaction between the iron oxides and the

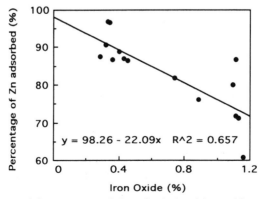

FIG. 7—*The correlation of the percentage of zinc adsorbed and iron oxide content for Ariake clay.*

clay surfaces. The isoelectric point of iron oxides is in the range of pH 7 to 9 irrespective of the difference in the type of iron minerals, leading to the positively charged iron oxide below the isoelectric point. Ohtsubo et al. (1991) indicated from determination of the zeta potential that the negative charge of illite decreased markedly by addition of only 2% poorly-ordered iron oxide. From these findings, a decrease in the amount of zinc adsorbed due to an increase in the iron oxide content (Fig. 7) can be attributed to a reduction in the negative charge of the clay by the precipitation of iron oxide produced from pyrite oxidation. Thus the decrease in the zinc adsorbed through the production of iron oxides, due to the reclamation of marine clays or the spreading of dredged clays on the ground, would release the zinc retained on the clay surfaces into the pore water.

Summary and Conclusions

The production of SO_4^{2-} and iron oxides due to pyrite oxidation was demonstrated based on an investigation of the chemistry of clay profiles and on incubation tests on remolded samples. For the clay profile at Nakaumi, the pH of the clays at the depth of 40 cm exposed to subaerial conditions decreased to below 4.0 within two years. The SO_4^{2-} concentration for the incubated Nakaumi clay samples increased from 3.2 to 13 mol/m^3 while the pH decreased from 7.6 to 4.6 for 12 months.

For Ariake and Nakaumi marine clays, the percentage of zinc adsorbed decreased from 60 to 37% and 53 to 11%, respectively, with decreasing pH at a pH range of 4.4 to 6.3. A negative correlation was observed between the percentage of zinc adsorbed and the iron oxide content for Ariake clay where the zinc adsorbed decreased 20% due to an increase in the iron oxide of only 1%. This suggests that the zinc adsorbed by marine clays would be released into the soluble phase due to exposure of the sediment surface to the atmosphere.

References

Adachi, T., Matsumoto, Y., and Hara, T., 1992, "Process of Acid Sulfate Soil Formation from the Viewpoint of Moisture Conditions on Coastal Muddy Soil," *Transactions of the Japanese Society of Irrigation, Drainage and Reclamation Engineering,* No. 62, pp. 89–96.

Bertolacini, R. J. and Barney, J. E., 1957, "Colorimetric Determination of Sulfate with Barium Chloranilate," *Analytical Chemistry,* Vol. 29, No. 2, pp. 281–283.

Dowdy, R. H. and Volk, V. V., 1983, "Movement of Heavy Metals in Soils," in *Chemical Mobility and Reactivity in Soil Systems,* Soil Science Society of America, American Society of Agronomy, pp. 229–240.

Harter, R. D., 1983, "Effect of Soil pH on Adsorption of Lead, Copper, Zinc, and Nickel," *Soil Science Society of America Journal,* Vol. 47, pp. 47–51.

Kawasaki, H., 1988, "Properties of Sediment of New Creeks in the Lower Chikugo River Basin," *The Bulletin of the Kyushu National Agricultural Experiment Station,* Vol. 25, No. 1, pp. 77–92.

Kodama, H. and Jaakkimatien, M., 1981, "A Comparative Study of Selective Chemical Dissolution Methods for Separating Non-Crystalline Components Produced by Grinding of Silicates," *Proceedings of International Clay Conference,* Bologna, Italy, pp. 399–410.

Ohtsubo, M., Takayama, M., and Egashira, K., 1983, "Relationships of Consistency Limits and Activity to Some Physical and Chemical Properties of Ariake Marine Clays," *Soils and Foundations,* Vol. 23, No. 1, pp. 38–46.

Ohtsubo, M., Yoshimura, A., Wada, S., and Yong, R. N., 1991, "Particle Interaction and Rheology of Illite-Iron Oxide Complexes," *Clays and Clay Minerals,* No. 39, pp. 347–354.

Ritsema, C. J. and Groenenberg, J. E., 1993, "Pyrite Oxidation, Carbonate Weathering, and Gypsum Formation in a Drained Potential Acid Sulfate Soil," *Soil Science Society of America Journal,* Vol. 57, pp. 968–976.

Stahl, R. S. and James, B. R., 1991a, "Zinc Sorption by Iron-Oxide-Coated Sand as a Function of pH," *Soil Science Society of America Journal,* Vol. 55, pp. 1287–1290.

Stahl, R. S. and James, B. R., 1991b, "Zinc Sorption by B Horizon Soils as a Function of pH," *Soil Science Society of America Journal,* Vol. 55, pp. 1592–1597.

Torrance, J. K., Hedges, S. W., and Bowen, L. H., 1986, "Mossbauer Spectroscopic Study of the Iron Mineralogy of Post-Glacial Marine Clays," *Clays and Clay Minerals,* Vol. 34, No. 3, pp. 314–322.

Yamamoto, K., 1983, "Adsorption Characteristics of Copper and Zinc by Allophane Soils," *Journal of Soil Science and Manure, Japan,* Vol. 54, No. 6, pp. 89–96.

Masaharu Fukue,[1] Yoshihisa Kato,[1] Takaaki Nakamura,[1] and Shoichi Yamasaki[2]

Heavy Metal Concentration in Bay Sediments of Japan

REFERENCE: Fukue, M., Kato, Y., Nakamura, T., and Yamasaki, S., "**Heavy Metal Concentration in Bay Sediments of Japan,**" *Dredging, Remediation, and Containment of Contaminated Sediments, ASTM STP 1293,* K. R. Demars, G. N. Richardson, R. N. Yong, and R. C. Chaney, Eds., American Society for Testing and Materials, Philadelphia, 1995, pp. 58–73.

ABSTRACT: Because industry discharge wastes into the sea, marine sediments can be contaminated with various kinds of hazardous and toxic substances. This study discusses how the "degree of pollution" of heavy metals affects the marine sediments from Osaka Bay and Tokyo Bay.
  In this study, the concentrations of various metals, such as manganese, iron, aluminum, titanium, vanadium, copper, phosphorus, etc., were measured from sediment samples obtained from different sites in the bays. However, the results had to be corrected because background concentrations for each metal differ with site location and grain size characteristics. The large difference between background and individual concentrations at various soil depths indicates that the surface layers of the seabed are significantly polluted with some species of heavy metal and other elements.

KEYWORDS: heavy metals, sediments, background value, degree of pollution

In some port and bay areas in Japan, the sediments have been polluted with various kinds of hazardous and toxic substances as a result of industrial development along the coast. If the sediments are dredged, the soils must be separately treated depending on the degree of contamination or pollution. If the sediments are seriously polluted, and if the pollutants can be redissolved easily in water, the soils cannot be directly used as construction materials, but must be treated as hazardous or toxic waste, i.e., treatment for detoxification, remediation, encapsulation, etc. This in turn requires that background levels of contamination be established.

Geochemists have extensively studied the chemical properties of the sea floor for academic reasons. However, the seriousness of the pollution problem has raised added concerns regarding the effects of heavy metals contained in sediments on human health and the environment. Matsumoto and Yokota (1977) measured the concentration of heavy metals and discussed the history of pollution in Tokyo Bay sediments. They found that the fluxes of heavy metals into the sediments after 1960 were three to eight times that of the natural fluxes before 1920. Matsumoto (1983) obtained similar results, and pointed out that Tokyo Bay sediments were seriously polluted with some species of heavy metals but not with iron,

---

[1] Professor and associate professors, respectively, Marine Science and Technology, Tokai University, 3-20-1 Orido, Shimizu, Japan 424.
[2] Aoki Marine Ltd., 6-13-7 Fukushima, Fukushima-Ku, Osaka, Japan 553.

aluminum, and magnesium. He also mentioned that the heavy metal concentration was lower in the top sediment layer as a result of the reduction of the discharged amount of heavy metals since 1970, assuming that background values are almost constant with depth. In his estimation, variation of heavy metal content with sediment mineral composition was not taken into account.

The "degree of contamination" can usually be determined by comparing the background values with measured concentrations of the elements under consideration, but an accurate estimation of background values is required.

General Description of Heavy Metals in Sediments

Concentration of Elements in Noncontaminated Soils

The concentration of elements contained in soils depends on:

(a) mineralogy, including crystal structure,
(b) adsorptive characteristics on the particle surfaces, and
(c) the amount of solutes in pore water.

If there are no contaminants discharged through human activities, the sediments are noncontaminated. Then, the concentration of each metal contained in the sediment is considered to be a background value. However, the amount of background contamination depends on the types of soil minerals and solutes in the pore water. The mineralogy governs the type and amount of metals forming the crystal structure and the adsorptive properties onto the particle surfaces. In general, as a result of isomorphous substitutions, clay minerals can contain greater amounts of some species of metals, such as aluminum, iron, magnesium, etc., in the crystal structure than primary minerals. The adsorption capacity is also greater for clay minerals than for primary minerals. Therefore, the background of some heavy metals is usually higher for clayey soils than sandy soil and gravel.

Definition of "Degree of Pollution"

The discharge of heavy metals increases the amount of heavy metals in the sediments, especially on the surfaces of particles and in pore water. The soil is then called "contaminated" or "polluted" with higher heavy metal concentration.

This study defines the degree of pollution (P_d) for a given heavy metal as

$$P_d = \frac{\text{concentration (Cm)} - \text{background (B)}}{\text{background (B)}} \qquad (1)$$

For sediments, the background for a heavy metal is usually taken as the mean concentration below the contaminated soil layer, as illustrated in Fig. 1. However, it should be noted that the background often depends on sediment mineralogy. If the mineralogy of a sediment varies with depth, it is obvious that the background may also vary with depth.

Method, Results, and Discussion

Measurement of Heavy Metal

Sediment samples were sieved through mesh with 105 μm openings and were dried at 110°C for 24 h. The dry samples were then ground and element concentrations were mea-

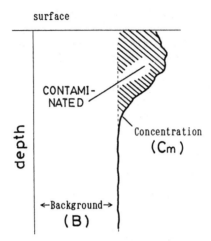

FIG. 1—*General profile of heavy metal concentration and definition of background.*

sured. Approximately 0.5 g of sediment samples were completely dissolved with procedures, as shown in Fig. 2. Then, the concentrations of the elements were determined using the Inductively Coupled Plasm Emission Spectroscopy method, i.e., by atomic absorption spectrometry. The method is commonly used in geochemistry.

Tokyo Bay Sediments

Metal concentrations were measured from four core samples obtained from the different sites in Tokyo Bay area, as shown in Fig. 3. At the A-1 site, the soil sample was obtained using a boring technique with a thin walled sampler. The sampling depth was 23 m. The other samples obtained from the bottom of Tokyo Bay (T-1, T-2, and T-4), were obtained using a piston core sampler. The sampler consists of a cylindrical tube with a diameter of 82 mm. The sampling soil depth at the sea sites was shallow, i.e., approximately 100 cm. With this technique, samples are little disturbed (Nakamura et al. 1993).

The concentration profile of each element measured at the A-1 site is shown in Fig. 4. The substrate is called the Yurakucho layer, which is an alluvial deposit that started depositing about 18 000 years ago, mostly under sea conditions.

Since the deeper soil layer was formed many thousands of years ago, it is considered that the concentration of elements at depth is the background which has not been influenced by human activities. However, as can be seen in Fig. 4, the background values vary with depth. The typical example can be seen in the profile of aluminum. Since aluminum is contained in the clay crystal structure, the concentration may primarily depend on the type of clay minerals.

Table 1 shows the classification of the soils in Site A-1. The soil is tentatively classified into eight layers, i.e., I to VIII. The characteristics of the soil particles can be represented by the grain size in terms of D_{60}, D_{30}, D_{10}, and liquid and plastic limits, as shown in Table 1. Figure 5 shows a correlation between D_{60}, grain size at 60% finer, and the concentration of aluminum for the Site A-1 soil. In Fig. 5, the open circles are considered to be background values and the solid circles show the concentration of aluminum for the top contaminated soil layer. The figure shows that the greater the grain size, the lower the background, and that the concentration is relatively high in the contaminated sediments. A similar trend is

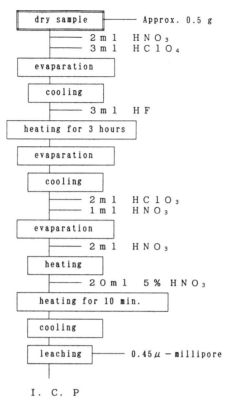

FIG. 2—*Procedure for the measurement of metal concentrations.*

obtained in terms of the correlation between the liquid limit and the background, as shown in Fig. 6. The background increases as the liquid limit increases. This is because the greater the grain size the lower the liquid limit. A similar trend can be obtained for the relationship between background and clay content.

From Fig. 4 and Table 1, it is seen that the concentration of aluminum varies from 60 g/kg for sandy silts to 90 g/kg for silt. This trend is also seen for the profile of iron. Thus, this trend might result from the variation in mineralogy with depth. On the other hand, we believe that the concentration of other elements is primarily dependent on adsorption onto the surfaces of particles and solutes in the pore water (Yong et al. 1992).

At a depth of 11 m, the soil contains relatively large sand fractions. This resulted in a reduced aluminum concentration because of a lower clay content, whereas the concentrations of titanium, copper, and zinc, considered to be adsorbed heavy metals, are relatively high. For example, the concentration of zinc is the highest at this elevation through the Yurakucho layer. We surmise that some heavy metals are high in this sandy layer because ground water seepage contributed to the transport of the heavy metals.

The variation in the background values can be obtained approximately as indicated by the broken lines in Fig. 4. Therefore, the degree of contamination or pollution can be evaluated by the comparison between the background indicated by the broken line and the measured concentration at the same elevation.

62 CONTAMINATED SEDIMENTS

FIG. 3—*Sampling sites in Tokyo Bay area.*

In Fig. 4, the concentration of all the elements except manganese is relatively high at the thin layer below the fill with a thickness of 1.35 m. This fact obviously indicates that the surface of the substrate is contaminated by discharge through human activities. The concentration of manganese is relatively low at the surface in comparison to the background. This may be due to relatively rapid chemical reactions, oxidation and/or reduction.

In the bay sediments (Figs. 7 and 8), the situation is similar to those shown in Fig. 4. Figure 7 shows the concentration profiles for the elements measured on the Site T-2 sediments. Background values are again indicated by broken lines which vary with depth. It is important to realize that the background values are lower near the top layer. This means that the top layer is considerably contaminated. If the mean value of the data for relatively deep soils is used as the background, the aspects of contamination are not clearly indicated, because the mean value may be high or low in comparison to the real background value at a given depth. From Fig. 7 we can surmise that the top layer, about 20 cm thick, is significantly contaminated. The background values at this site are more or less similar to those for Site A-1.

Figures 9 and 8 show the concentration profiles for Sites T-1 and T-4, respectively. The profiles for both the sites show a similar trend. The background values vary with depth and grain size distributions. In Fig. 8, the profile of aluminum concentration obviously shows that in the top layer, with a thickness of about 20 cm, the background varies rapidly with depth. The degree of contamination is lower in comparison to the Sites A-1 and T-2 sediments. Thus, background values vary with sediment type, i.e., with depth and site.

Osaka Bay Sediments

Core samples were obtained from two sites in Osaka Bay, as shown in Fig. 10. The concentration profiles for the measured elements are shown in Figs. 11 and 12. Background

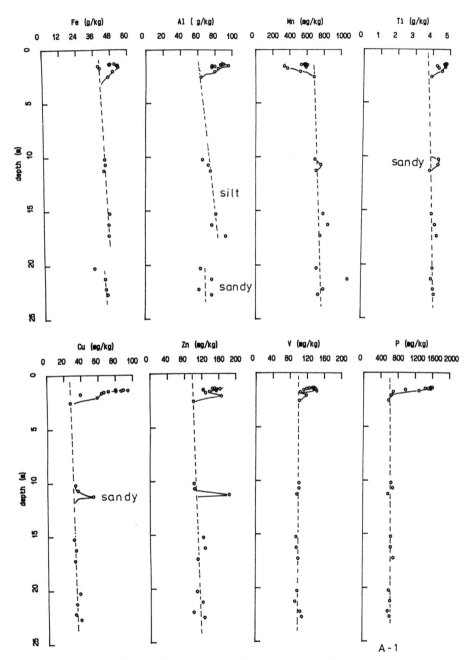

FIG. 4—*Concentration profiles of elements at Site A-1.*

TABLE 1—Soil classification at Yurakucho layer.

No.	Depth, m	Classification	W.C., %	Liquid Limit, %	Plastic Limit, %	D_{10}, mm	D_{30}, mm	D_{60}, mm
I	0–1.3	Artificial fill: sand, gravel, asphalt water table: 1.20 m	—	—	—	—	—	—
II	1.3–2.1	Silt to sandy silt	45–52	—	—	—	0.022	0.05
III	2.1–3.5	Sandy silt	42	39.6	21	0.003	0.032	0.097
IV	3.5–6.7	Fine sand with silt	—	—	—	—	—	—
V	6.7–11.0	Soft silt with sand	56–65	70–81	29–35	—	0.0034	0.018
VI	11.0–12.4	Sandy silt	40–54	48–57	26	—	0.016	0.065
VII	12.4–18.7	Soft silt	60–70	60–92	26–46	—	0.0015	0.018
VIII	18.7–24.0	Silty sand or sandy silt	—	—	—	—	0.013	0.045

W.C.: water content, D_{10}: grain size at 10% finer, D_{30}: grain size at 30% finer, D_{60}: grain size at 60% finer.

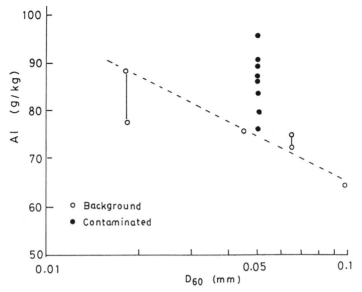

FIG. 5—*Relationship between aluminum background value and grain size at Site A-1.*

values for each element are relatively low in comparison to the Tokyo Bay sediments. For example, surface sediment from Sites O-1 and O-2 revealed copper background values of approximately 8 and 16 mg/kg, respectively, whereas the values in the Tokyo Bay sediments varied from 15 to 50 mg/kg. Each background is different for Sites O-1 and O-2. This may be due to the mineralogical differences between sediments. Evidence for this theory is that the background of aluminum is considerably lower at Site O-1 compared to the Site O-2 and Tokyo Bay. The contaminated layer thickness at Site O-2 is approximately 40 to 50 cm. On the other hand, the contaminated layer is only 20 cm thick at Site O-1.

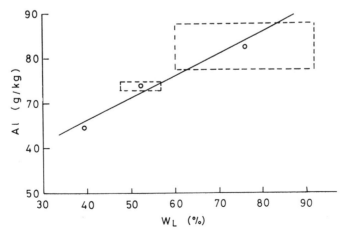

FIG. 6—*Relationship between aluminum background value and liquid limit at Site A-1. Boxes indicates ranges in liquid limit and corresponding background values.*

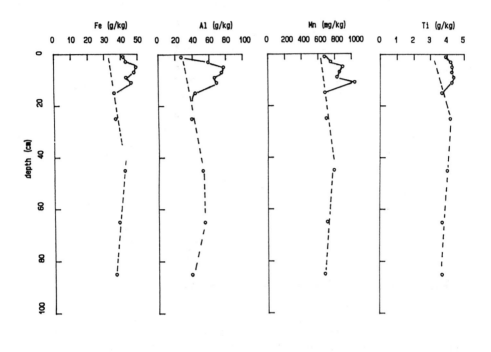
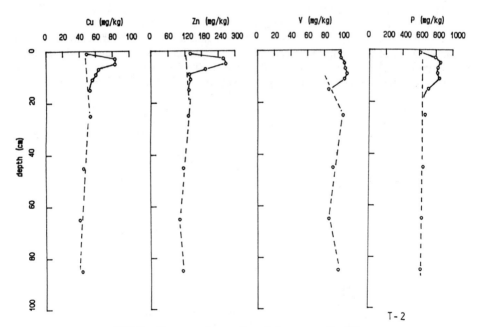

FIG. 7—*Concentration profiles of elements at Site T-2.*

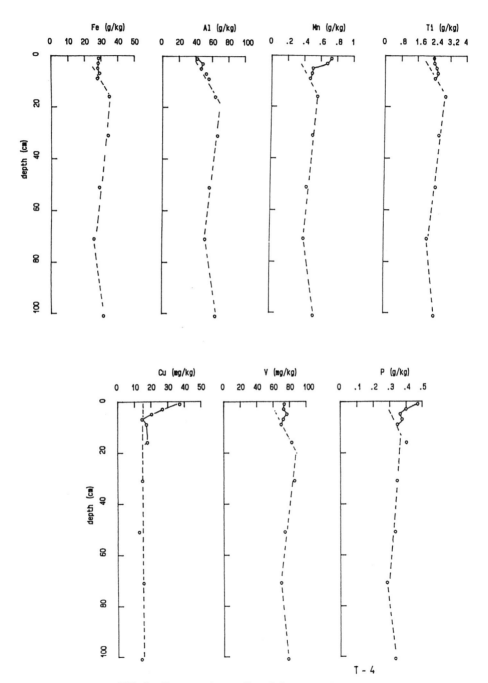

FIG. 8—*Concentration profiles of elements at Site T-4.*

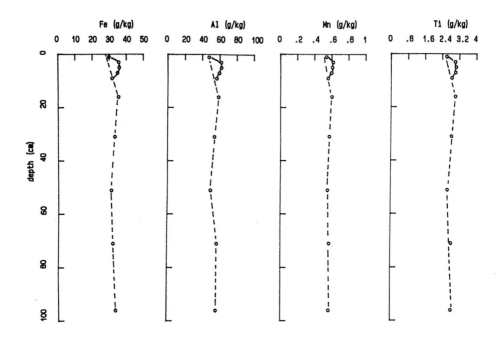
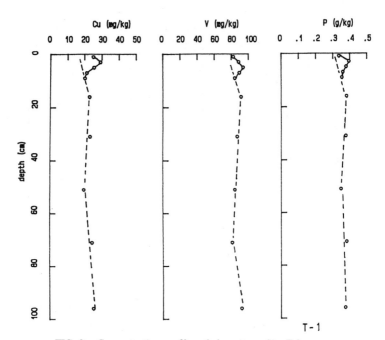

FIG. 9—*Concentration profiles of elements at Site T-1.*

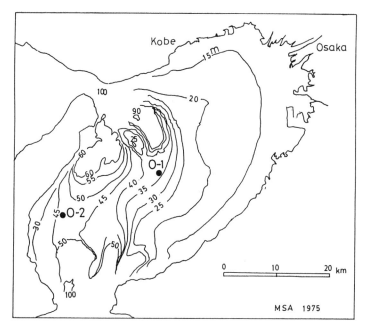

FIG. 10—*Sampling sites in Osaka Bay area. The contours show water depth* (m).

Degree of Pollution

At the same site, the heavy metal concentration in noncontaminated layers vary with soil type, especially grain size characteristics. This may be attributed to the mineralogical aspects as noted previously. In general, for noncontaminated soils, smaller particle sizes are related to higher concentrations. If the variation in depth is neglected and if the mean value is taken as the background, the degree of pollution usually becomes lower.

This study indicates that the background varies with the element species and sampling site, as shown in Table 2. The backgrounds for aluminum and iron are relatively high in comparison to other metals, because these elements are contained in the mineral crystals. From the table, the range of backgrounds for each element is large for Site A-1 sediments. This is due to the large length of the sample, i.e., more variation in minerals due to the deposition during a relatively long duration. Some variation is also due to the amount of metals adsorbed on the particle surfaces, depending on the adsorptive characteristics of the charged particles, under the natural environment.

It may be important to note that the concentration itself does not mean "safety or danger" for human health and environment, unless desorption of heavy metals from the soil particles can occur. Some heavy metals can be strongly retained on the particle surfaces (Yong et al. 1992). Therefore, one can conclude that the background is only the minimum concentration for safety, but not in regard to limiting or allowable concentration.

In Table 2, the degree of pollution for Tokyo Bay and Osaka Bay sedimens is also presented. As can be seen in the table, contamination of some metal species is extremely high. For example, the degree of pollution for copper in Site O-2 exceeds 7. On the other hand, at the same site, the background of aluminum and iron is relatively high but the degree of pollution is relatively low. This indicates that the sediments contain a large amount of clay minerals that have adsorbed a large amount of heavy metals, such as copper. Consequently,

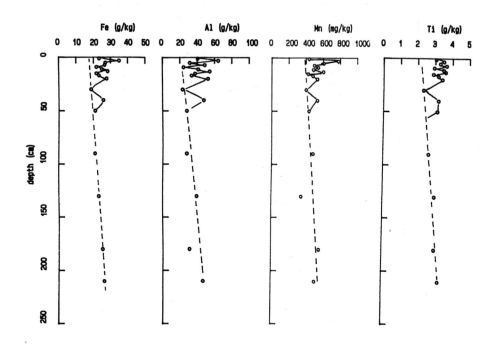
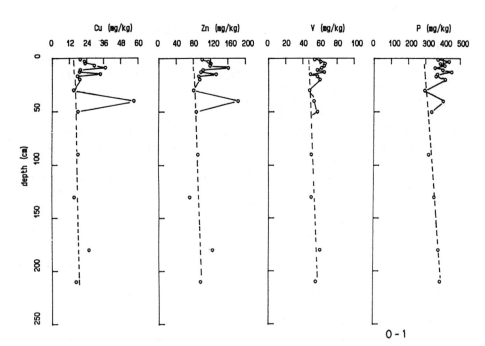

FIG. 11—*Concentration profiles of elements at Site O-1.*

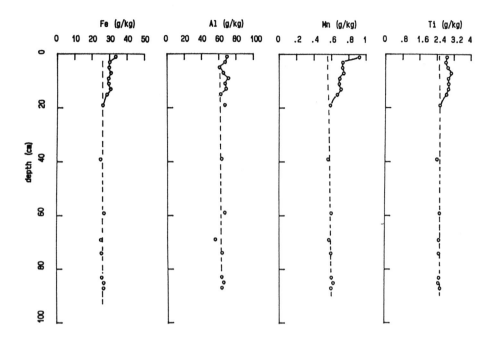
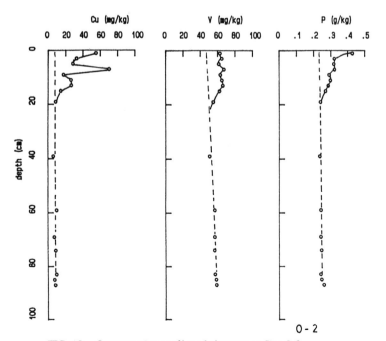

FIG. 12—*Concentration profiles of elements at Site O-2.*

TABLE 2—Background values and degree of pollution (P_d) for various sediments.

Site	Core Length, cm	Fe, g/kg	Al, g/kg	Mn, mg/kg	Ti, g/kg	Cu, mg/kg	Zn, mg/kg	V, mg/kg	P, mg/kg
A-1	2500	38–46	55–90	680–860	3.7–4.4	28–42	95–130	90–100	600–700
	P_d	0.35	0.52	—	0.26	2.39	0.6	0.4	1.67
T-1	100	28–36	45–58	500–610	2.6–3.0	15–27	—	80–90	300–380
	P_d	0.2	0.2	0.15	0.15	0.67	—	0.13	0.25
T-2	85	32–43	26–58	620–790	3.1–4.2	40–53	100–150	95–100	600–650
	P_d	0.43	1.6	0.52	0.23	0.71	1.16	0.32	0.34
T-4	100	22–35	38–65	300–550	1.9–2.9	12–15	—	60–85	260–400
	P_d	0.4	0.4	1.3	0.33	1.46	—	0.15	0.71
O-1	220	17–28	20–50	320–560	2.2–3.3	15–28	80–130	45–60	290–400
	P_d	1.05	1.9	1.0	0.56	2.3	1.1	0.46	0.5
O-2	85	25–28	57–68	550–620	2.4–2.5	7–9	—	45–60	230–250
	P_d	0.31	0.18	0.67	0.20	7.8	—	0.38	0.9

the degree of pollution for aluminum and iron becomes low. However, the amount of these elements discharged by human activities is relatively large, i.e., at most 10 g/kg, whereas the amount of copper at the maximum degree of pollution is only 0.06 g/kg. This fact suggests that we should distinguish the metals into two categories, elements constituting the mineral crystals, such as aluminum, iron, etc., and others representing contaminants.

Conclusion

In order to estimate the degree of pollution, the proper estimation of the background is important. The background values vary with the mineralogy of soil particles which can be represented by the grain size characteristics, liquid limit, clay content, etc. The study shows that coarser particles correlate with lower background values, and that higher liquid limits are related to larger background values.

The top layers of the bay sediments in Osaka Bay and Tokyo Bay are contaminated. Since it is considered that the pollution is not only of heavy metals but also other toxic and hazardous materials, proper treatment of the sediments is required for the cleanup of the bottom.

References

Matsumoto, E. and Yokota, S., 1977, "Records of Pollution in Tokyo Bay Sediments," *Journal of Geochemistry*, Vol. 11, pp. 51–57 (in Japanese).

Matsumoto, E., 1983, "The Sedimentary Environment in the Tokyo Bay," *Journal of Geochemistry*, Vol. 17, pp. 27–32 (in Japanese).

Nakamura, T., Fukue, M., and Naoe, K., 1993, "Effects of Calcium Carbonate on Geotechnical Properties of Sediments in Tokyo Bay," The 3rd International Offshore and Polar Engineering Conference, Vol. 1, pp. 647–651.

Yong, R. N., Mohamed, A. M. O., and Warkentin, B. B., 1992, "Principles of Contaminant Transport in Soils," *Developments in Geotechnical Engineering*, Elsevier, The Netherlands, pp. 143–180.

Dredging Transport and Handling

John B. Herbich[1]

Removal of Contaminated Sediments: Equipment and Recent Field Studies

REFERENCE: Herbich, J. B., "**Removal of Contaminated Sediments: Equipment and Recent Field Studies,**" *Dredging, Remediation, and Containment of Contaminated Sediments, ASTM STP 1293,* K. R. Demars, G. N. Richardson, R. N. Yong, and R. C. Chaney, Eds., American Society for Testing and Materials, Philadelphia, 1995, pp. 77–111.

ABSTRACT: The problem of contaminated marine sediments has emerged as an environmental issue of national importance. Harbor areas in particular have been found to contain high levels of contaminants in bottom sediments due to wastes from municipal, industrial, and riverine sources. This paper briefly examines the extent and significance of marine sediment contamination in the United States, and reviews the state of the art of contaminated sediment cleanup and recent field studies or demonstrations.

KEYWORDS: sediments, contaminants, dredging, specialized equipment, case studies

Contamination of marine sediment in all areas of the world, particularly in shallow water areas, poses a potential threat to marine resources and human health. Improving the capability to assess, manage, and remediate these contaminated sediments is critical not only to the well-being of the marine environment but as well as to its use for navigation, commerce, fishing, and recreation.

There are many definitions for contamination of sediments. The Marine Board of the National Research Council (1989) defined contaminated sediments as follows:

"Contaminated sediments are those that contain chemical substances at concentrations which pose a known or suspected environmental or human health threat."

Many contaminated marine sediments are found along all coasts of the contiguous United States and Canada, both in local "hot spots" and distributed over large areas. There is a wide variety of contaminants including: heavy metals, polychlorinated biphenols (PCBs), DDT, and polynuclear aromatic hydrocarbons (PAHs). At present there are no generally accepted sampling techniques or testing protocols; however, a variety of classification methods are available:

(a) Sediment bioassays-sediment toxicity on a crustacean, infaunal bivalve, and infaunal polychaetes. Essentially marine life is subjected to various levels of toxicity and their survival noted.

---

[1] W. H. Bauer Professor Emeritus and Director Emeritus, Center for Dredging Studies, Texas A&M University, Civil and Ocean Engineering, College Station, TX 77843-3136; vice president, Consulting & Research Services, Inc., P.O. Box 10295, College Station, TX 77842-0295 and 721, South Alu Road, Wailuku, HI 96793-1569.

(b) Sediment quality triad:
 1. contamination quantified by chemical analysis,
 2. toxicity determined by laboratory bioassays, and
 3. benthos community structure determined by taxonomic analysis of biofauna.
(c) Apparent effects threshold technique equilibrium partitioning (AET), a tool for deriving sediment quality values for a range of biological indicators to assess contaminated sediments. The AET is the contaminant concentration in sediment above which adverse effects are always expected for a particular biological indicator.

To ensure that decision making is informed and scientifically based, continued research and use of assessment methodologies should provide information to determine:

- a range of concentrations of chemicals in sediments that will result in biological effects, and
- whether in-place sediments are causing biological impacts.

A tiered approach to the assessment of contaminated sediments should be used. The approach would progress from relatively easy and less expensive (but perhaps less definitive) tests to more sensitive methods as needed (Herbich 1990).

Legislative authority for the management of contaminated marine sediments falls largely under three statutes: (1) the Comprehensive Environmental Response, Compensation, and Liability Act of 1980 (CERCLA), (2) the Marine Protection Research and Sanctuaries Act (MPRSA), and (3) the Clean Water Act (CWA). MPRSA was amended by the Superfund Amendments and Reauthorization Act (SARA) of 1986 and is principally aimed at the cleanup and remediation of inactive or abandoned hazardous waste sites, regardless of location. Reauthorization of the Clean Water Act is being considered by Congress. Superfund sites have been ranked by the U.S. Environmental Protection Agency (EPA) based on the hazard that the contaminants may pose to human health and the environment through releases of contaminants or toxic materials to ground water, surface water, and air.

As a result of legislative responsibility and programmatic interests, a wide variety of federal agencies has shown active interest in this subject. EPA's responsibilities under Superfund and the CWA are the source of its interests in water quality concerns and remediation of uncontrolled hazardous waste sites. The U.S. Army Corps of Engineers (COE) is involved because of its responsibility to dredge and maintain navigable rivers and harbors. The COE also assists in the design and implementation of remedial cleanup actions under Superfund. NOAA has responsibility for assessing the potential threat of Superfund sites to coastal marine resources as a natural resources trustee as well as under its NS&T program. The U.S. Fish and Wildlife Service has legal authority for various endangered coastal species, food chain relationships, and habitat considerations, all of which are potentially impacted by contaminated sediments. The U.S. Navy has had experience in assessing contaminated sediments and now must grapple with such problems in maintaining homeports for Navy vessels.

Although the dredged material management strategy developed by the Corps of Engineers may be relevant to severely contaminated sediments, it is important from a management standpoint to differentiate them from less contaminated sediments. In particular, the most highly sophisticated remedial technologies (i.e., those involving treatment or destruction of associated contaminants) are likely to be cost-effective only in small areas and for sediments with relatively high contamination levels. Sediment contamination problems often involve large volumes of sediments with relatively low contamination levels. As a result, some highly sophisticated technologies may be inapplicable or inefficient for remediating contaminated sediments.

"No action" may be the preferred alternative in cases in which the remedy may be worse than the disease; e.g., where dredging or stabilizing contaminated sediments results in more biological damage than leaving the material in place. Contaminants generally accumulate in depositional zones and, if the source is controlled, new clean sediments will deposit and cap the contaminated material over time. In effect, no action alternatives in such cases may result in natural capping.

The procedures may be summarized as follows:

1. "No action" may be an acceptable option if the contamination degrades or is buried by natural deposition of clean sediment in a relatively short period of time (6 to 12 months). In using the no action strategy as a form of natural capping of contaminated material, consideration should be given to the length of time it takes for contaminants to be isolated from the food chain.

2. In-place capping may be a useful option if the sediments are not in a navigation channel or if ground water is not flowing through the site.

3. Removal and subaqueous burial off-site may be a viable option, although experience with this technique is limited to relatively shallow water [less than 30.5 m (100 ft)].

4. Incineration seems to be viable only for sites with relatively small amounts of sediments containing high concentrations of combustible contaminants.

5. Other techniques to assist in remediation of contaminated sediment may be appropriate in special cases. Examples include a variety of sediment stabilization or solidification techniques, and biological and/or chemical treatment.

6. Additional evaluation should be conducted to determine the applicability of the Corps of Engineers' dredged material management strategy to more severely contaminated sediments.

The purpose of the paper is to provide a review of equipment and methods employed in the removal of contaminated sediments.

Remedial Technologies

From a remediation standpoint, the most important factors are likely to be a definition of the cleanup target, technical and cost feasibility, natural recovery estimates, and ability to distinguish and/or control continuing sources of contaminants.

Dredging technology exists that is capable of greatly reducing turbidity and resuspension during the dredging of bottom sediments; however, special equipment has to be deployed and modified operational methods must be used (Herbich and Brahme 1990; Herbich 1993).

Dredging Equipment

Contaminated marine sediments have been a hot topic during the last five or six years and as cleanup was ordered by the EPA, it became obvious that the most efficient and cost-effective way of cleaning up sediment would be by dredging. In selecting appropriate equipment for the removal of contaminated sediment, mechanical, hydraulic, and pneumatic dredges can be considered (Herbich 1992, 1993).

Mechanical Dredges

Mechanical dredging equipment (such as clamshell dredges, dipper dredges, draglines, and backhoes), offers the advantage of removing the sediment at near its *in situ* density since the minimal amount of water is retained in the bucket containing the dredged material. This advantage benefits disposal operations because less volume is required for the initial storage and less effluent (potentially requiring treatment) is produced. On the other hand, the operating characteristics of mechanical dredges produce low ratings for many of the factors listed herein including sediment resuspension, cleanup precision, costs, and production. Thus mechanical dredges are somewhat limited because of the secondary environmental effects. One piece of equipment developed in recent years is the Cable Arm bucket which reduces the resuspension of the sediment.

Hydraulic Dredges

Hydraulic dredges include the cutterhead, dustpan, sidecast, and hopper dredges. The cutterhead dredge is suitable for the removal of contaminated sediments because of its high production rates, flexibility, operational characteristics, and costs. The disadvantages include:

(a) a large amount of water is added, which is an important factor in the cleanup of contaminated water,
(b) sediment is resuspended by the rotating cutter or auger, and
(c) production rates are decreased when an attempt is made to reduce resuspension of sediments.

Specialty Dredges

A number of specialty dredges have been developed, principally in Italy, the Netherlands, and Japan for the removal of contaminated sediments. These include the Pneuma pump (available in Italy), the Oozer dredge, the Cleanup dredge and Refresher system (available in Japan), the Dutch-designed "Matchbox-head" dredge (now available in the United States) and the horizontal-auger dredges, bucketwheel dredge, and positive displacement pump (either under development or available in the United States). A new combination dredge employing mechanical, hydraulic, and pneumatic concepts has been developed in Japan. This high-density pneumatic conveying system has been used in Kumamoto Port, Japan.

In this paper several "special purpose" dredges are described and their capabilities discussed.

To achieve an efficient operation, the selection of proper dredging equipment for any project is important. In the case of contaminated sediments, it is even more important, because any additional contamination generated during dredging must be avoided. The selection depends on a number of factors, including:

(a) characteristics of sediments,
(b) quantity of sediments to be removed,
(c) degree of contamination,
(d) toxicity of contaminants,
(e) location,
(f) environmental conditions at the site (waves, currents, tides, etc.),
(g) distance to the disposal site,
(h) type of disposal,
(i) availability of particular equipment,
(j) resuspension rates during dredging, and
(k) production rate.

There are several types of dredges that may be placed in a "special purpose" category or are specially designed for dredging contaminated sediments (Herbich 1990–1992). These are listed below under mechanical, mechanical-hydraulic, hydraulic, and pneumatic type:

(a) Mechanical type
  1. Enclosed clamshell bucket
  2. Cable Arm clamshell bucket
(b) Mechanical-hydraulic type
  1. Cutterhead dredge
  2. Mud Cat
  3. "Cleanup" system
  4. High-density dredge
  5. Scraper dredge
(c) Hydraulic type
  1. "Refresher"
  2. "Matchbox"
  3. "Wide Sweeper" cutter-less dredge
  4. Vibrating auger-positive displacement pump
(d) Pneumatic
  1. "Pneuma"
  2. "Oozer"
  3. Agitation and emulsion dredge

In addition, an encapsulation method may be considered which provides for a dredge to inject material into silt deposits.

Mechanical-Type Dredges

Enclosed Clamshell—The Japanese have developed a "watertight" clamshell bucket for use with grab bucket dredges. An evaluation of the watertight bucket was made by the U.S. Army Engineer Waterways Experiment Station in 1982 (Fig. 1) (Hayes et al. 1984). Experiments conducted at the Jacksonville District indicated that the watertight bucket significantly reduced water column turbidity and did not decrease production.

Figure 2 shows the benefit of using an enclosed bucket. Operation of the dredge can be modified slightly to reduce sediment resuspension by slowing the raising and lowering of

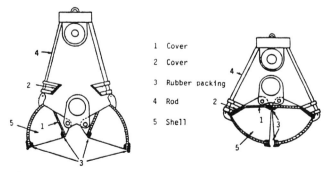

FIG. 1—*Open and closed positions of the watertight clamshell bucket.*

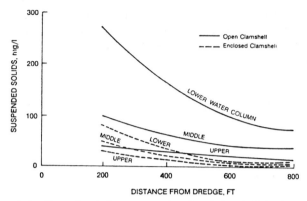

FIG. 2—*Resuspended sediment levels from open and enclosed clamshell dredge operations in the St. Johns River.*

the bucket through the water column. It must be noted that this operation modification reduces the production rate of the dredge; generally high unit costs are associated with this type of mechanical dredging.

Field tests were conducted by the U.S. Army Corps of Engineers in the St. Johns River, Jacksonville, FL; the James River, VA; Black Rock Harbor, CT; and the Thames River, CT.

Results obtained in the St. Johns River indicate that the enclosed bucket offers a marked advantage over the open clamshell dredge in the upper water column, but not near the bottom. Plume characteristics at the St. Johns River indicate the highest sediment concentration for the enclosed clamshell was 360 mg/L. At Black Rock Harbor the maximum sediment concentrations ranged from 111 to 761 mg/L; the highest value was measured 30.5 m (100 ft) from the source. Concentrations between 0 and 111 mg/L were observed as far as 244 m (800 ft) from the source. Further details are presented in a later section.

Cable Arm Clamshell Bucket—The design of the cable-arm clamshell bucket differs from the conventional clamshell buckets, as the sweep of the bucket is controlled by the employment of cables (Fig. 3). One main cable controls the descent of the bucket, four spreader cables control the opening of the shell, and another cable closes the shell and lifts the bucket.

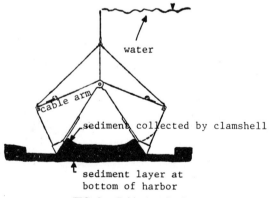

FIG. 3—*Cable Arm bucket.*

To reduce resuspension of the sediment, a venting system is constructed to allow water and air to pass through the bucket during its descent. The Cable Arm bucket was deployed by the Environment of Canada (Buchberger et al. 1993) during the Toronto Harbor demonstration. An improvement in the design was made possible through the installation of an inner deflecting plate that restricted lateral movement and a greater quantity of sediment was retained within the bucket as it closed.

Mechanical-Hydraulic Type Dredges

Mud Cat—The Mud Cat has a horizontal cutterhead equipped with knives and spiral augers that cut the material and move it laterally toward the center of the augers where it is picked up by the suction. The dredge can remove sediments in a 2.4 m (8 ft) width and up to water depths of 4.6 m (15 ft). The dredge operates on anchors and cables; the manufacturer claims that the dredge leaves the bottom of the dredged area flat and free of windrows that are characteristic of a typical cutterhead and hopper dredge operation (Fig. 4).

By covering the cutter-auger combination with a retractable mud shield, the amount of turbidity generated by the Mud Cat's operation can be reduced.

The Mud Cat was developed in the United States and has been used extensively in removing sediments from reservoirs and disposal ponds. Production is relatively low and depth limitation is 4.6 m (15 ft). There are units available capable of dredging from a 7.6 m (25 ft) depth. Resuspension of sediment may be classified as low to moderate; field test results are presented in a later section. A remotely controlled unit is also available.

Cleanup System—To reduce or minimize resuspension of the sediment, Toa Harbor Works, Japan, has developed a unique "Cleanup" system for dredging highly contaminated sediment (Sato 1984). The Cleanup head consists of a shielded auger that collects sediment when the dredge swings back and forth; the auger guides the sediment toward the suction of a submerged centrifugal pump (Fig. 5). To minimize sediment resuspension, the auger is shielded and a movable wing covers the sediment as it is being collected by the auger. Sonar devices indicate the elevation of the bottom. An underwater TV also indicates the amount of material being resuspended during a particular operation. Figure 6 shows details of a shielded auger. Fairly large volumes (2 200 000 m^3 up to 1981) have been excavated by Cleanup dredges in soft muds and sand containing various contaminants such as mercury, cadmium, PCB, oily, and organic substances.

Cleanup dredges have been used extensively in Japan. In 1981 there were four such dredges having centrifugal dredge pumps; one dredge was equipped with a Pneuma pump. The resuspension of the sediment may be classified between low and moderate. Several dredges that have been developed specifically for cleanup of contaminated sediments are manufactured outside of the United States and thus are unavailable under the Jones Act

FIG. 4—*Mud Cat dredge (shield over the auger head raised to show the augers).*

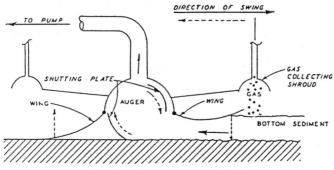

FIG. 5—The "Cleanup" system.

restrictions (46 U.S.C. §883). (The Jones Act specifies that hulls of the vessels must be constructed in the United States if they are to operate in U.S. waters.)

High-Density Dredge—There are five types as follows:

Screw and impeller dredging system—The screw and impeller dredging system consists of a vertical Archimedes' screw and a centrifugal pump (a mechanical action combined with hydraulic suction). The dredged material is conveyed pneumatically through a discharge pipe to a disposal area (Fig. 7) (Iwaskai et al. 1992).

Airtight bucket wheel dredging system—The bucket wheel is covered with an airtight hood. The fact that the dredge head is driven into the sediment, and since the discharge pipeline is totally enclosed, there is very little resuspension of sediment at the dredge head and no contaminated sediment is released to the water column (Fig. 8). The field tests were conducted at Shin-Maji Port.

Over a period from August 1987 to December 1991, the airtight bucket wheel system dredged about 700 000 m^3 (915 565 yd^3) of sludge, at a rate between 85.2 and 103.0 m^3/h (111 and 135 yd^3/h).

High-density pneumatic conveyance system—This system is based on the design of the screw conveyor type dredge and has been developed for long-distance transport of dredged material. The system is composed of a mud tank equipped with primary and secondary

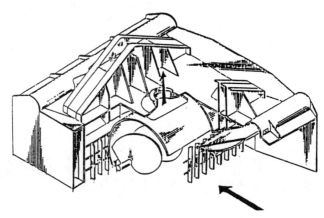

FIG. 6—"Cleanup" shielded auger head.

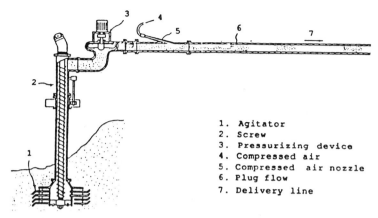

FIG. 7—*Operating principle of the screw and impeller dredging system.*

1. Agitator
2. Screw
3. Pressurizing device
4. Compressed air
5. Compressed air nozzle
6. Plug flow
7. Delivery line

screens and an agitator (Fig. 9). The sludge intake and the pneumatic feed system is similar to that of the screw and impeller system. Its performance is governed by the specific gravity of mud, water content, grain size distribution, liquid limit, plastic limit, and other characteristics.

The system was tested with mud in Kumamoto Port. A 100 m (328.1 ft) navigation channel was dredged to a water depth of −4.5 m (−14.8 ft) and to −5.0 m (−16.4 ft), −5.5 m (−18.0 ft), and −7.5 m (−24.6 ft) at anchorage sites. Silt content of the *in situ* soil was approximately 50%, and sand content about 30%. *In situ* water content of the soil was 65%. The *in situ* material was excavated with grab dredges, transported about 1000 m (3281 ft) by barges to a pressure-transport dredge. Dredged material was then placed in a hopper by means of a backhoe and transported under pressure through a 400 m (1312 ft) pipe to a disposal site (Fig. 10).

Resuspension of sediment was said to be below 10 mg/L. The excess water in the pneumatic system was controlled by valves installed in the pipe near discharge points in the disposal area. The measured water content was 86.6% at site A and 90.9% at site B. Since

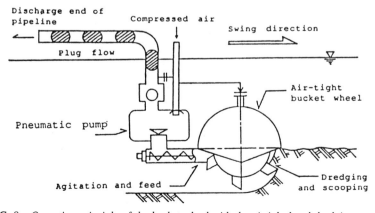

FIG. 8—*Operating principle of the bucket wheel with the airtight hood dredging system.*

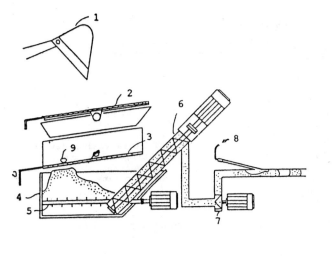

1. Backhoe
2. Primary screen exciter
3. Secondary screen (vibrating screen)
4. Mud tank
5. Agitator
6. Screw
7. Pressurizing device
8. Compressed air
9. Foreign objects

FIG. 9—*Operating principle of screw-type pneumatic conveying system.*

the *in situ* water content was approximately 65%, the sediment gained about 24% in water content. Water had to be added to wash the cutoff valves, and some of the water adhered to the grab buckets. Difficulties encountered were due to trash, stones, bamboo stems, and iron scrap accumulating in the hoppers.

In summary, the system worked reasonably well; the slurry water content was kept low, considerably lower than in a conventional hydraulic dredging operation. There was less

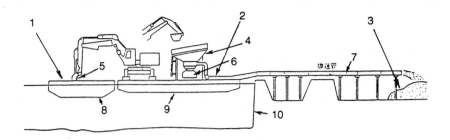

1	unloading/pneumatic transport system	6	air pressure
2	sea/overland pipeline	7	sand discharge tube
3	reclamation	8	soil transporting ship
4	hopper	9	pneumatic transport ship
5	backhoe	10	wharf

FIG. 10—*Flow chart of dredging, pressure transport, and reclamation.*

contaminated water to be treated; also, with a lower water content of dredged material, the settlement and consolidation in the disposal area proceeded at a much faster pace.

The pneumatic conveyance system shows promise; however, additional research is needed.

Hopper pneumatic conveyance system—This system uses three hoppers instead of a screw conveyor (Fig. 11). It has been successfully tested at Kumamoto Port.

A vessel named TOTRA-1 is equipped with three such units and it is said to have a total capacity of 600 m^3/h (785 yd^3/h).

Transfer tank pneumatic conveyance system—Dredged material is transferred to a barge into a large hopper which has two discharge ports, shown in Fig. 12. The material is discharged automatically into one of the two pressurized transfer tanks. Dredged material can be transported continuously through a pipeline by operating two tanks alternately.

The vessel MATS-600 is equipped with three such systems and has a nominal capacity of 600 m^3/h (785 yd^3/h).

Scraper Dredge—A new type of dredge was developed in the Netherlands satisfying the following requirements:

(a) no turbidity in the surrounding water,
(b) no transfer of contaminants to the adjacent sediments,
(c) no spill of dredged material, and
(d) accurate positioning of the dredge.

Figures 13 to 15 depict a Scraper Dredge which removes the top layer of contaminated sediment. Flexible polyurethane scraper blades are mounted on endless chains in a totally enclosed unit. The scraped material is fed into a collector bin; a special screw-centrifugal pump is used to pump the material through a floating pipeline.

The dredge moves at about 0.3 m per s (1 ft per s) with almost no resuspension of sediments. The draft of the vessel is controlled by four ballast tanks. In very shallow water the vessel can be fitted with floats. The dredging depth is between 0.3 and 1.1 m (1 and 3.6 ft). Prototype tests were conducted in the spring of 1990 with good results.

Future designs may include: (1) a scraper dredge with a ladder, and (2) a scraper dredge equipped with a ladder and scraper wheel. The scraper dredge is a novel and promising idea;

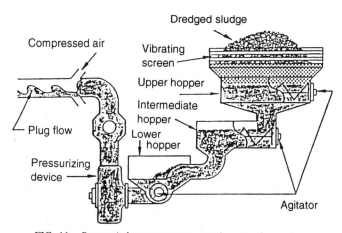

FIG. 11—*Reservoir hopper-type pneumatic conveying system.*

88 CONTAMINATED SEDIMENTS

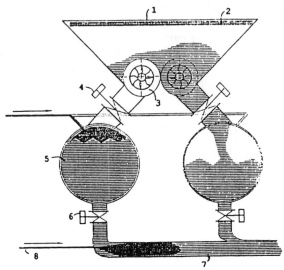

1. hopper
2. vibrating screen
3. screw feeder
4. inlet valve
5. transfer tank
6. discharge valve
7. discharge pipe
8. compressed air line

FIG. 12—*System flow of transfer tank type pneumatic conveying system.*

FIG. 13—*Scraper dredge.*

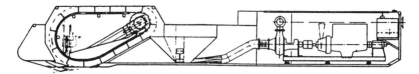

FIG. 14—*Cross-section of the Scraper dredge.*

the prototype dredge has been deployed in several projects. Apparently there is one such dredge in existence in the Netherlands.

Hydraulic-Type Dredges

"Refresher"—The Refresher dredge was developed by Penta-Ocean Construction Company, Ltd. purposely for the removal of contaminated materials.[2] The dredged material is confined by a specially designed flexible enclosure that completely covers the cutter, preventing escape of sediments to the outside of the immediate dredging area (Fig. 16). The working open section is always on the swing side of the cutterhead. A gas removal system is also installed and can be activated as needed to prevent gas from moving up the suction pipe. The flexible enclosure of the cutterhead is automatically adjusted to the bottom contours.

The Refresher dredge is equipped with a main pump and an additional pump on the ladder to provide a high level of protection. Automatic valves in the suction pipe prevent sediment-water mixture from flowing back in case of power failure (Fig. 17). The resuspension of the sediment may be classified as low.

"Matchbox-Head" Dredge—A special suction head was developed by a dredging contractor in the Netherlands to replace the traditional cutterhead. The main design points are as follows (Fig. 18):

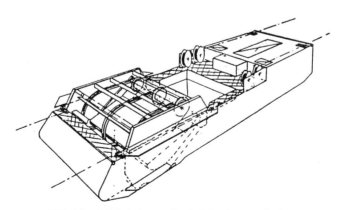

FIG. 15—*"Artistic impression" of the Scraper dredge.*

---

[2] Shinsha, H., Penta Ocean Construction Limited, Japan, Personal communication with author, 1988.

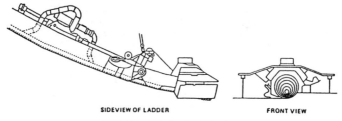

FIG. 16—*"Refresher" dredge.*

1. A large plate covers the top of the dredge head to avoid inflow of water and escape of gas bubbles.
2. Adjustable angle between the drag head and the ladder to create an optimum position of the drag head independent of the dredging depth.
3. There are openings on both sides of the drag head to improve dredging efficiency. During swinging action the leeward side is closed to prevent water inflow.
4. Dimensions of the head must be carefully designed for the average flow rate and swing rate.

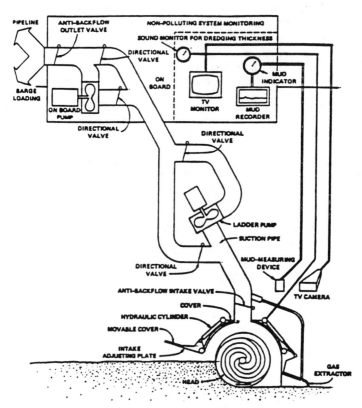

FIG. 17—*Description of a "Refresher" dredge.*

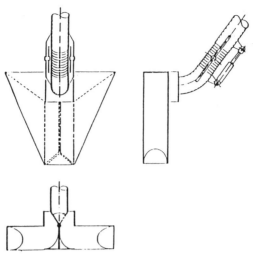

FIG. 18—*Matchbox suction head.*

A diffuser may be installed at the submerged end of the discharge pipe to reduce the dispersion of fine sediment in the water column. By its gradually widening cross-section, the flow can decelerate to an acceptable velocity to reduce turbulence. Outflow velocities are designed to be between 0.2 to 0.3 m/s (0.66 to 0.98 ft/s); however, it is questionable that contaminated material would be allowed by the regulatory agencies to be discharged in open water. A possible application may be to employ such a diffuser in a containment area. A degassing system is also installed to reduce the amount of gas moving up the suction pipe.

A direct comparison between a matchbox suction head and conventional cutterhead was made by the Waterways Experiment Station in Calumet Harbor (Hayes et al. 1988). The matchbox was specifically designed to be fitted on the ladder of the U.S. Army Corps of Engineers dredge *Dubuque*. The Calumet Harbor demonstration project indicated that the clamshell dredge generated the largest suspended sediment plume affecting the entire water column. The cutterhead (when operated properly) and the matchbox dredges were able to limit the sediment resuspension to the lower portion of the water column. The cutterhead slightly outperformed the matchbox dredge, shown in Table 1.

Wide Sweeper Cutterless Dredge—This hydraulic suction dredge does not have a cutter and is principally employed for removal of contaminated materials without resuspending the sediment particles. The main features of Wide Sweeper are:

TABLE 1—*Plume area for 10 mg/L contour for the Cutterhead, Clamshell, and Matchbox dredges (Hayes et al. 1988).*

Depth, %	Cutterhead		Clamshell		Matchbox	
	hectares	acres	hectares	acres	hectares	acres
5	0	0	0.7	1.7	0	0
50	0	0	0.7	1.8	0	0
80	0	0	—	—	0.2	0.4
95	0.5	1.2	1.4	3.5	1.2	2.95

1. Bottom sediments can be removed essentially without resuspension of particles.
2. Acoustic sensors determine the characteristics of sediment to be dredged.
3. Suction head can follow sea bed configuration to some extent or can be kept in a horizontal position.
4. Turbidity generated is monitored by a TV camera.
5. The dredge is equipped with a ladder pump and a main pump (Fig. 19).

Vibrating Auger and Positive Displacement Pump—The vibrating auger cutterhead was developed by Ellicott Machine Corporation, Baltimore, MD, for removal of chromium, lead, and nickel from the Welland River, Ontario, Canada. The dredge (Fig. 20) is a modification of the Mud Cat dredge.

Pneumatic-Type Dredge

Pneuma System—The Pneuma system developed by SIRSI, Florence, Italy, was the first dredging system to use compressed air instead of centrifugal motion to excavate and pump slurry through a pipeline. This system has been used in European and Japanese dredging projects. The Pneuma system consists of a pump body (composed of three cylinders), compressor, shovel, and a distribution system that automatically controls the supply of compressed air to the cylinders. When the pump is submerged, sediment and water are forced into one of the empty cylinders through an inlet valve. After the cylinder is filled, compressed air is forced into the cylinder, closing the inlet valve and simultaneously forcing the material out of an outlet valve and into a discharge line. When the cylinder is empty, the air pressure is reduced to atmospheric pressure, the outlet valve closes, and the inlet valve opens (Fig. 21) (Richardson et al. 1982). The two-stroke cycle is then repeated. The distribution system controls the cycling phases of all three cylinders, thus there is always one cylinder operating in the discharge mode. The system is in use in water depths of 50 m. According to available unpublished data, the amount of suspended sediment generated by the Pneuma system is minimal. Pneuma systems have been employed under license in the United States and have recently been used in Canada. The pump is usually supported by a crane.

"Oozer" Dredge—The Oozer pump was developed by Toyo Construction Company, Japan. The pump operates in a manner similar to that of the Pneuma pump system; however, there are two cylinders instead of three and a vacuum is applied during the cylinder-filling stage

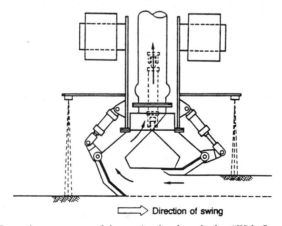

FIG. 19—*General arrangement of the suction head on dredge "Wide Sweeper No. 6."*

FIG. 20a—*Modified Mud Cat dredge.*

FIG. 20b—*Vibrating auger closeup.*

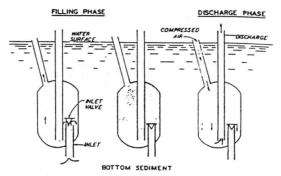

FIG. 21—*The pneumatic pump system (courtesy of Dredge Material Research Program, USAE Waterways Experiment Station).*

when the hydrostatic pressure is not sufficient to rapidly fill the cylinders. The pump is usually mounted at the end of a ladder and equipped with special suction heads and cutter units, depending on the type of material being dredged. The conditions around the dredging system, such as the thickness of the sediment being dredged, the bottom elevation after dredging, as well as the amount of resuspension, are monitored by high-frequency acoustic sensors and an underwater television camera. A large Oozer pump has a dredging capacity ranging from 300 to 500 m^3/h (392 to 654 yd^3/h). During one dredging operation suspended solids levels within 3 m (9.8 ft) of the dredging head were all within background concentrations of less than 6 mg/L. Figure 22 is a sketch describing the Oozer dredge.

The main features of the "Oozer" dredge are as follows:

1. The dredge can effectively remove contaminated sediments from a maximum depth of 18 m (59 ft).

2. Because the swing speed can be adjusted from 0 to 20 m/min (0 to 65.5 ft/min), the dredge can be effective in removing suspended sediments.

3. Five acoustic sediment sensors can measure the bearing pressure of sediment to be removed and the thickness of various sediment layers.

4. Underwater TV cameras monitor the presence of turbidity near the suction intake.

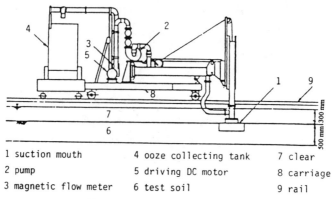

1 suction mouth	4 ooze collecting tank	7 clear
2 pump	5 driving DC motor	8 carriage
3 magnetic flow meter	6 test soil	9 rail

FIG. 22—*Outline of Oozer dredge.*

5. Toxic gases released during dredging pass through gas scrubbers to remove toxic content before gases are released to the atmosphere.

6. Screen is provided at the suction mouth to prevent large objects from entering. Double-suction valves and electrically-controlled check valves provide secondary protection.

7. Oozer dredges can, under ideal conditions, pump sediments at *in situ* density.

8. Different cutters and suction heads are available for dredging sediments ranging from clay to sand.

Agitation and Emulsion (AgEm) Dredging

The idea for agitation and emulsion dredging was developed in response to environmental constraints of dredging in harbors, navigational channels, and rivers (D'Olier, 1993).

The AgEm dredge consists of a pontoon barge with a crane supporting a Toyo submersible pump capable of handling up to 1000 m^3/h (1308 yd^3/h) of water-solids mixture. The mixture is pumped to an onboard liquidizer/emulsifier equipped with an overboard discharge valve trapping any large particles. The slurry then enters the liquidizer unit where four rotating agitators break down the solid particles to their original particle size. Air under pressure is added to discharge the emulsified mixture through a double exit diffuser discharging behind the dredge into the turbulent thrust from the two main engines. The suspension cloud then moves away from the dredge and the sediment is disposed thinly over a wide area.

Encapsulation

The feasibility of injecting contaminated silty material into a layer of clean silt (particles smaller than 0.06 mm) has recently been proposed (Davids 1992). Rheological properties of silt should be evaluated before encapsulation of contaminated sediments is successful.

Many contaminants are adsorbed by the fine fractions of silt; thus, silt is the best medium to be used for encapsulation. Adsorbed substances can only be transported by erosion or through uptake by biological life. Since many silts exhibit characteristics of Bingham Body, the contaminated silt body will remain intact, as if it were a solid body.

Before encapsulation is contemplated, the characteristics of silt layer must be determined. If a large enough silt layer is not available, a pit or a depression may be excavated and filled with clean silt of desirable characteristics. Dredges may be employed to pump the contaminated material from the site directly into a silt layer for encapsulation (Fig. 23).

Other Specialty Dredges

IMS Versi Dredge—Versi dredges have a pump installed at the bottom of the ladder to increase the dredging depth and minimize cavitation (Fig. 24).

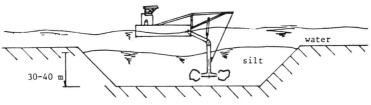

FIG. 23—*Encapsulation concept.*

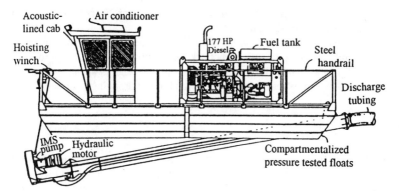

FIG. 24—*IMS versi-dredge.*

Worm-Wheel Dredge—This dredge developed by VOW in cooperation with HAM Research (the Netherlands) consists of a rotating screw-shaped auger (or worm) that cuts the material and transports it toward the center of the worm where an oval-shaped suction mouth is mounted to pump the material through a discharge pipeline. The dredge, named HAM 254, has been used in the restoration of Lac Nord near Tunis, Tunisia, and in Geulhaven, the Netherlands. It was stated that the maximum layer that can be removed by the dredge is equal to 50 to 70% of the outer diameter of the worm. A hood can be installed over the dredge head forming a closed box (in which a slight underpressure is maintained) to reduce sediment resuspension by the dredge. The Worm-Wheel dredge appears to be similar in design to the Mud Cat dredges available in this country.

Nessie Dredge—The Nessie Dredge Model 8DX has a bucketwheel dredge head and a standard dredge pump positioned at water level for easier priming. A suction line connects the bucketwheel with the pump (Fig. 25). Instead of a conventional ladder, an articulated

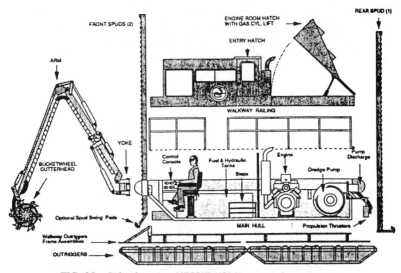

FIG. 25—*Side plan view NESSIE N8DX—general components.*

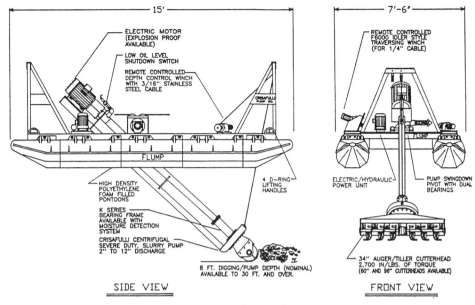

FIG. 26—*Flump dredge.*

arm is employed to lower and raise the bucketwheel dredge head. The main hull is supported by outrigger pontoons.

Crisafulli Dredges—The Crisafulli Pump Company manufactures a series of small auger dredge head pumps for removal of sediment from ponds, docks, lagoons, lakes, marinas, etc. The dredges include a Flump (Fig. 26) and Rotomite (Fig. 27).

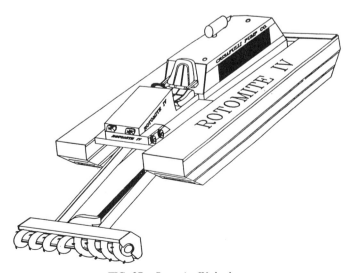

FIG. 27—*Rotomite IV dredge.*

Crisafulli dredges are small and are designed for a variety of excavating projects. No information on resuspension of sediments by a dredge head is available. The resuspension rates may be comparable to Mud Cat dredges. Mudshields are available for the Sludge dredge.

Field Studies, Demonstrations

James River, Virginia

The James River extends from Norfolk and Newport News, VA to West Virginia (WV). It is tidal to the City of Richmond, and is a major tributary to Chesapeake Bay. During the period 1967 to 1975 chlorinated hydrocarbon pesticide (Kepone) was discharged into the river at Hopewell. The Kepone became adsorbed onto the fine-grained, organic-rich sediments of the river, with the bulk accumulating in the area of maximum turbidity in the middle estuary. In this area, the Kepone is stored in the sites of high deposition, i.e., in dredged ship channels, tributary mouths, and reaches of wide cross-section where tidal currents are reduced. U.S. Army Engineer Division, Norfolk, VA conducted a dredging demonstration project as part of normal maintenance of the James River Channel. The areas dredged during the project were Goose Hill Flats and Dancing Point-Swam Point Shoal. The removed sediments consisted of an underconsolidated, very soft, saturated silty clay (CH). The Kepone concentration in the sediment averaged 0.045 ppm.

The dredge employed in one phase of the project was the cutterhead dredge ESSEX. The average cutter speed was reduced to 16 rpm to lessen resuspension of the bottom material. During the second phase of the demonstration project, the cutterhead was removed and replaced with a modified dustpan dredge head. The normally used water jets were not employed during this demonstration to reduce the sediment resuspension.

Samples of suspended sediment concentration indicated values between 42 mg/L near the surface and 86 mg/L near the bottom of the water column for the cutterhead dredge, and between 53 mg/L near the surface and 90 mg/L near the bottom. Table 2 shows a comparison between the cutterhead dredge and the dustpan dredge. These results suggest that the modified dustpan reduced the size of the suspended sediment plume significantly, although higher concentrations were observed near the dredge. Table 3 shows the range of resuspended solids for the dustpan dredge field test (McLellan et al. 1989).

Major design modifications in the dustpan dredge head included the following:

(a) curved plate mounted above the head opening to improve hydraulic entry condition,
(b) winglets or splitter added to either end and center to stabilize the dustpan head, prevent spillage from the sides of the head, seal the head, and improve the suction entry conditions, and

TABLE 2—*Comparison between Cutterhead and Dustpan dredges (McLellan et al. 1989).*

Dredge	Length of Plume, m (ft)	Max TSS[a] Divided by Background TSS
Cutterhead	610 (2000)	2.2
Dustpan	290 (950)	3.8

[a] Total suspended solids.

TABLE 3—*James River modified dustpan dredge field test (McLellan et al. 1989).*

Day	No. of Samples	Suspended Solids at Dredgehead, mg/L		
		Maximum	Minimum	Average
1	20	147	0	67
2	27	302	6	101
3	20	130	0	42
4	28	122	0	35

(c) trailing plate hinged below the head to act as a sealing strip, reducing sediment loss beneath the head and increase suction efficiency.

The modified dustpan dredgehead experienced repeated clogging.

Using the 80 mg/L contour (twice the ambient surface total suspended sediment level) shows that the cutterhead resuspended sediment plume affects 80% of the water column at least 304.8 m (1000 ft) in both the ebb and flood directions.

If the 100 mg/L contour is used as the plume's boundary in the dustpan demonstration test, the plume affects 40% of the water column during a tidal cycle.

The most important lesson from the James River Study was that the EPA determined that the contaminated sediments should be left in place and allow nature to clean the river over the years.

Tylers Beach, Virginia

The U.S. Army Corps of Engineers Dredging Research Program conducted a field study in the fall of 1991 to monitor the movement of dredged material during placement operations in James River, off Tylers Beach, VA. The main objective of the study was to collect wide-area measurements of the current, and resuspended sediment prior to and during maintenance dredging operations at Tylers Beach. A vertical pipe single-point discharge was used in this study to reduce resuspension of sediment. The dredge material placement site is a depression with a maximum depth of about 8.2 m (27 ft); the placement site has an area of approximately 20.2 hectares (50 acres). Approximately 13 761 m^3 (18 000 yd^3) of sediments were discharged at the placement site.

The sediment movement in the study area is dominated by tide and wind-generated waves. During peak periods of flood and ebb, current velocities reached 0.6 m per s (2 ft per s).

Background concentrations of suspended sediments had minimum concentrations of 10 to 30 mg/L during the slack water. Concentrations of 60 to 70 mg/L were observed during peak flood and ebb velocities. Near the bottom of the channel the concentrations were greater than 100 mg/L.

During dredging and placement operations, concentrations of resuspended sediment were similar to the background conditions, except at the bottom in the immediate vicinity and down current of the discharge point, where concentrations reached above 15 000 mg/L. Clouds of suspended material were observed in the shallow water on Point of Shoals.

Cold Spring, New York

The Marathon Battery Company site is situated in the Village of Cold Spring in Putnam County, NY approximately 65 km (40 mi) north of New York City.

The site is comprised of three study areas: Area I, which consists of East Foundry Cover Marsh and Constitution Marsh; Area II, which encompasses the former nickel-cadmium battery plant, the surrounding grounds, a vault with cadmium contaminated sediments dredged from East Foundry Cover in the 1970s, and adjacent residential yards; and Area III, which includes East Foundry Cover, West Foundry Cove, and the Hudson River in the vicinity of the Cold Spring pier. The eastern boundary of the sites includes the Old Foundry, a national historic site.[3]

Water depths in the West Foundry Cove and the Hudson River in the vicinity of the Cold Spring pier range from 0 to about 6 m (0 to 20 ft), increasing dramatically within several hundred metres of the riverbank. The main channel of the Hudson River in this area averages between 20 and 80 m (65 to 262 ft) in depth. The Cold Spring pier area is in an eddy zone created by the pier at the south end of this area and encompasses an area of 110 m^2 (361 ft^2) from the pier. Similarly, West Foundry Cove is in an eddy area created by Constitution Island. These slow-flow eddy areas have a significantly higher deposition of contaminants. Water circulation between Foundry Cove and the Hudson River is mainly influenced by a tide of 1 to 1.5 m (3.3 to 5 ft), exposing a considerable portion of the East Foundry Cove bottom at low tide. Because of the shallow water depths in the Cove, almost one third of the Cove bottom is covered with aquatic plant growth and is considered an emergent wetland.

The remediation plan calls for dredging 0.3 to 1.2 m (1 to 4 ft) of sediment in the marsh area east of East Foundry Cove, 0.3 m (1 ft) of sediment in the East Foundry Cove, 0.3 (1 ft) of sediment on the Constitution Island east of Metro-North Railroad line and 0.3 to 0.9 m (1 to 3 ft) of sediment in Hudson River next to the Cold Spring pier.

The remediation work commenced in June 1993 and was suspended early in January 1994 for the winter season. The scope of work includes the removal, on-site treatment, and off-site disposal of contaminated soils and sediments.[3] This remedial action is quite unique because of the variability in the situations and site conditions, including:

(a) removal by controlled dredging of 51 990 *in situ* m^3 (68 000 *in situ* yd^3) of contaminated sediments in shallow estuaries adjacent to a wildlife sanctuary and along the shoreline of the Hudson River adjacent to a recreational area,
(b) removal by controlled excavation of 25 995 *in-situ* m^3 (34 000 *in-situ* yd^3) of very soft marsh deposits using soft ground excavation techniques, and
(c) removal of 19 878 *in-situ* m^3 (26 000 *in-situ* yd^3) of plant ground and storage vault material adjacent to residential properties.

A modified dredge similar to a Mud Cat was used to remove 0.3 to 0.9 m (12 in. to 36 in.) of sediment under close monitoring of turbidity and suspended solids to control and minimize the disposal of cadmium contaminated sediment in the water column. Quality control includes horizontal and vertical positioning [3 cm (0.1 ft) accuracy is claimed by using on-board gages, skill of the operator, and proprietary procedures]. Resuspension of sediments is controlled by silt curtains, reduction in auger rotational speed, and prevention of back-flushing of slurry in the suction line.

---

[3] Russell, T. W. and Marano, P. F., Advanced GeoServices Corp., Chadds Ford, PA, Personal communication with author, 1993.

Major problems encountered include variability in sediment types (production rates varying from 15 to 61 m³/h (20 to 80 yd³/h) and debris and boulders that must be removed before dredging. Remedial activities are expected to continue through early 1995 with reconstruction and restoration of the East Foundry Cove Marsh.[3]

St. Johns River, Florida

This field study was conducted in the St. Johns River near Jacksonville, FL in 1982, to directly compare sediment resuspension from conventional and enclosed clamshell dredges. The U.S. Army Corps of Engineers, Jacksonville Division, was performing maintenance dredging at Pier Basin 139, U.S. Naval Air Station, Jacksonville.

The dredging work involved deepening of the pier basin to 4.6 m (15 ft). A clamshell dredge operated about 10 h/day which allowed a certain amount of flushing during "non-dredging" hours.

The sediment was characterized as silt (MH) with specific gravity of 2.40. Ninety-eight percent of the sediment was finer than 0.062 mm.

The enclosed bucket was a modified Yawn-Williams 9.9 m³ (13 yd³) clamshell-type bucket. The modification consisted of welding the side and top plates onto a standard bucket. The edge of each half of the bucket was lined with rubber to assure a watertight seal. The open clamshell bucket was a 9.2 m³ (12 yd³) Yawn-Williams bucket.

Average suspended sediment levels were 47 mg/L near the surface and 72 mg/L near the bottom; current measurements remained below 6.1 cm per s (0.2 ft per s). The highest sustained resuspended sediment contour for the open clamshell was 480 mg/L (or 6.7 times background) and 360 mg/L (or 5.0 times background) for the enclosed clamshell. Table 4 shows that near the bottom the enclosed bucket influenced an area 2.4 hectares (5.9 ac) [24%] greater than the open bucket. The enclosed bucket had lower levels of suspended sediment near the point of dredging (Hayes et al. 1984).

Savannah, Georgia Field Study

This field study was conducted on the dredge *Clinton* while dredging in the Back River near Savannah, GA. This quiescent reach was formed in 1969 by placement of tide control

TABLE 4—*Impacted area (acres) for the St. Johns River study (Hayes et al. 1984).*

Bucket Type	Concentration[a]	Area, hectares (acres)			
		25% depth	50% depth	75% depth	100% depth
Open	2	2.52 (6.23)	1.69 (4.17)	5.83 (14.4)	7.65 (18.9)
	4	0 (0)	0 (0)	0 (0)	0.21 (0.515)
	6	0 (0)	0 (0)	0 (0)	0 (0)
Enclosed	2	[b]	3.74 (9.25)	0.19 (0.47)	10.03 (24.8)
	4	0 (0)	0 (0)	0 (0)	0.08 (2.0)
	6	0 (0)	0 (0)	0 (0)	0 (0)

[a] Background multiplier indicating relative plume TSS concentration.
[b] Unable to obtain area for contour which did not close.

[3] Russell, T. W. and Marano, P. F., Advanced GeoServices Corp., Chadds Ford, PA, Personal communication with author, 1993.

gates across the Back River. The Back River lies parallel to the Savannah River along this reach and interconnects with the river at both ends. The tide gates are normally open during the flood tide allowing the sediment-laden water to flow into the Sediment Basin. During the ebb tide the gates are closed to increase flow through the Savannah River and decrease sediment deposition there, while decreasing the flow from the Sediment Basin and increasing sedimentation. The material dredged was silty clay with an average moisture content of 44.3%. The softness of the material deposited in the Sediment Basin along with the absence of traffic created almost ideal dredging conditions.

Equipment Description—The *Clinton* is a 0.46 m (18 in.) hydraulic cutterhead dredge. The Savannah District imposes operational restrictions on dredging within the Sediment Basin to reduce the resuspension of the light, soft material. The restrictions are specified in each dredging contract for work in the Sediment Basin. These limitations are usually outlined as:

(a) tangential swing speed must not exceed 0.3 m/s (1 ft/s),
(b) tangential tip speed of the cutterhead must not exceed 0.6 m/s (2 ft/s),
(c) cutterhead may not be buried more than 50% of its diameter below the mudline.

The restrictions were temporarily lifted during the testing period so that a wider range of speeds could be evaluated.

The average background suspended solids concentrations during the Savannah River cutterhead study ranged from 17 mg/L near the surface to 67 mg/L near the bottom. Tides and gates influenced the currents, and the speeds ranged from 0.07 to 0.34 m/s (0.24 to 1.1 ft/s) for the ebb tide and 0.2 to 0.48 m/s (0.67 to 1.56 ft/s) for the flood tide. Figure 28 depicts the average suspended solids concentrations collected during the field study. The 120 mg/L isopleths located near the bottom represents approximately 1.7 times the background level and extends for 45 m (150 ft). Applying the respective background concentrations for each depth increment, the 50% depth would have a background concentration of 45 mg/L. Therefore, the 40 mg/L contour represents a conservative estimate of the plume's boundaries and indicates that the plume remains below mid-depth of the water column. Figure 28 indicates that the lower speed ebb currents may retain more material in suspension than the higher speed floor currents. Using the 40 mg/L contour as the plume's boundary, the dredge-induced plume covers approximately 366 m^2 (1200 ft^2) during the tidal cycle (McLellan et al. 1989).

Sheboygan River, Wisconsin

The Sheboygan River is 278 km (173 mi) long and has a drainage area of 271 km^2 (105 mi^2) discharging into Lake Michigan. The Superfund site encompasses the lowermost 22.6 km (14 mi) of the river and a 40.5 hectares (100 acres) harbor. The sediments are contaminated with PCBs and metals.

As part of a pilot study, a total of 1911 m^3 (2500 yd^3) of sediments with a PCB concentration greater than 700 ppm was targeted for removal. Areas to be dredged were completely enclosed by two separate curtains. Chains along the bottom of the curtains conformed to the river bed and anchored the curtains. A large barge-mounted clamshell, modified to minimize spillage, and a backhoe were employed to remove the sediments. Two complete passes of the clamshell were effective in reducing residual contamination to less than 10 ppm. The backhoe was less successful in reducing PCB concentrations to acceptable levels. Four areas that were dredged had to be armored to contain the residual PCB concentrations.

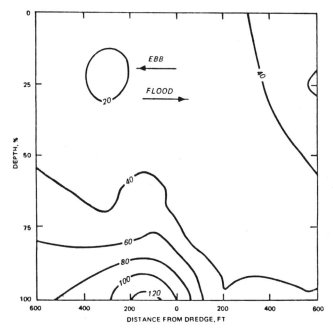

FIG. 28—*Concentration profiles, in mg/L, in a vertical section of the water column taken along a horizontal transect, describe a resuspended sediment plume resulting from conventional cutterhead dredging in the Back River near Savannah, GA.*

A total of 1394 m² (15 000 ft²) of river bed in five separate areas was slated for armoring. Average water depths in areas for dredging and armoring were in the range of 0.6 to 0.9 m (2 to 3 ft). Capping and armoring consisted of placing a geotextile over the contaminated sediments, covering it with run of the bank material (gravel), placing a geotextile over the run of the bank material, and covering it with cobbles and limestone. The layered system was anchored to the river bed and river banks by rock-filled steel cages called gabions.

In-situ anaerobic PCB dechlorination has detoxified the mixtures found in Sheboygan River sediment by more than 50% since armoring was completed. In addition, the armor has been covered with sediment and aquatic plants have taken root.

The second phase of dredging began in June 1991 and about 1911 m³ (2500 yd³) of sediment were removed from the river.

Waukegan Harbor, Illinois

Discharges of PCBs into Slip No. 3 in Waukegan Harbor continued for about 20 years, until the early 1970s, resulting in highly contaminated sediments and contaminated soil in the parking lot north of the plant and the "North Ditch," a tributary to Lake Michigan.

The remedy for the site was based on site specific modeling of the PCB contaminated sediment transport from the Harbor to Lake Michigan. The results showed that if the PCB concentrations were reduced to 100 ppm then the transportation of the contaminated sediments would be eliminated. A 50 ppm cleanup level was selected by the EPA to ensure the requirements were met and to protect the aquatic organisms against resuspension due to dredging. The EPA determined that the 50 ppm level would result in a greater than 95% removal rate of the PCBs from the entire harbor area.

Contaminated sediments were removed by dredging. Specific actions included:

(a) Replacing Slip No. 3 with the construction of a new slip on the east side of the Harbor; relocating Larsen Marine to the new slip;
(b) Constructing a double-walled pile cutoff wall to permanently isolate Slip No. 3;
(c) Dredging to remove highly contaminated sediments greater than 500 ppm from Slip No. 3 and sediments greater than 50 ppm in the Upper Harbor;
(d) Constructing two additional containment cells on OMC property for subsequent disposal of excavated soils and sediments from the OMC property; excavating soils with levels exceeding 10 000 ppm;
(e) Designating soils and sediment removed from Slip No. 3 and the OMC property for a treatment process that removes 97% of the PCBs;
(f) During remedial construction activities, treating and discharging the water, through a short-term water treatment facility, to a sanitary district or to an on-site location approved by the EPA; and
(g) After excavation, treatment (if applicable), and disposal in containment cells, capping the cells with a high-density polyethylene liner and soil cover.

A temporary cofferdam was placed at the end of Slip No. 3 to prevent the resuspended sediments from reaching the Upper Harbor. Dredging was performed within the cofferdam which effectively isolated that part of Slip No. 3 from the Upper Harbor. That part was effectively separated from the rest of Slip No. 3 and the heavy contamination of PCBs was confined to a small area. A small cutterhead dredge was deployed to remove contaminated sediment all the way to the clean substratum. The dredge was lowered to the "hot spot" area by a crane. The dredging proceeded during December 1991 and January 1992. Ice formed during that time but did not interfere with the dredging operations.

Dredging of the rest of Slip No. 3 and the Upper Harbor was performed with a cutterhead dredge. To prevent resuspended sediment from reaching the Lower Harbor, an impermeable silt curtain (all the way to the bottom) was placed in at the lower part of Upper Harbor. A silt curtain was also placed at the entrance to new Slip No. 4 to prevent resuspended sediment from reaching the new slip. The curtain failed three times during the operation and dredging had to be stopped until the silt curtain was restored. The failure of the curtain was due to weather conditions (wind and wind-generated currents). The volume of dredged material placed in the enclosed area of Slip No. 3 was 24 465 m^3 (32 000 yd^3). The contaminated sediments have been temporarily capped with clean sand, and after consolidation of silty, sandy material, a permanent cap will be placed over the disposal area.

New Bedford Harbor, Massachusetts

The City of New Bedford is located in Bristol County, MA. The harbor area is comprised of a broad outer bay and an inner harbor. A hurricane barrier was constructed at the harbor entrance to protect the area from storm surge flooding.

PCBs were used in the New Bedford area in the production of electronic capacitors. Discharges of waste effluent of not only PCBs but also heavy metals resulted in the contamination of sediments covering an area of 398 hectares (985 acres) north of the hurricane barrier. The PCB concentrations ranged from a few ppm to over 100 000 ppm. The water column in New Bedford Harbor has been measured to contain PCBs in the parts per billion (ppb) range. New Bedford Harbor was designated a Superfund Site in 1982.

A pilot study was conducted in New Bedford Harbor from May 1988 to February 1989. The pilot study was conducted to answer some of the questions that could not be adequately addressed by the Engineering Feasibility Study:

1. What is the efficiency of dredging for contaminant removal?
2. What are contaminant release rates from dredging?
3. Can the contaminated material be contained in a contained aquatic disposal (CAD)?

Three hydraulic pipeline dredges were selected for the pilot study:

(a) a cutterhead dredge,
(b) a horizontal auger (Mud Cat) dredge, and
(c) a matchbox-head dredge.

The dredges were employed to remove approximately 7645 m^3 (10 000 yd^3) of sediment of which 2217 m^3 (2900 yd^3) were contaminated. The dredged material was obtained from two separate sites designated as dredging areas 1 and 2.

Two disposal methods were used during the study:

(a) confined disposal facility (CDF), and
(b) contained aquatic disposal (CAD).

All dredges used during the pilot study were able to remove the contaminated sediment. The cutterhead and Matchbox-head dredges were able to remove 0.3 to 0.5 m (1.1 to 1.5 ft) of sediments (making two passes), leaving PCB levels of less than 10 ppm in the remaining sediments.

Resuspension of the sediment was minimized by slowing down operating parameters such as the swing speed, rate of advance, cutterhead rotation, etc. Resuspension rate and contaminant release at the point of dredging varied between different dredges, but impacts 125 m (500 ft) away were minimal for all dredges. The average rates of suspended sediments and PCB release are summarized in Table 5.

It should be pointed out that the pilot study was carefully designed, employed the best of equipment available, and was closely observed and monitored. In a project requiring removal of large volumes of contaminated sediment, higher resuspension of sediment rates can be expected (Herbich 1993) (Table 6).

In summary:

1. The cutterhead dredge generated the lowest amount of resuspended sediment. However, to reduce sediment resuspension at the cutter, several measures had to be taken including a

TABLE 5—*Sediment resuspension and contaminant release during dredging operations (Herbich 1993).*

Dredge	Resuspension Rate, g/s	Total Suspended Solids, mg/L	PCBs (ppb)		
			Dissolved	Particulate	Total
Cutterhead	12	82	0.6	22.3	7.0
Horizontal Auger	329	1610	10.1	200.3	2.6
Matchbox Head	46	319	0.5	56.9	54.9

TABLE 6—*Resuspended sediments by special purpose dredges (Herbich and Brahme 1991).*

Name of Dredge	Reported Suspended Sediment Concentrations[a]
Pneuma Pump	48 mg/L 3 ft above bottom
	4 mg/L 23 ft above bottom [4.9 m (16 ft) in front of pump]
Cleanup System	1.1 to 7.0 mg/L above suction
	1.7 to 3.5 mg/L at surface
Oozer Pump	Background level (6 mg/L) 3.0 m (10 ft) from head
Refresher System	4 to 23 mg/L 3 m (10 ft) from head

[a] Suspended solids concentrations were adjusted for background concentrations.

reduction in swing speed, reduction in cutter rotation, and only a 0.6 m (2 ft) advance per swing. These measures caused greater than expected reduction in production.

2. Dredging operations were delayed because of problems deploying and moving heavy swing anchors in shallow water. Other problems were related to a considerable amount of debris plugging the dredgeheads.

3. Two passes of the cutterhead dredge were required to reduce the average PCB level in remaining sediments to 8.6 ppm (after one pass the PCB level was 80.5 ppm).

4. Resuspension rate of sediments at the dredgehead varied from 12.0 to 329 g per s. This means that in a 6-month, 10-h/day, dredging operation between 79 713 kg (175 739 lb) to 2 167 424 kg (4 778 357 lb) of sediment will be resuspended.

5. The average suspended solids concentrations measured around the contained aquatic disposal dredging operations were 32.5 and 175.8 mg/L. Samples taken some 243 m (800 ft) from the point of discharge indicated levels of suspended sediments from 12.0 to 98 mg/L.

6. The capping of the confined aquatic cell was unsuccessful as the PCBs' contents in the upper 0.6 m (2 ft) of sediment measured four months later were between 0.1 and 95.9 ppm.

7. The silt curtains deployed during the pilot study sustained substantial damage during severe weather. When the silt curtains were not deployed, high suspended sediments were observed up to 305 m (1000 ft) away from the dredging operation at the contained aquatic disposal site.

8. The unit cost of dredging 7645 m^3 (10 000 yd^3) from a "hot spot" is estimated to be about $85.01 per m^3 ($65.00 per yd^3). The actual dredging began in April 1994.

Welland River, Ontario, Canada

The Welland River dredging site is located in Ontario, Canada near the U.S. border between Lake Erie and Lake Ontario. The contaminants included chromium, lead, and nickel.

A specially modified Mud Cat dredge (MC 915 ENV) was deployed to remove some 150 m^3 of industrial deposits and contaminated clayey silt sediments (Great Lakes Action Plan 1993). Modifications included:

(a) a multiflight variable pitch dual convergence horizontal auger, to minimize windrows,
(b) a truss boom, to allow the auger head to swivel and tilt to adjust for sloped bottom conditions,
(c) an increase in the size of the shroud encompassing the auger, to reduce sediment resuspension, and
(d) hydraulic vibrators, to increase the digging capability of the augers (Fig. 20).

There were no sustained plumes of resuspended material propagating away from the dredge head and measurements indicated that the total suspended solid levels within 25 m from the dredge were less than the Environment Canada criteria (Buchberger et al. 1993).

Collingwood Harbor, Ontario, Canada

Collingwood Harbor is located approximately 150 km north of Toronto, Canada along the southern shore of Georgian Bay. The toxic sediments in the harbor are a result of the harbor's ship building industry dating back over 100 years. In 1987, surficial sediment surveys and core sediment samples indicated heavy metals and chemicals. High concentrations exceeding MOE guidelines were found for PCBs, chromium, copper, lead, zinc, iron, and oil and grease.

A Pneuma pump was selected for this demonstration project to examine new technologies that remove sediment with minimal disturbances and adverse environmental impacts. Pneuma Pump No. 150/30 dredged approximately 4430 m^3 (5794 yd^3) of marginally contaminated sediments and pumped the sediments 1.2 km (0.75 mi) to a confined disposal facility. Total suspended solids level of 28 mg/L was measured 10 m (2.4 ft) from the dredge in the East Slip, which was higher than the background level by 18 mg/L. At another location, the level was 24 mg/L higher than the background level (Pelletier 1992).

Hamilton Harbor, Ontario, Canada

In October 1992, the Hamilton Harbor Removal Demonstration was undertaken by Environment Canada to evaluate the Cable Arm 100E closed clamshell bucket, manufactured by Cable Arm (Canada) Inc. of Pickering, Ontario, Canada. The performance of the bucket had been evaluated previously in Toronto Harbor (Orchard 1992) and was determined to be an effective remediation technology. Based on this evaluation, the bucket was revised to meet the site specific conditions in Hamilton Harbor. Approximately 160 m^3 (209 yd^3) of highly contaminated sediment were removed from the harbor, transported and off-loaded to a treatment demonstration facility. Once again, the bucket was successful in removing contaminated sediment with minimal disturbance to the aquatic environment. Since these demonstrations, the commercial value of this bucket has been recognized for its environmental use and for its ability to decrease operational costs (Pelletier and Leaney-East 1992).

Gibraltar Lake, near Santa Barbara, California

Gibraltar Lake has served as a principal source of domestic water for the City of Santa Barbara since 1920. The reservoir formed by the dam had an initial capacity of 1788 hectare-m (14 500 acres-ft), the dam was raised 23 ft (7 m) and increased the capacity to 2713 hectare-m (22 000 acres-ft). The May 1979 survey showed the capacity reduced to 1134 hectare-m (9200 acres-ft) (Gibralter Lake Restoration Project 1981).

A Pneuma pump was selected to dredge the accumulated sediments from the lake. In all, about 879 237 m^3 (1 150 000 yd^3) of saturated silt were removed from the lake.

Field Studies, Demonstrations, Summary Tables

Table 7 summarizes pertinent data from recent projects or demonstrations. It should be emphasized that many of the projects have been pilot studies or demonstrations. Consequently the production rates are based, in some cases, on only a few days' operations. Nevertheless, the production rates observed have been low because of concern for resuspen-

108　CONTAMINATED SEDIMENTS

TABLE 7—*Recent projects or demonstrations.*

Equipment	Sediment Type	Water Depth, m (ft)	Contaminated Sediment Depth, m (ft)	Solids Content, %	Average Production, m³/h (yd³/h)	TSS, mg/L	Location, m (ft)
Cable Arm (100E bucket)	very fine sand to clay	0.1 (0.3)	5.7 (18.7) 0.2–0.4 cuts (0.6–1.3 cuts)	44 to 48	3–7 min/cycle (positioning difficulty; not an accurate evaluation for commercial use)	<25	10 (32) from dredge Hamilton Harbor, Ontario[a]
Pneuma 150/30	sand 26% silt 64% clay 10%	4.1–6.3 (13.5–20.6)	0.05–1.8 (0.16–5.9)	17.8–25.7 (avg = 18.2)	avg = 25.4 (33.2)	2–128 (surface) 2–509 (bottom) (avg 20.7–37.0)	10 (32) from dredge Collingwood Harbor, Ontario
Pneuma 450/8	sand, silt, and clay	18.3 max (60) max	18.3 (60)	40–50 (max 70%)	240–300 (314–392) 1.2–530 overall (1.6–693 overall)	≈2.0	3.0 (10) Gibraltar Lake, CA[b]
Pneuma				25.7		minimal	Demonstration in the Netherlands
Horizontal Auger (Mud Cat 915 ENV)	primarily silt with clay, sand mixture and some gravel		≤1.5 ≤(4.9)	8 to 22 (dependent on dredging mode)	not available due to modifications made throughout demonstration	avg 6.0 max 18.0	10 (32) from dredge Welland River, Ontario, Canada
Horizontal Auger (custom design)			0.30[c] (1.0) 0.18 first pass (0.6) 0.12 second pass (0.4)		15.3–61.0 (20–80)	Site A 7.5 (avg) Site B 6.0 (avg) Site C 50.9 (avg) Site A 21.0 (max) Site B 18.0 (max)	Cold Spring, NY Marathon Superfund
Cutterhead	black organic sandy silt to silt	shallow	0.6 (2.0)		12.3 (16.0)	82.0	New Bedford Harbor, MA
Horizontal Auger (Mud Cat)	black organic sandy silt to silt	shallow	0.6 (2.0)		up to 31.3 (up to 41.0)	1610.0	New Bedford Harbor, MA
Matchbox-head	black organic sand silt to silt	shallow	0.6 (2.0)		18.7 (24.5)	319.0	New Bedford Harbor, MA
Pneuma	sand 26% silt 64% clay 10%		0.03 avg (0.1 avg)	30.7	45 (58.9)	results still pending[a]	(commercial use) Collingwood Harbor, Ontario

[a] Buchberger 1993.
[b] City of Santa Barbara, 1987.
[c] See footnote 3 on page 100.

TABLE 8—*Summary of dredge capabilities used for removing contaminated sediment.*

Type	Production	Depth Limitation, m (ft)	Resuspension of Sediment	Comments
Mechanical				
Open clamshell watertight	low	9.1–12.2 (30–40)	high	
Watertight clamshell bucket	low	9.1–12.2 (30–40)	low	Experiments conducted in St. Johns River
Cable-arm bucket	low	9.1–12.2 (30–40)		Experiments conducted at Hamilton Harbor, Ontario, Canada
Mechanical-Hydraulic				
"Mud Cat"	moderate	4.6–7.6 (15–25)	low to moderate	Extensively used
"Mud Cat ENV"	moderate		low	Experiments conducted in Sydney, Nova Scotia, Canada
Remotely controlled "Mud Cat"	low	4.6 (15)	low to moderate	New development
"Cleanup" system	moderate	21.3 (70)	low to moderate	Extensively used in Japan
Cutterhead	moderate to high	12.2 (40)	low	Pilot study in New Bedford, Massachusetts
Hydraulic-Suction				
"Refresher"	moderate to high	18.2–35.0 (60–115)	low	Extensively used in Japan
"Matchbox"	moderate to high	25.9 (85)	low to moderate	Experiments conducted at Calumet Harbor
"Wide Sweeper"	moderate	30.5 (100)	low	Used in Japan
Pneumatic				
"Pneuma"	low to moderate	60.9 (200)	low	Evaluated by USAE Waterways Experiment Station
"Oozer"	moderate to high	18.0 (59)	low	Used extensively in Japan
Mechanical-Hydraulic-Pneumatic				
Screw-impeller	low to moderate	6.1 (20)	low	Used in Japan (high-density)
Airtight bucketwheel	low to moderate	4.6 (15)	low	Used in Japan (high-density)

sion of sediments. The resuspension of sediments while dredging varied from 2 to 1610 mg/L total suspended solids (TSS).

A variety of small-sized equipment was employed in the pilot, demonstration projects; the dredging equipment ranged from hydraulic suction (cutterhead), mechanical-hydraulic (augerhead), pneumatic, and mechanical clamshell (Cable Arm). Figure 8 summarizes the dredging equipment capabilities in removal of contaminated sediment.

Since there are many sites where contaminated sediment will have to be removed in the future, refinement in design of equipment and removal methods will have to improve. Perhaps ideas developed overseas will be incorporated in equipment development in the United States and Canada.

Summary and Conclusions

In some cases, modified conventional dredging equipment may be used to handle contaminated material; however, in most cases special-purpose dredges are employed. These special-purpose dredges have been developed principally overseas and have successfully removed contaminated sediments.

The following projects have been reviewed in preparation of this paper: Hamilton Harbor, Ontario, Canada; Collingwood Harbor, Ontario, Canada; Gibraltar Lake, Santa Barbara, CA; Demonstration in the Netherlands; Welland River, Ontario, Canada; Cold Spring, NY; and New Bedford Harbor, MA.

Two important factors must be observed in selection of equipment for removal of contaminated sediments:

(a) Low rate of sediment resuspension by dredging equipment to prevent secondary contamination, (Andrassy and Herbich 1988; Hayes 1986).
(b) Low water content in dredged sediments to reduce the volume of slurry to be treated and to reduce the volume of water that will have to be treated to remove contaminants.

Many contaminated sites await remediation; equipment and methods will have to be improved to speed up the removal process and to reduce the associated costs.

References

Andrassy, C. and Herbich, J. B., 1988, "Generation of Suspended Sediment at the Cutterhead," *The Dock and Harbour Authority*, Vol. 68, No. 797, pp. 207–216.
Buchberger, C., January 1993. "Environment Canada Demonstrations," *Terra et Aqua*.
Buchberger, C., Santiago, R., and Orchard, I., May 1993, "Contaminated Sediment Removal Program Status with a View to the Future," *Proceedings, 26th Dredging Seminar*, CDS Report No. 327, Texas A&M University, College Station, TX.
Davids, S. W., de Koning, J., Miedema, S. A., and Rosenbrand, W. R., April 1992, "Encapsulation: A New Concept for the Disposal of Contaminated Sediment. A Feasibility Study," *Proceedings WODCON XIII*, Bombay, India, pp. 140–154.
D'Olier, B., 25 May 1993, "Agitation and Emulsion Dredging in Resuspension," *Proceedings, 26th Annual Dredging Seminar*, Atlantic City, NJ, CDS Report No. 327, pp. 267–273.
Gibraltar Lake Restoration Project, 1981, Final Report submitted to Environmental Protection Agency by the City of Santa Barbara, CA.
Great Lakes Action Plan, 21 January 1993, "Workshop on the Removal and Treatment of Contaminated Sediments," Etobicoke, Ontario, Canada.
Hayes, D. F., 1986, "Guide to Selecting a Dredge for Minimizing Resuspension of Sediment," Environmental Effects of Dredging, Technical Notes, EEDP-09-1, U.S. Army Engineer Waterways Experiment Station, Vicksburg, MS.
Hayes, D. F., Raymond, G. L., and McLellan, T. N., 1984, "Sediment Resuspension from Dredging Activities," Dredging 1984, Clearwater, FL, American Society of Civil Engineers.
Hayes, D. F., McLellan, T. N., and Truitt, C. L., 1988, "Demonstration of Innovative and Conventional Dredging Equipment at Calumet Harbor, Illinois," MP EL-88-01, U.S. Army Engineer Waterways Experiment Station, Vicksburg, MS.
Herbich, J. B., July 1990, "Extent of Contaminated Marine Sediments and Cleanup Methodology," *Chapter 219, 22nd International Coastal Engineering Conference*, ASCE, Delft, the Netherlands.
Handbook of Coastal and Ocean Engineering, J. B. Herbich, Ed., 1990–1992, Vols. I–III, Gulf Publishing Company, Houston, TX.
Herbich, J. B. and Brahme, S. B., 1991, "A Literature Review and Technical Evaluation of Sediment Resuspension during Dredging," Contract Report HL-91-1, U.S. Army Engineer Waterways Experiment Station, Vicksburg, MS.
Herbich, J. B., 1992, *Handbook of Dredging Engineering*, McGraw-Hill, Inc., New York.

Herbich, J. B., 5–8 December 1993, "Dredging Equipment for the Removal of Contaminated Sediment State-of-the-Art," 6th International Symposium on the Interactions Between Sediments and Water, Santa Barbara, CA.

Herbich, J. B., May 1993, "Review of New Bedford Harbor Pilot Study—Removal of Contaminated Sediments," *Proceedings, 26th Dredging Seminar,* Texas A&M University, pp. 25–43.

Iwaskai, M., Kuioka, K., Izumi, S., and Miyata, N., 7–10 April 1992, "High Density Dredging and Pneumatic Conveying System," *Proceedings, XIIIth World Dredging Congress, WODCON XIII,* Bombay, pp. 773–792.

McLellan, T. N., Havis, R. N., Hayes, D. F., and Raymond, G. L., 1989, "Field Studies of Sediment Resuspension Characteristics of Selected Dredges," Technical Report HL-89-9, U.S. Army Engineer Waterways Experiment Station, Vicksburg, MS.

National Research Council, 1989, Marine Board, *Contaminated Marine Sediments,* National Academy Press.

Orchard, I., 30 September 1992, "Toronto Harbour Contaminated Sediment Removal Demonstration," Environment Canada.

Pelletier, J. P. and Leaney, A., 1992, "Hamilton Harbor Removal and Treatment Demonstration," Environmental Screening Document, Environmental Protection, Environment Canada, Toronto, Ontario, Canada.

Pelletier, J. P., 1992, "Collingwood Harbour Sediment Removal Demonstration," Preliminary Report on the Water Quality Monitoring Program, Environmental Protection, Environment Canada, Toronto, Ontario, Canada.

Richardson, T. W., Hite, J. E., Shafer, R. A., and Ethridge, J. D., 1982, "Pumping Performance and Turbidity Generation of Model 600/100 Pneuma Pump," TR HL-82-8, U.S. Army Engineer Waterways Experiment Station, Vicksburg, MS.

Sato, E., 1984, "Bottom Sediment Dredge CLEAN UP," *Proceedings, 8th U.S. Japan Experts Meeting, Principles and Results, Management of Bottom Sediments Containing Toxic Substances,* T. T. Patin, Ed., U.S. Army Engineer Waterways Experiment Station, Vicksburg, MS, pp. 403–418.

van Drimmelen, C. and Schut, T., April 1992, "New and Adapted Small Dredges for Remedial Dredging Operations," WODCON XIII, Bombay, India, pp. 156–169.

Jean-Pierre Pelletier[1]

Demonstrations and Commercial Applications of Innovative Sediment Removal Technologies

REFERENCE: Pelletier, J.-P., **"Demonstrations and Commercial Applications of Innovative Sediment Removal Technologies,"** *Dredging, Remediation, and Containment of Contaminated Sediments, ASTM STP 1293,* K. R. Demars, G. N. Richardson, R. N. Yong, and R. C. Chaney, Eds., American Society for Testing and Materials, Philadelphia, 1995, pp. 112–127.

ABSTRACT: The Contaminated Sediment Removal Program (CSRP) of Environment Canada was founded in November 1990 following a request from the Great Lakes Cleanup Fund to the Environmental Protection Service-Ontario Region to provide leadership in the identification of removal technologies and procedures for contaminated sediments in the Great Lakes.

Following a request for proposal issued by the CSRP, proposals were received from vendors of innovative sediment removal technologies to conduct contaminated sediment removal demonstrations in different Areas of Concern (AOCs) on the Canadian side of the Great Lakes.

In 1992, the CSRP conducted the demonstration of two innovative sediment removal technologies at three different sites. The Cable Arm 100E clamshell bucket was demonstrated in Toronto and Hamilton Harbours, while the Pneuma Pump was demonstrated in Collingwood Harbour.

Those three demonstrations led to the first Canadian commercial applications of the Cable Arm 100E clamshell bucket in Pickering, Ontario, and of the Pneuma Pump in Collingwood, Ontario.

KEYWORDS: sediment, sediment removal, removal technology, clamshell bucket, Pneuma Pump

The International Joint Commission has identified 43 Areas of Concern (AOCs) throughout the Great Lakes where water quality impairments prevent full beneficial uses of the port and/or harbor. Of those 43 areas, 12 are on the Canadian side, and 5 are interconnecting channels and are therefore joint U.S.-Canada AOCs (Fig. 1). In those AOCs, Remedial Action Plans (RAPs) were put together to assess and remediate the contamination problems.

The Canadian Government, in keeping with the Canada-U.S. Water Quality Agreement, launched in 1989 its five-year, $125 million Great Lakes Action Plan. From the Action Plan, $55 million was allocated to the Great Lakes Cleanup Fund. The main objectives of the Cleanup Fund are to assist the Canadian Remedial Action Plan teams in assessing and remediating contamination problems in the different AOCs. With many others, three distinct programs have been established by the Cleanup Fund to concentrate on assessment and remediation of contaminated sediment: the Contaminated Sediment Removal Program, the

---

[1] Chemist and senior environmental project officer, Contaminated Sediment Removal Program, Environmental Protection Service-Ontario Region, Environment Canada, Toronto, Ontario, Canada, M4T 1M2.

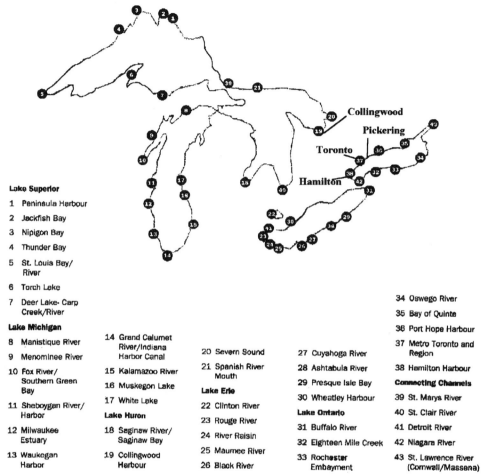

FIG. 1—*Great Lakes Areas of Concern (AOCs) and study areas.*

Contaminated Sediment Treatment Technology Program, and the Contaminated Sediment Assessment Program.

The Contaminated Sediment Removal Program (CSRP) was formed in 1990 following a request from the Great Lakes Cleanup Fund to the Environmental Protection Branch-Ontario Region (Environment Canada), to provide leadership in the identification of innovative removal technologies and procedures for contaminated sediment in the Great Lakes. The objectives of the CSRP are to identify and demonstrate innovative contaminated sediment removal technologies that could eventually be used for full-scale cleanups in AOCs. Operational and performance standards have been developed as means to evaluate proposals and assist during the on-site technology assessment. It is expected that, once demonstrated, these technologies will have applicability to contaminated sediment in non-AOCs as well as to routine navigational and recreational dredging projects.

In June 1990, the CSRP decided upon an approach that encompassed the basic elements of a request for proposal from vendors. A notice to vendors who wished to submit proposals or expressions of interest for the demonstration of removal technologies was published in

November 1990. A list of suitable firms was prepared from those expressing interest. The firms on the list were sent a request for proposal document (RFP) in May 1991.

Responses to the RFP document were submitted in a workshop held in June 1991. About 25 proposals were received from 12 companies for 3 chosen demonstration locations (Collingwood, Hamilton, and Port Hope Harbours). Approximately nine proposals were found to have merit.

Two of those proposals led to the demonstration of the Cable Arm Environmental Clamshell bucket in Toronto and Hamilton Harbours and the Pneuma Pump #150/30 in Collingwood Harbour. The second main objective of the CSRP was also met when the Cable Arm Environmental bucket was used commercially for the first time by Ontario Hydro to remove marginally contaminated sediment from their water intake channel, and in Collingwood Harbour where the Pneuma Pump #150/30 was also used commercially for the first time by Transport Canada to remove marginally contaminated sediment from a section of the shipping channel.

The Toronto Harbour Demonstration

With a population of over three million (including the Greater Metro area), and about one third of Ontario's population, Metropolitan Toronto is the commercial, industrial, and administrative center of the Province of Ontario (Fig. 1). Port activities and development have reshaped Toronto's waterfront through dredging, land reclamation, and lake-filling over the past 150 years (Central Waterfront Planning Committee, 1974). Extensive filling operations resulted in creation of the port lands which transformed much of the waterfront. A 5 km headland to provide an outer harbor for port expansion is now one of Toronto's most prominent shoreline features.

Sediment contamination in Toronto Harbour consists mainly of elevated levels of total phosphorus, total Kjeldahl nitrogen, oil and grease, copper, lead, zinc, and PCBs.

In April 1992, the Great Lakes Cleanup Fund of Environment Canada requested the CSRP to conduct a removal demonstration in Toronto Harbour in June 1992. The CSRP, after assessing all the proposals from vendors of innovative removal technologies from its information base, decided that the Cable Arm Environmental clamshell bucket was the most suitable technology for the proposed demonstration. This innovative clamshell bucket could be able to reduce sediment resuspension. Its moveable top plates allow surface water to pass through as it submerges and a sensor indicates complete closure of the bucket, therefore decreasing sediment resuspension on contact and on closure and lifting. Its overall weight, approximately 40% lower than conventional clamshell bucket, could reduce the regular cycle time while increasing the percentage of solids in the sludge. A schematic of the bucket operating procedures is presented in Fig. 2.

After conducting a sediment sampling survey across the waterfront slips in April 1992, the CSRP decided (based partly on the results of that survey) that Parliament Street Slip (Fig. 3) was the most suitable site for the demonstration due to the low sediment contamination encountered, the particle size of the sediment, the low usage of the slip, and, of most importance, the absence of storm sewer outflow that would complicate postdemonstration site assessment.

Mobilization of the equipment started on Friday, 29 May 1992, and ended on Tuesday, 2 June 1992. A silt curtain was deployed at the mouth of the slip to prevent release of resuspended sediment to the rest of the inner harbor. During mobilization, large patches of oil and grease were noticed on the water surface. In anticipation of important oil release during sediment removal, an apparatus called "Working Cell" or "Confined Work Area" was put in place. This cell was made of four pieces of 18 m by 8 m Terrafix 400 that were sewn

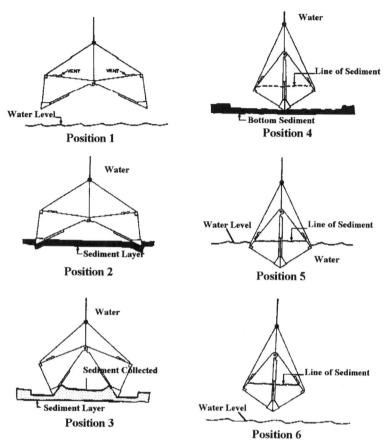

FIG. 2—*Cable Arm Clamshell Bucket dredge cycle.*

together and extended 1 m above the water surface and 1.5 m above the slip bottom. A basic pretreatment process was also put in place in order to decrease the volume of water included in the sediment that was transported to the treatment facility. The containment area of the transport barge was divided into two sections by 20 ft (6.1 m) containers welded in place to span the entire width of the boat. The front section had a total capacity of 360 m³. Five tons (4536 kg) of 2 cm gravel formed a cove on the inside wall of the partition. Since the ship was ballasted to effect a 15° degree slope to the stern, the excess water from the sludge was able to filter through the gravel and find its way under the partition.

The removal operations started on 2 June 1992 and ended on 4 June 1992 when 250 m³ of marginally contaminated sediment were removed. During that period, the CSRP audited the dredge on many points: sediment resuspension, cycle time, percentage of solids in the sludge, etc.

Two water quality monitoring programs were put in place in order to measure the rate of resuspension of sediment and the extent of resuspended sediment release to the surrounding slip and the adjacent harbor. In addition to the concentration of contaminants in water samples, turbidity, total suspended solids, and total organic carbon were thoroughly monitored. Figure 4 presents the plant layout and the water sampling stations as used during the dem-

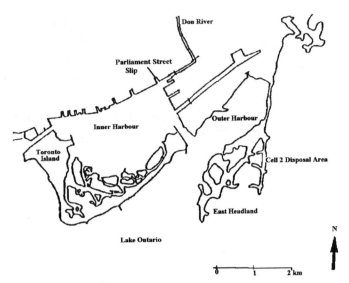

FIG. 3—*Location of Parliament Street Slip in Toronto Harbour.*

onstration, while Fig. 5 presents the total suspended solids concentration measured during the demonstration. It was noticed through total suspended solids and turbidity measurements that the level of resuspension of sediment was related to the level of familiarity of the operator with the specific operating procedures related to the Cable Arm bucket. The level of turbidity never exceeded 11 NTU throughout the demonstration.

Prior to dewatering, the percentage of solids in the dredged material was 49% (dry basis), which was very close to the *in-situ* percentage of solids ranging between 50% and 62% (dry basis). These percentages suggest that the Cable Arm bucket can remove sediment very near or at its *in-situ* percentage of solids (dry basis).

The basic pretreatment process used during this demonstration proved to be very efficient since approximately 75 m^3 of water was removed from the 250 m^3 of sediment dredged. This volume represents approximately 60% of the total volume of water that was mixed with the sediment as interstitial or excess water. After this dewatering process, the percentage of solids (dry basis) in the sludge was approximately 70%.

Most cycles performed on 3 June 1992 were recorded with entry and removal time. The shorter cycle time was 1 min and 30 s, while the longest was 8 min and 5 s. The average cycle time on 3 June 1992 was approximately 3 min and 30 s. It is believed that, with proper operator training and positioning system, the cycle time could be reduced to anywhere between 1 min to 1 min and 30 s (depending on water depth).

The success of this demonstration highlighted the need for further testing of the Cable Arm Environmental Clamshell bucket in a more contaminated area located in open waters.

The Hamilton Harbour Demonstration

Hamilton Harbour is located at the west end of Lake Ontario (Fig. 1). It is separated from Lake Ontario by a sand bar, but connected to the lake through the Burlington Ship Canal.

The highest concentration of heavy industry in Canada is located on the south shore of Hamilton Harbour and represents a major portion of the city's economy. Past discharge from

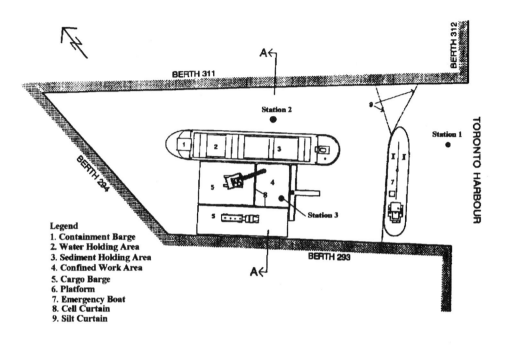

FIG. 4—*Plant layout and water quality sampling station locations.*

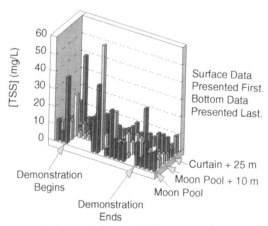

FIG. 5—*Variation of TSS concentration.*

the iron, steel, and other heavy industries in conjunction with municipal sewage and different types of urban and rural runoffs (mainly associated with an important system of highways) have adversely impacted on Hamilton Harbour water and sediment quality.

In the summer of 1992, a sediment sampling survey was undertaken in the study area. During sampling, the sampling crew noticed an odor from the sediment collected and an oily sheen on the water surface, meaning that oil was being released from the sediment. They also noticed that black tarry material was present in visible amounts as a separate phase. Past studies (Murphy et al. 1990) have shown that, in this area, the tarry/oily sediment can go as deep as 30 to 40 cm as per 1990.

Particle size analyses performed on those samples have shown that approximately 65% of the material was of a clayey particle size. Chromium, iron, and zinc were the only metals with concentrations above the Ontario Ministry of the Environment and Energy's severe effects level guidelines of 110 ppm, 40 000 ppm, and 820 ppm, respectively. The total PAH concentration was in the high hundreds ppm. The high molecular weight PAH (4 to 6 ring compounds) made up over half of the total PAHs.

Whole sediments from near Pier 16 greatly depressed the oxygen consumption of the bacteria *Photobacterium phosphoreum*, whereas other sediments were much less toxic. Sediment near Pier 16 in the removal area is anoxic and contains an extremely high concentration of hydrogen sulphide (100 mg/L). Hydrogen sulphide is believed to be the cause of the high toxicity of the sediment located in the area (Murphy et al. 1990). Coal tar cannot be biodegraded in anoxic sediments (Heitkamp and Cerniglia 1987; Mihelcic and Luthy 1988).

A demonstration of an innovative *in-situ* sediment treatment process was performed by Environment Canada in the early summer of 1992. Calcium nitrate was injected in the sediment to a depth of 20 cm. It was believed that, by injecting calcium nitrate in the top 20 cm, the production of hydrogen sulfide would be decreased, and that the sediment would have an increased and more stable oxygen level. This would allow the natural organisms, present in the sediment, to biodegrade the PAH molecules contaminating the bottom material.

The CSRP managed to coordinate removal of only the top 20 cm of pretreated sediment, during a demonstration of the Cable Arm Environmental Clamshell bucket. This demonstration had therefore two purposes: (1) assessing the efficiency of the Cable Arm bucket to perform "surgical dredging" in very contaminated sediment located in open water, and (2) evaluating the efficiency of the *in-situ* treatment process by performing chemical analyses on the removed sediment.

The total volume of contaminated sediment planned to be removed during this demonstration was 150 m^3; i.e., 60 m^3 of pretreated material, and 90 m^3 of untreated material in an area adjacent to the pretreatment area (Fig. 6). Once removed, the material was taken to a bioremediation treatment plant located nearby in Hamilton Harbour. The treatment process required that the removed material be free of debris and have a percent solids of approximately 50% (dry basis).

Mobilization started on 6 October and ended on 8 October 1992. An 82 m by 15 m deck barge was anchored to Pier 15 in order to provide a stable anchoring point to the flat deck barges that would be used to place the crane, the containers, and the rest of the required equipment (Fig. 7). Due to the success of the working cell used in the Toronto Harbour demonstration, the same system setup was put in place for this demonstration. This time, four pieces of 10.9 m by 10.3 m Terrafix 400 were sewn together making up the working cell and hung from 1 m above the water surface and down to 1.5 m above the harbor bottom. A floating oil boom was secured inside the working cell in order to prevent gross contamination of the curtain. In order to ensure that the removed material would meet all the requirements for immediate treatment, a 3 m by 3 m classifying screen with a sieve of

PELLETIER ON REMOVAL TECHNOLOGIES   119

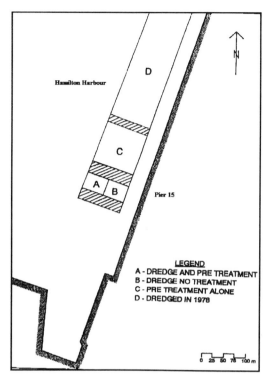

FIG. 6—*Hamilton Harbour demonstration area.*

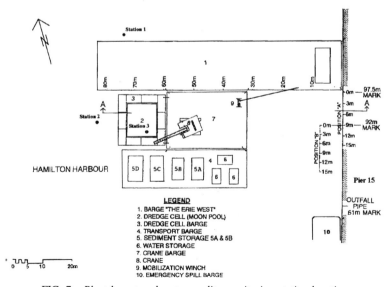

FIG. 7—*Plant layout and water quality monitoring station locations.*

approximately 5 cm was placed on top of the containers to remove any debris that would be dredged with the sediment.

The removal demonstration started on 8 October 1992 at approximately 4:00 p.m. and ended on 9 October 1992 at approximately 1:00 p.m. Again, during that period, the CSRP audited the dredge on many points: sediment resuspension, cycle time, percentage of solids in the sludge, etc.

One water quality monitoring program was established in order to assess the sediment resuspension rate caused by the removal operations. Figure 7 shows the location of the sampling stations in relation to the removal area. Figure 8 shows the turbidity levels measured throughout the monitoring program. Those results indicate that the resuspension of material was kept to a minimum.

During the demonstration, approximately 50 m^3 of pretreated material to a depth averaging 34 cm and 100 m^3 of untreated material were removed. The results of sludge samples indicated percentages of solids between 44% and 48% (dry basis).

The cycle time averaged approximately 2 min and 30 s. The increased length of time related to positioning of the bucket (slower descent and precision placement) and the pause at the surface for decantation purposes added as much as 1 min. It is again believed that, with proper operator training and use of a positioning system, the cycle time could be reduced to anywhere between 1 min to 1 min and 30 s (depending on water depth).

Some problems were encountered during this demonstration. The first one was related to the classifying screens. It was believed that since the material was fine, gravitation would be sufficient and vibration would not be necessary. Debris and large patches of clay made the used of the screens impossible. The second problem encountered was related to the size of the working cell. The cell was not large enough, therefore restricting the full use of the crane boom and the clamshell footprint. A great amount of time was spent in moving the working cell, since only six cycles could be performed with the size of the cell used in Hamilton Harbour.

This demonstration, in conjunction with the Toronto Harbour demonstration, highlighted the fact that the Cable Arm Environmental Clamshell bucket was marketable for environmental use, and could compete with conventional dredging technologies for navigation and recreational dredging projects.

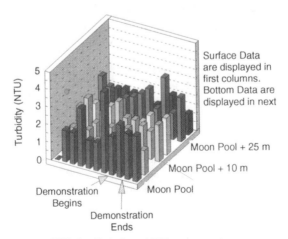

FIG. 8—*Variation of TSS concentration.*

Partnership with the Provincial and municipal governments as well as private industries was established during this demonstration. In fact, the Hamilton Harbour demonstration could not have been realized without the financial, technical, and "in-kind" support from those valuable partners.

The Collingwood Harbour Demonstration

Collingwood Harbour is situated in Georgian Bay on the south shore of Nottawasaga Bay (Fig. 1). The adjacent Town of Collingwood, first incorporated in 1858, has a permanent population base of 12 000 (1985 statistics) and a sizeable population of seasonal residents.

In the 1980s, sediment was found to have metals (mainly copper, lead, and zinc) and polychlorinated biphenyls (PCBs) in excess of the Ontario Ministry of the Environment and Energy's severe effect guidelines. The localized contamination of harbor sediments with PCBs and some metals was attributed mainly to historic use of the harbor as a center for the repair and construction of Great Lakes vessels. These activities have ceased and do not contribute to sediment contamination.

In the summer of 1991, the Collingwood Harbour Remedial Action Plan Team approached the CSRP to conduct a sediment removal demonstration/cleanup in Collingwood Harbour. The purpose of the demonstration/cleanup was to demonstrate an innovative sediment removal technology and to assist the RAP in delisting Collingwood Harbour as an Area of Concern through removal of contaminated sediment in a small area of the harbor. In response to the RAP's request, the CSRP selected the Pneuma Pump #150/30 for the demonstration/cleanup of Collingwood Harbour.

The Pneuma Pump uses static water head and compressed air inside special cylinders in a manner comparable to a piston. The hydrostatic head and induced vacuum causes each cylinder to sequentially fill with sediment slurry. As each cylinder is filled, compressed air supplied by a distributer linked to a compressor, acts as a piston and the slurry is forced through a valve to the discharge pipeline. As the cylinder empties, the compressed air is discharged thus releasing the internal pressure of the cylinder. The vacuum is once again applied causing the cylinder to refill with the slurry as the procedure starts over again. Figure 9 shows the operating cycle of the Pneuma Pump. The pumping rate of this particular Pneuma Pump could be as high as 150 m^3/h, with a reflow distance of up to 2 km. Since this pump is more efficient as the water depth is increased (due to the increase of the hydrostatic pressure), and the water depth in the working areas being only 2 to 4 m, a lower pumping rate than 150 m^3/h was anticipated.

Mobilization of the equipment started on 1 November 1992 and ended on 5 November 1992. A silt curtain was deployed at the mouth of the west slip in order to confine the work area, and to prevent release of potentially resuspended material to the outer harbor.

The demonstration started on 5 November 1992 and ended on 23 November 1992. The Pneuma Pump was attached to a crane that was loaded on a flat deck barge. The pump being used on a trailing mode, a cable winch was used to pull the barge from the beginning of a sweep to the end of that sweep. A tugboat was then used to bring the barge back to the beginning of the following sweep. The material was transported through a 15 cm diameter pipeline for a distance of 1.2 km, to a confined disposal facility (CDF) located at the mouth of Collingwood Harbour.

A water quality monitoring program was established in order to assess the sediment resuspension rate caused by the removal operations. Figure 10 shows the location of the sampling stations in relation to the removal area. Figure 11 shows the total suspended solids concentration measured throughout the monitoring program. This figure indicates that the

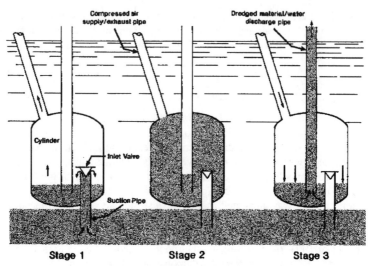

FIG. 9—*Pneuma Pump operating cycle.*

average concentration of total suspended solids recorded at 10 m in from of the pump was approximately 30 mg/L, with a background of approximately 15 mg/L. High levels of total suspended solids were recorded at this station on three occasions. These high levels (which also matched high levels of turbidity) were due to the fact that the pump was bubbling vigorously when sampling occurred. This was the result of debris being caught in the inlet valve, thereby preventing the valve to close when compressed air was discharged into the chambers. If the inlet valve is not totally closed, the air exits the chambers and is released in the sediment, resulting in bubbling and high resuspension of sediment. During the whole demonstration the level of turbidity averaged approximately 20 NTU at 10 m in front of the

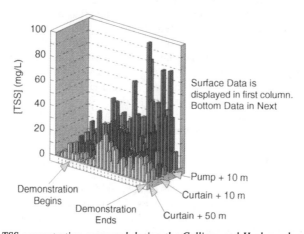

FIG. 10—*TSS concentration measured during the Collingwood Harbour demonstration.*

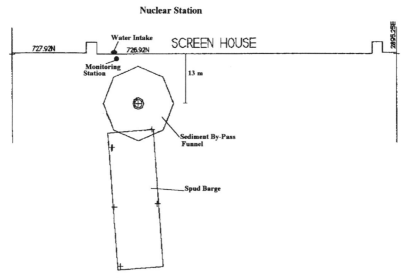

FIG. 11—*Project location (Pickering B NGS)*.

pump, except at the same three occasions mentioned above when turbidity reached levels between 50 and 110 NTU. The background turbidity averaged approximately 10 NTU.

During the 14 days of the demonstration, the volume of sediment removed from the west slip was estimated to be approximately 1800 m^3. The pumping rate varied from 20 to 35 m^3/h of slurry, while the percentage of solids in the slurry varied from 15 to 30% (dry basis).

Medium- and large-size debris, remnants of ship building activities, were abundant in the bottom sediment of the working area. Due to those debris, downtimes were numerous and sometimes lengthy since cylinder cleanups were required in order to improve pumping efficiency. In fact, as the concentration of debris in the cylinders increased, it was found that the turbidity around the pump was increasing while the production was decreasing.

Even though some minor problems were encountered during the demonstration, the project was allowed to become a full-scale cleanup activity since three other areas required sediment removal: (1) the remaining of the west slip, (2) the east slip, and (3) a portion of the outer harbor at the mouth of the west slip.

On 29 November 1992, the equipment was transferred from the west slip to the east slip, where approximately 500 m^3 of contaminated sediment was pumped to the CDF. During this whole phase, turbidity was marginally increased (2 to 3 NTU) above the background levels, while total suspended solids increased by 5 to 10 mg/L above background, at 10 m in front of the pump. The pumping rate was between 25 and 45 m^3/h, while the percentage of solids in the slurry was between 20 and 40%.

On 11 December 1992, the removal activities being completed in the east slip, the equipment was transferred to the outer harbor. Removal of sediment at the mouth of the west slip started on 12 December 1992 and ended on 19 December 1992. During this phase, a volume of contaminated sediment of approximately 1800 m^3 was removed. The turbidity and the total suspended solids concentration measured at 10 m from the pump never exceeded 10 NTU and 30 mg/L, respectively. The pumping rate and the percentage of solids in the slurry were comparable to those measured during the east slip portion.

This project showed the great potential of the Pneuma Pump #150/30 at removing contaminated sediment in areas where debris are in low abundance. The next step for this innovative technology was to compete with conventional dredging technologies for navigational dredging projects.

Pickering B Nuclear Generating Station

Since commissioning, Pickering B Nuclear Generating Station has experienced sediment entrainment problems within the cooling water intake channel resulting in frequent unit derating. A solution to the station's sedimentation problem was to install a sediment bypass system in 1990. The system is comprised of precast concrete slabs creating a funnel shape approximately 19 m in diameter. The sediment will be pumped through a 2 m diameter fiberglass pipe to the tempering screenhouse and then into the discharge channel. Conditional approval for the sediment bypass system to become operational was granted by the Ontario Ministry of the Environment in the spring of 1992. The terms and conditions associated with this approval stated that the 150 m^3 of sediment that had accumulated on top of the funnel area was to be removed before beginning operation of the system. Due to the amount of sediment resuspension, including problems with the containment and treatment of discharge water, conventional mechanical and hydraulic dredging technologies were not chosen by Ontario Hydro. The Cable Arm Environmental Clamshell bucket was selected in order to meet both Environmental Regulatory standards and Ontario Hydro's operational requirements.

Low resuspension of bottom material was a prerequisite since most of the water intake would still be in operation during dredging. It was therefore crucial that turbidity and total suspended solids levels be kept to a minimum throughout the dredging operations. An increase of 10 mg/L above ambient levels, at 25 m from the dredge, was set as the limit not to be exceeded. The closest water intake was located between 15 and 25 m from the dredge (depending where on the funnel dredging was taking place) (Fig. 11), with a primary current pushing the material toward the intake and a secondary current pushing the resuspended material the opposite direction. Increasing suspended particle concentration could lead to unit derating and possible shutdown of the nuclear generator, since a great amount of the resuspended material would be sucked in by the water intake.

Mobilization of the equipment started on 17 May 1993. The main equipment consisted of: one deck barge with spuds on which the crane was placed; one floating barge attached to the spud barge; one 35 m^3 and one 46 m^3 container; one tugboat; and one 3.2 m^3 Cable Arm Environmental Clamshell bucket weighing 2950 kg equipped with an underwater camera, a closure confirmation system, an air-operated vents dewatering system, and a depth transducer.

The actual removal operations started on 18 May and ended on 25 May 1993. In total, the two containers were filled to capacity twice, leading to a volume of sludge of approximately 160 m^3 of bottom material consisting mainly of organic debris (leaves, branches, dead fish, etc.). Sample analyses showed that the percentage of solids in the sludge was approximately 60% (dry basis), therefore leading to a volume of *dry* bottom material removed of approximately 96 m^3. The total time required to remove the 160 m^3 of bottom material was approximately 7 h. The average production rate was therefore approximately 23 m^3/h, with an average cycle time of 8 min. At the beginning of the project, Pickering NGS requested a cycle time of 10 min to reduce the amount of resuspended material between dredging cycles.

A water quality monitoring program was established in order to assess the quantity of resuspended material generated by the removal activities. Total suspended solids and turbidity levels were monitored at the closest water intake (13 to 22 m from the bucket). Total suspended solids concentrations showed a relationship with the operator's lack/increase of knowledge of the specific operating procedures related to the Cable Arm Environmental Clamshell bucket. During the first day, several bucket loads were overfilled resulting in important increase of total suspended solids concentrations. During the 158 min of dredging, the total suspended solids concentrations were increased to levels as high as 110 mg/L on approximately five occasions, although the average total suspended solids concentration was kept near the preset limit of 22 mg/L. All five occasions when the concentration of TSS reached levels as high as 110 mg/L were related to overfilling events. As the day went by, the operator became more familiar with the operating procedures related to this "unusual" bucket, and recorded TSS concentrations decreased significantly down to levels close to the preset limit, and most of the time below the 22 mg/L limit. Turbidity measurements indicated very low levels (below 7 NTU). This phenomenon was explained by the fact that most of the bottom material was composed of small debris that could not be detected by the nephelometer but were detected by the on-line monitor used for TSS measurements. During the following days of operation, the concentration of TSS was kept near or below the preset limit.

The Cable Arm Environmental Clamshell bucket proved to be not only environmentally sound for dredging activities, but also proved to be economically viable. The most important objective of this project was successfully met, i.e., no effect on station operation and reactor safety was noticed by Ontario Hydro personnel.

Collingwood Harbour

Following the 1992 contaminated sediment removal demonstration/cleanup performed by the CSRP and his partners, Transport Canada, the Collingwood Harbour Remedial Action Plan (RAP) team, and the CSRP jointly decided that an area of approximately 10 000 m^2 required removal of sediment. This action was necessary since levels of lead, exceeding the Ontario Ministry of the Environment and Energy severe effect level guideline of 250 ppm, showed toxicity to some benthic invertebrates. Earlier studies performed by the RAP had indicated that the contaminated silt was overlaying a clean clay layer on top of bedrock. This suggested that the clay layer was from the preindustrial activity era. It was then decided that removal of the silt (with thickness varying from 0.2 to 0.5 m) only was sufficient to remediate the area.

In October 1993, Transport Canada started a public tender process. Six companies responded with proposals. The contract was awarded to Voyageurs Marine Construction Co. Ltd. of Dorion, Québec. This contractor is the same one that performed the sediment removal demonstration/cleanup with the Pneuma Pump in Collingwood Harbour in 1992. The same equipment and personnel were used during this navigational dredging project as were used during the demonstration/cleanup of 1992.

Mobilization started on 22 November and ended on 24 November 1993. Due to the results of the demonstration/cleanup activities of 1992, no silt curtain nor any other type of siltation control mean was felt necessary for this project.

The actual operations started in late afternoon, on 24 November and ended on 8 December 1993. In total, approximately 3000 m^3 of contaminated silt was pumped to the CDF through the same 15 cm reflow pipeline. Postdredging soundings have indicated that sediment was removed to an average depth of approximately 0.3 m. Since 66 h were required to remove approximately 3000 m^3 of sediment, the average pumping rate could be estimated to be 45 m^3/h, with an average percentage of solids in the slurry of 30%, leading to a volume of dry solids removed from the bottom of the harbor of 900 m^3.

The CSRP monitored the water quality around the dredging activities periodically. The results for TSS at 25 m from the pump have indicated that the average concentration was approximately 5 mg/L on surface, and 10 mg/L at 1 m from the bottom. The maxima reached were 10 mg/L on surface and 17 mg/L at the bottom. The ambient total suspended solids concentration averaged approximately 5 mg/L, both on surface and bottom.

Turbidity measured at 25 m from the pump averaged approximately 6 NTU, with a maximum value recorded of 20 NTU on surface; and approximately 8 NTU, with a maximum turbidity level recorded of 36 NTU at 1 m from the bottom. Ambient levels of turbidity were approximately 5 NTU, on both surface and bottom.

From these turbidity results and the above TSS concentrations, it can be seen that very little disturbance was caused on surface waters, and only minimal disturbance occurred near the bottom.

Both the performance and the water quality monitoring results indicated that the Pneuma Pump is very effective at removing contaminated sediment in areas where very little debris are encountered. This project also proved that the Pneuma Pump can compete with conventional technologies for navigational and recreation dredging projects.

Conclusion

It is a proven fact that some areas where sediment contamination is important are not allowed to be dredged. This problem sometimes leads to important loss of revenues to major city ports. With the type of development and testing performed by Environment Canada and other agencies, it is believed that, eventually, a whole range of technologies will be available for any type of dredging project requiring special care due to sediment contamination, the presence of major water intakes and spawning beds nearby, or the importance of keeping the volume of excess water to a minimum.

Both the Cable Arm Environmental Clamshell bucket and the Pneuma Pump #150/30 can be used for some projects, with the special measures mentioned above. It is believed that these two technologies are economically viable due to competitive costs, and are both adaptable to specifications related to dredging contaminated sediments.

The most important objectives of the Contaminated Sediment Removal Program of Environment Canada are to identify and demonstrate innovative dredging technologies that could eventually help modernize the dredging industry. The CSRP succeeded when the Cable Arm Environmental Clamshell bucket and the Pneuma Pump both became accepted as proven technologies by some dredging companies. It is now expected that these two innovative technologies will be strongly considered to remediate areas like Hamilton Harbour and Thunder Bay, where sediment contamination is a threat to the environment and where sediment removal has now become an option.

Bibliography

C. B. Fairn & Associates Ltd., 1993, *Report on Harbour Dredging, Collingwood Harbour, Ontario* (prepared for Public Works-Government Services Canada), December.

―――, 1993, *Report on Funnel Dredging of Sediments, Ontario Hydro Pickering NGS B, Cooling Water Intake Channel*, report for the Contaminated Sediment Removal Program of Environment Canada, June.

Cable Arm Inc., L. B. Tanker Inc., 1993, Ontario Hydro-Nuclear Plant Cooling Water Intake Channel Dredging.

H.S.P. Inc., 1993, *Demonstration of Pneuma Dredging Technology at Collingwood Harbour, Ontario*, prepared for the Contaminated Sediment Removal Program of Environment Canada, May.

Heitkamp, M. A. and C. E. Cerniglia, 1987, "Effects of Chemical Structure and Exposure on Microbial Degradation of Polycyclic Aromatic Hydrocarbons in Freshwater and Estuarine Ecosystems," *Environmental Toxicology Chem.*, Vol. 6, pp. 535–546.

L. B. Tanker Inc., 1992, *Contaminated Sediment Removal Demonstration: Hamilton Harbour*, report for the Contaminated Sediment Removal Program, Environmental Protection-Ontario Region, Environment Canada.

Lajeunesse, J., 1992, *Contaminated Sediment Removal Demonstration of Cable Arm Clamshell in Toronto Harbour, L. B. Tanker*, report prepared for the Contaminated Sediment Removal Program of Environment Canada, November.

Leaney-East, A. and Pelletier, J.-P., 1992, *Toronto Harbour Removal Demonstration: Environmental Screening Document*, report for the Contaminated Sediment Removal Program, May.

McQuest Marine Research and Development Company Ltd., 1993, *Evaluation of an Acoustic Scanner for Bucket Positioning During Dredging Operations, Pickering Nuclear Station*.

Metropolitan Toronto Remedial Action Plan (RAP), 1989, Metropolitan Toronto Remedial Action Plan, report prepared jointly by Environment Canada, Environment Ontario, Ministry of Natural Resources and Metropolitan Toronto and Region Conservation Authority.

Mihelcic, J. R. and Luthy, R. G., 1988, "Microbial Degradation of Acenaphthalene and Naphthalene under Denitrification Conditions in Soil-Water Systems," *Applied Environ. Microbial.*, Vol. 54, pp. 1188–1198.

Murphy, T. P., Brouwer, H., Fox, M. E., et al., 1990, "Coal Tar Contamination Near Randle Reef, Hamilton Harbour," NWRI Contribution No. 90-17, Lakes Research Branch, National Water Research Institute, Environment Canada.

Pelletier, J.-P., 1993, Ports de Collingwood, Hamilton et Toronto, Rivière Welland et Station Nucléaire Pickering: cinq sites d'extraction de sédiments contaminés utilisant des techniques innovatrices. Articles pour la 4e Conférence de l'Association Internationale Villes et Ports. Programme d'extraction des sédiments contaminés. Service de protection de l'environnement-région de l'Ontario. Environnement Canada. Octobre.

Pelletier, J.-P., 1994, *Collingwood Harbour Contaminated Sediment Removal Demonstration/Cleanup: Report on the Water Quality Monitoring Program*, Restoration Techniques Development Program, Environmental Protection Branch-Ontario Region, Environment Canada, January.

Pelletier, J.-P., 1994, *Toronto Harbour Contaminated Sediment Removal Demonstration: The Bio-Assay Results and the Chemical and Physical Characteristics of Parliament Street Slip Sediment*, Restoration Techniques Development Program. Environmental Protection Branch–Ontario Region, Environment Canada, January.

Pelletier, J.-P., 1994, *Toronto Harbour Contaminated Sediment Removal Demonstration: Report on the Water Quality Monitoring Program*, Restoration Techniques Development Program, Environmental Protection Branch–Ontario Region, Environment Canada, January.

Pelletier, J.-P., 1994, *Hamilton Harbour Contaminated Sediment Removal Demonstration: Report on the Water Quality Monitoring Program*, Restoration Techniques Development Program, Environmental Protection Branch–Ontario Region, Environment Canada, January.

Persaud, D., Jaagumagi, R., and Hayton, A., 1992, *Guidelines for the Protection and Management of Aquatic Sediment Quality in Ontario*, Water Resources Branch, Ontario Ministry of the Environment and Energy, June.

Public Works Canada-Ontario Region, 1992, *Contaminated Sediment Removal Technology Demonstration, Hamilton Harbour, Hamilton, Ontario*, Project Field Report, Architectural and Environmental Services, Project No. 649678, October.

Public Works Canada-Ontario Region, 1993, *Contaminated Sediment Removal Technology Demonstration, Collingwood West Dry Dock, Collingwood, Ontario*, Project Field Report, Architectural and Environmental Services, Project No. 649678, June.

Royal Commission on the Future of the Toronto Waterfront (Canada), 1990, *East Bayfront and Port Industrial Area: Environment in Transition. A Report on Phase I of and Environmental Audit of Toronto's East Bayfront and Port Industrial Area*, The Honorable David Crombie, Commissioner, April.

Daniel E. Averett[1]

Data Requirements for Advancing Techniques to Predict Dredge-Induced Sediment and Contaminant Releases— A Review

REFERENCE: Averett, D. E., "**Data Requirements for Advancing Techniques to Predict Dredge-Induced Sediment and Contaminant Releases—A Review,**" *Dredging, Remediation, and Containment of Contaminated Sediments, ASTM STP 1293*, K. R. Demars, G. N. Richardson, R. N. Yong, and R. C. Chaney, Eds., American Society for Testing and Materials, Philadelphia, 1995, pp. 128–135.

ABSTRACT: In many areas of the world, contaminated sediments are being considered a major factor in the redistribution of toxic chemicals in the environment. While removal of contaminated sediments from the aquatic environment is often the preferred alternative for reducing the potential impacts of contaminated sediment, regulatory agencies and the public often express concern about contaminant releases during dredging operations. The U.S. Army Corps of Engineers continues to develop techniques for making *a priori* estimates of the sediment resuspension rates and contaminant releases during hydraulic and mechanical dredging activities. However, appropriate field data to verify and refine these techniques for a wide range of conditions are currently limited. Data needs include physical and operational characteristics of the dredge, waterway characteristics, sediment characteristics, sediment contaminant data, and water quality data collected during the dredging activity. This paper discusses key parameters required to improve the current predictive techniques and outlines the type of monitoring program needed to improve the comparability of the techniques to measured releases. The recommended monitoring program is derived from experiences with previous monitoring efforts. Planners of future dredging demonstrations are encouraged to collect similar data in order to advance the state of the art for predicting sediment and contaminant releases associated with dredging.

KEYWORDS: sediment, contaminant, dredging, sediment resuspension, monitoring, water quality, contaminant release, dredged material, suspended sediment

Background

Needs to remove contaminated sediments from waterways to maintain navigation depths or to remediate sites impacted by contaminated sediments (environmental dredging) have led to increased concerns about the release of contaminants to the water column during the dredging activity. Dredging refers to removal or excavation of subaqueous sediments from a stream, lake, harbor, or other waterway. Most dredging technologies use tools that penetrate, grab, rake, cut, or hydraulically scour the bottom of the waterway to loosen or dislodge the sediment. Once loosened, the sediment or dredged material is lifted to the surface of the waterway by mechanical devices such as buckets or by hydraulic suction into a pipeline

---

[1] Environmental engineer, U.S. Army Corps of Engineers, Waterways Experiment Station, Vicksburg, MS 39180-6199.

system facilitated by centrifugal, pneumatic, or positive displacement pumps. Sediment mechanically disturbed by the dredge and not removed from the waterway may be released to the water column. Sediment "lost" during the dredging operation to the water column is referred to as resuspended sediment. In areas where sediments have become contaminated by point and/or nonpoint pollution sources, contaminants associated with the sediment particles or desorbed from the sediment particles to the water are an environmental concern. This concern arises from the potential to spread contamination to cleaner areas of the waterway and because of potential toxicity or bioaccumulation in aquatic organisms.

Planning dredging projects where sediments are contaminated requires an assessment of the impact of the project on the environment, particularly with respect to water quality. Commonly asked questions are how much sediment will be resuspended during dredging and what is the associated contaminant loss from the project area. Knowing the answer to these questions helps to determine if the environmental risks associated with an environmental dredging project or a navigation dredging project are acceptable. This knowledge also addresses the need for control measures during dredging. These control measures may include selection of appropriate dredging equipment for the site, inclusion of turbidity containment technologies (silt curtains), selection of operational procedures, and instrumentation for control and optimization of the process.

Rudimentary predictive techniques for sediment resuspension by the more commonly used dredge types, i.e., hydraulic cutterhead dredge and clamshell bucket dredge, have been developed. These techniques are based on field studies of dredging operations for a limited number of sites and a limited range of operating conditions and are not fully validated. Laboratory procedures for estimating contaminant releases during dredging have been investigated, but field comparisons of the laboratory predictions for a variety of contaminants are rare, and the laboratory protocols are not yet considered field verified.

Monitoring dredging operations in the field requires field personnel, sampling equipment, analytical services, and a well-designed sampling and analysis plan. Satisfying these needs often requires significant financial expenditures for monitoring. Regulatory agencies often specify monitoring requirements to assess compliance with permit conditions or water quality standards for a dredging project. While compliance monitoring may not meet all of the data needs for furthering research on predictive techniques for sediment resuspension or contaminant releases, it can contribute useful data and can help to overcome the general inadequacy of research funding in this area. By carefully designing the compliance monitoring program for the project and by expanding the scope of the compliance monitoring program slightly at a relatively small cost, these projects can collect data that will also advance the state of the art for *a priori* estimates of sediment resuspension and contaminant releases. The Corps of Engineers and port authorities will continue to encounter these questions in future projects and should take advantage of available data collection opportunities.

Purpose

The purpose of this paper is to outline the types of data required to refine existing techniques or to develop improved predictive techniques for sediment resuspension and contaminant releases at the point of dredging.

Available Predictive Techniques for Sediment Resuspension

The Corps of Engineers has conducted a number of field studies aimed at defining sediment resuspension characteristics of dredges operating under various conditions. The results of these studies and others have been summarized by McLellan et al. (1989) and Herbich and Brahme (1991). Most results have been reported in terms of suspended sediment concentrations in the water column near the point of dredging. Where sufficient data are available, spatial concentration contours for suspended sediment concentration at one or more depths have been drawn to illustrate the plume development around the point of dredging.

A mathematical model to describe the plume induced by a cutterhead dredge has been reported by Kuo et al. (1985), and a mathematical model to describe the turbidity plume induced by a bucket dredge has been reported by Kuo and Hayes (1991). In addition to plume geometry, stream velocity, and particle settling data, the source strength or rate of sediment resuspended at the point of dredging is a critical parameter for application of these models. Information for estimating the source strength for various dredge operating conditions and sediment characteristics is the largest knowledge gap for procedures to estimate sediment and contaminant losses during dredging. In most of the literature dealing with this subject, models to define the source strength for sediment resuspension are considered near field models, and models to describe plume development are considered far field models.

Source strength models have been proposed by Hayes (1986), Collins (1989), Crockett (1993), and Hayes et al. (1994). Operating conditions considered in the development of cutterhead models include depth and thickness of cut, ladder angle, cutter size (length and diameter), cutter rotation velocity, swing speed and direction, and suction intake velocity. Sediment characteristics considered include moisture content, specific gravity, grain size distribution, particle settling velocity, and Atterberg limits. Hayes et al. (1994) and Crockett (1993) developed predictive equations based on regression techniques using dimensional analysis to reduce the number of relationships in conducting regression and correlation analyses with respect to collected field data. Variables found to be significant to the estimation of the mass rate of resuspension were the mass rate of sediment approached by the cutter, fraction of cutter surface exposed to free water, tangential velocity of cutter blades relative to the water, and suction intake velocity.

Collins (1989) developed a model for bucket dredging by correlating operational parameters for the dredge to near field suspended solids concentrations measured as close as practical to the point of dredging. Because of the danger to sampling personnel near a lifting, swinging crane and bucket, measurement of suspended solids concentrations precisely at the point of dredging is difficult. Pump samplers or turbidity probes mounted on the bucket may be used, but they must be quite rugged to withstand bucket operations. Collins extrapolated sediment concentrations measured at various radial distances from the dredge to extrapolate an approximate concentration at the zero radial position. Depth averaged concentrations were found to correlate to particle settling velocity, bucket size and geometry, and bucket cycle time. Bucket cycle time is defined as the total time to make a complete bucket lift, recovery swing, bucket opening and release, return swing, and bucket drop and return to the channel bottom. Three sites were included in Collins' analysis, representing only a narrow range of the operating variables.

Estimating Source Strengths from Field Measurements

The principal field measurement of sediment resuspension is the suspended solids concentration in the water column at a point as close to the point of dredging as possible, i.e., immediately above the cutterhead or as close as possible to the bucket. In order to arrive at a source strength value in terms of mass sediment resuspended per unit time, a time and volume component must be related to the concentration. Figure 1 illustrates the conceptual approach to the problem (Myers et al. 1995). The sediment resuspension rate R_p in g/s is given by

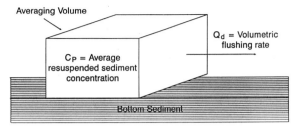

FIG. 1—*Definition sketch for measuring sediment resuspension at the point of dredging.*

$$R_p = C_p Q_d \quad (1)$$

where

C_p = characteristic suspended sediment concentration near the dredge head, g/cm³, and
Q_d = average volumetric flow of water through the control volume around the dredge head or bucket, cm³/s.

The volumetric flow is related to movement of the dredge head or bucket through the water column, which may be described by the velocity of the dredge head and the area of influence represented by the concentration measurement. Thus

$$R_p = C_p V A \quad (2)$$

where

V = velocity of the dredge head with respect to the water columns, cm/s, and
A = area of influence of the dredge head, cm².

How and where to measure these variables is subject to engineering judgment. Most field data suggest that suspended sediment concentrations decrease rapidly as the distance from the dredge head increases. Therefore, the area of influence has been defined as a small area around the cutter or bucket in a plane perpendicular to the direction of the movement. For the cutterhead dredge, suspended sediment samples are usually pumped to the deck of the dredge from sampling ports on the dredge ladder located 10 to 100 cm above the cutterhead. Therefore, the swing velocity of the ladder at the point of sample collection through the water column may be used for V in Eq 2. Hayes (1994) arbitrarily assumed that the area influenced was the product of the length of the cutterhead and a height above the cutterhead equivalent to twice the diameter of the cutter.

The geometry and velocity components for the bucket dredge are conceptually more difficult. Collins (1989) defined the control volume for the bucket as a cylinder about the axis of the bucket's rise and fall. Cross-sectional area in the horizontal plane was estimated to be a factor of 2 to 3 times the area of the bucket in the horizontal plane based on observations by Bohlen (1978) who used a wake formulation to estimate the size of the initial mixing zone for a bucket dredge. The velocity is related to the speed of movement of the bucket through the water column, which is turn-dependent on bucket cycle time and the fraction of the cycle time during which the bucket is falling or rising through the water column. Thus, with concentration, area, water depth, and velocity of the bucket, a mass rate of resuspension can be estimated.

Estimating sediment resuspension source strengths based on field measurements requires, as a minimum, sample collection for suspended sediment analysis in the immediate vicinity of the cutter or bucket, geometry of the cutterhead or bucket, and records of the velocity component either as ladder swing speed or bucket cycle time. Suspended sediment concentrations for water column samples display significant temporal and spatial variability. This complicates sampling and suggests the use of integrating samplers or high-frequency response instruments rather than point sampling alone.

In addition, operational parameters for the dredging operation that have been shown to correlate to sediment resuspension rates should be collected. These parameters are not routinely available; therefore, provisions for collecting these data should be coordinated with the dredge operator. Additional personnel are often required to collect and record the needed information. The examples discussed in this paper are for the commonly used cutterhead and clamshell dredges. However, the concept of concentration and control volume with a time component are applicable to other types of dredges with modification specific to the geometry and operating procedure for the alternate dredging technology.

Contaminant Releases

Contaminant releases associated with dredging are dominated by the contaminants associated with suspended sediment particles. A common way of estimating the contaminant release is to use the bulk sediment contaminant concentration and the suspended sediment concentration to estimate the contaminant release during dredging. However, the bulk sediment concentration may underestimate the contaminant concentration on the finer particles that have a greater affinity for contaminants and are more likely to be transported away from the dredging project area. Dissolved contaminants represent a small fraction of the potential total contaminant release, but may be of greater environmental significance. Equilibrium partitioning theory may be used to estimate an upper bound for dissolved organic contaminant concentrations at the point of dredging (Myers et al. 1994). Both dissolved and particulate contaminant releases are of interest for highly contaminated sites.

The Corps of Engineers has developed several test procedures with the objective of using laboratory data to predict the behavior of dredged material and associated contaminants in the field. The Corps of Engineers' "standard elutriate test" was developed to indicate the dissolved contaminant concentration for open water disposal of dredged material (Jones and Lee 1978). Ludwig et al. (1989) compared results of the standard elutriate test to water column concentrations in the vicinity of dredging. Palermo and Thackston (1988) described the test and field verification of a "modified elutriate test" for predicting the particulate and dissolved contaminant concentrations for effluent from a confined disposal facility for dredged material. DiGiano et al. (1993) took basic concepts and laboratory procedures from the standard and modified elutriate methods to develop a dredging elutriate test (DRET) that would predict the concentration of contaminants, dissolved and particulate, in the water column at the point of dredging.

The recommended procedure for the DRET begins with the collection of bottom sediment samples from the site to be dredged, as well as water samples from the site. Sediment and water are mixed at a concentration of 10 g per L for one hour. After allowing the suspension to settle for one hour, samples of the elutriate or supernatant are collected, separated into soluble and particulate fractions, and analyzed for the contaminants of concern.

DiGiano et al. (1993) compared the DRET to field data for a pilot dredging project at the New Bedford Harbor Superfund Site. Test results compared well to PCB concentrations measured in the field. Field measurements were based on samples collected from the plume approximately 30 m from the point of dredging. Plume samples rather than dredge head samples were used because they incorporated the effects of settling times used in the test. Further field verification of the DRET or alternate elutriate testing procedure is certainly needed before it is routinely applied to environmental dredging operations.

Field Monitoring Programs

An effective field monitoring program must be based on a well-designed and coordinated sampling and analysis plan. The plan should address the specific objectives of the monitoring program which may include estimating sediment resuspension rate for the dredge. The natural tendency is to include a large number of objectives. This paper encourages taking advantage of field monitoring opportunities to close the knowledge gaps for sediment resuspension and contaminant release. However, because of time and budget constraints for all projects, the sampling plan should address higher priority objectives completely rather than try to cover a wide range of objectives potentially yielding inconclusive results.

An integral part of the sampling and analysis plan is the quality assurance plan for the program. Quality assurance addresses a number of techniques to ensure that the data collected accurately represents the conditions measured not only in the laboratory but also in the field. Sampling protocols should be well-documented and understood by field personnel. Cross-contamination among samples is a potential problem that can be avoided by identifying proper procedures and specifying appropriate equipment and materials.

The sampling and analysis plan should explain the design of the experiment. Frequency, locations, numbers of measurements, and types of analyses should be described. Important field measurements for evaluating dredging are stream velocities, tidal changes, stream geometry, and stream temperatures. A large number of suspended sediment samples will be required, but costs will require that the number of samples for contaminant analyses be a small fraction of the suspended sediment samples. Contaminant analyses should include both dissolved and particulate fractions, or dissolved and total samples. Because of potential chemical changes with time, the filtration procedure for preparing the dissolved fraction must be performed in the field. Some analyses require that preservatives be added in the field.

Samples most critical to the sediment resuspension effort are those collected at the point of dredging. Because of the large temporal and spatial variability associated with these measurements, a large data set is required for a good statistical evaluation. Operating characteristics of the dredge should be recorded concurrently with sample collection. Critical to advancing predictive techniques is to evaluate a range of operating conditions, which may not be compatible with production. An agreement with the dredging contractor with regard to control of operating parameters must be reached prior to implementing the sampling program. During dredging, a log of changes in climate, dredge movement, operations, and other factors that may explain variability in the data should be meticulously recorded.

Definition of the plume geometry is important to defining the escape of sediment and contaminants from the point of dredging, and help to estimate the suspended sediment concentration at the point of dredging for a bucket dredge. To more efficiently use resources, the plume sampling area should be first defined by using towed or ship mounted optical or acoustic turbidity monitoring systems. These systems are becoming more available and provide instant readings that allow decisions regarding sampling locations. For some sites, aerial observation may be useful in defining the area of the plume and in locating appropriate sampling stations. Pumped water samples for suspended sediment analysis may be collected at a number of stations on radials extending outward from the point of dredging to the plume area. Station spacing on each radial should increase geometrically from the point of dredging. Samples should be collected at various depths, including one near the bottom of the water column. One or more relevant reference stations outside of the zone of dredging influence, typically upstream, should be sampled throughout the monitoring program to account for background conditions.

Logistics for implementation of the monitoring program should be planned in detail. Resources that must be identified include the numbers of personnel, boats, sampling equipment, sample containers, shipping containers and services, and analytical support. During the sampling operation, a field chief who understands the sampling plan should be designated and should be on site at all times to coordinate the sampling effort and to make decisions on adjustments to the sampling plan dictated by unforeseen changes in field conditions.

Summary

Predictive techniques to estimate sediment resuspension rates and contaminant releases while dredging contaminated sediments have been developed based evaluations at a limited number of sites. Contaminated sediment dredging projects are expected to increase in the future. Monitoring programs for these projects offer the opportunity to collect appropriate field data that will advance the state of knowledge for predicting sediment resuspension and contaminant losses induced by dredging.

Acknowledgments

This paper is based on research sponsored by the U.S. Army Corps of Engineers. Permission to publish this material was granted by the Chief of Engineers.

References

Bohlen, W. F., 1978, "Factors Governing the Distribution of Dredge-Resuspended Sediments," *Proceedings of the 16th Coastal Engineering Conference,* American Society of Civil Engineers, pp. 2001–2019.
Collins, M. A., 1989, "Dredging Induced Near Field Resuspended Concentrations and Source Strengths," *Contract Report* (unpublished), U.S. Army Engineer Waterways Experiment Station, Vicksburg, MS.
Crockett, T. R., 1993, "Modeling Near Field Sediment Resuspension in Cutterhead Suction Dredging Operations," M.S. thesis, University of Nebraska-Lincoln, Lincoln, NE.
DiGiano, F. A., Miller, C. T., and Yoon, J., 1993, "Predicting Release of PCBs at Point of Dredging," *Journal of Environmental Engineering,* American Society of Civil Engineers, Vol. 119, No. 1, pp. 72–89.
Hayes, D. F., 1986, "Development of a Near Field Source Strength Model to Predict Sediment Resuspension from Cutter Suction Dredges," M.S. thesis, Mississippi State University, Mississippi State, MS.
Hayes, D. F. and Crockett, T. R., 1994, "Modeling Near Field Resuspension in Cutterhead Suction Dredging Operations," Sixth International Symposium on the Interactions Between Sediments and Water, 5–8 December 1993, submitted for publication to *Australian Journal of Marine and Freshwater Research.*
Herbich, J. B. and Brahme, S. B., 1991, "Literature Review and Technical Evaluation of Sediment Resuspension during Dredging," *Contract Report HL-91-1,* U.S. Army Engineer Waterways Experiment Station, Vicksburg, MS.
Jones, R. A. and Lee, G. F., 1978. "Evaluation of the Elutriate Test as a Method of Predicting Contaminant Release during Open-Water Disposal of Dredged Sediments and Environmental Impact of Open-Water Dredged Material Disposal," *Technical Report D-78-45,* U.S. Army Engineer Waterways Experiment Station, Vicksburg, MS.
Kuo, A. Y., Welch, C. S., and Lukens, R. J., 1985, "Dredge Induced Turbidity Plume Model," *Journal of Waterway, Port, Coastal, and Ocean Engineering,* Vol. 111, No. 3, pp. 476–494.
Kuo, A. Y. and Hayes, D. F., 1991, "Model for Turbidity Plume Induced by Bucket Dredge," *Journal of Waterway, Port, Coastal, and Ocean Engineering,* Vol. 117, No. 6, pp. 610–622.

Ludwig, D. D., Sherrard, J. H., and Amende, R. A., 1989, "Evaluation of the Standard Elutriate Test as an Estimator of Contaminant Release at Dredging Sites," *Journal of the Water Pollution Control Federation,* Vol. 61, No. 11/12, pp. 1666–1672.

McLellan, T. N., Havis, R. N., Hayes, D. F., and Raymond, G. L., 1989, "Field Studies of Sediment Resuspension Characteristics of Selected Dredges," *Technical Report HL-89-9,* U.S. Army Engineer Waterways Experiment Station, Vicksburg, MS.

Myers, T. E., Palermo, M. R., Olin, T. J., and Averett, D. E., 1995, "Estimating Contaminant Losses from Components of Remediation Alternatives for Contaminated Sediments," U.S. Environmental Protection Agency Great Lakes National Program Office, Chicago, IL.

Palermo, M. R. and Thackston, E. L., 1988, "Test for Dredged Material Effluent Quality," *Journal of Environmental Engineering, American Society of Civil Engineers,* Vol. 114, No. 6, pp. 1310–1330.

Shoichi Yamasaki,[1] *Hiroshi Yasui,*[2] *and Masaharu Fukue*[3]

Development of Solidification Technique for Dredged Sediments

REFERENCE: Yamasaki, S., Yasui, H., and Fukue, M., **"Development of Solidification Technique for Dredged Sediments,"** *Dredging, Remediation, and Containment of Contaminated Sediments, ASTM STP 1293,* K. R. Demars, G. N. Richardson, R. N. Yong, and R. C. Chaney, Eds., American Society for Testing and Materials, Philadelphia, 1995, pp. 136–142.

ABSTRACT: The sediments deposited on the bottoms of seas, lakes, and rivers can be contaminated with hazardous and toxic substances as a result of the discharge of human activities. Therefore, since the natural remediation process cannot be expected, contaminated or polluted as well as highly organic sediments must be treated as waste and be properly disposed for human health and environmental protection.

One method of disposal may be to remove the sediments by dredging and to treat them with a proper technique. The main problems in the dredging method are as follows: 1) Since sediments usually have very high water content, it is necessary to decrease the volume and solidify them for the next procedure, e.g., landfill. (2) The leachates from the sediments should be treated also. It is required that the water to be discharged be kept at a quality satisfying "the level of standards."

This paper describes an experimental study using a solidification system performed for the cleanup of the bottom of a river. To promote the solidification of the system, several agents, such as lime, cement, polymer, resin, etc., were used. The results show that these agents strongly influence the solidification characteristics of the sediments and the quality of the leachate from the sediments.

KEYWORDS: sediments, dredging, solidification, unconfined compressive strength, biochemical oxygen demand, dissolved oxygen, suspended solid, pH

Sediments at the bottoms of seas, rivers, and lakes are more or less contaminated or polluted with hazardous and toxic substances (Matsumoto and Yokota 1977). Near coasts, the top surfaces of sediments are covered with the organic matter, e.g., wastes from the pulp and paper industry. This has changed the environment for the worse. Since the hazardous and toxic substances contained in the sediments are possibly ingested by aquatic living things, i.e., benthos, fish, shell, etc., it is urgent that we remedy the problem.

In Japan, another problem arising in sea regions is the deposit of feed from the raising farms for fish, where fry will be grown up inside of the surrounding net. In some cases, the feed-sediments have accumulated and the thickness exceeds 1 m. In such regions, it is also urgent to clean the bottoms. For the cleanup of bottoms, there are some techniques being developed, such as the use of bioremediation, dredging of the sediments, or covering the bottom with clean sand or gravel. The practical application of the bioremediation process may be the most ideal because it is due to a natural process. However, it requires some time

---

[1] Executive manager, Aoki Marine Co., Ltd., Fukushima-Ku, Osaka, Japan, 553.
[2] President, San O Co., Ltd., Kaide-cho, Mukou, Kyoto-fu, Japan.
[3] Professor, Marine Science and Technology, Tokai University, Shimizu, Japan, 424.

to clean up the sediments. The method of cleanup by covering the bottom with sand cannot always be useful for some reasons. One of the reasons is that the sand cover will be eroded by the stream current or waves, or both. Another reason is the serious shortage of sand materials in Japan.

Therefore, cleanup by dredging of contaminated sediments is one of the most effective ways. As well, dredging techniques have been used to excavate for ship navigation and for obtaining construction materials for land reclamation from the sea. However, the dredged materials are usually too soft to transport and construct the landfill, unless the sediments are solidified. This requires development of a solidification technique for sediment disposal or for use as construction materials.

This paper describes the solidification technique developed and shows results of the field experiments of solidification on river sediments.

Summary of Solidification System

To make dredged sediments plastic or solid, the water content must be decreased. In this study, the decrease in water content is achieved by extracting the pore water and compressing the sediments.

The solidification plant system developed is schematically shown in Fig. 1. The plant consists of mainly three components, (1) mixer, (2) presser with water extractor, and (3) pH controller for water release. These components are installed on vessels if needed. Slurry-like sediments dredged using any type of dredging techniques are mixed with a specified type of agent to provide for the solidification in the mixing chamber, as shown in Fig. 1. In this study, various types of agent were examined for evaluating the solidification characteristics of the sediments. Although various sizes of the mixing and press chamber are available, medium sizes of components were used in this study. The selection of the size will be depend on the amount of sediments to be treated and construction period desired. The press chamber used has 120 unit cells (press rooms) separated by filters. Each cell has an extracting wall area of 20 m² for a sediment volume of 0.34 m³ at one cycle of treatment. As an example, a larger press chamber has 145 unit cells separated by 144 filters.

After mixing, the mixture is transferred into the press chamber where the water is extracted through filters with applied pressure. After a 30 min period, i.e., time required for transfer, the mixture is compressed by the presser with a compressive pressure of 630 kPa during the extraction of water. The extracted water is transferred into the pH control chamber, where

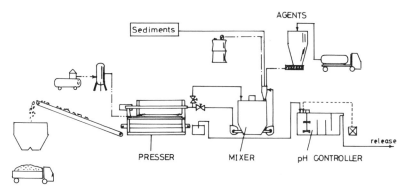

FIG. 1—*Solidification plant system developed.*

TABLE 1—*Properties of dredged sediments.*

Water Content, %	243–279
Sand Fraction, %	19–28
Silt/Clay Fraction, %	72–81
Ignition Loss, %	6.1–7.2
pH	6.3
Bulk Density, g/cm^3	1.2

the water was neutralized with acid or alkaline. When the quality of water is satisfied, it is released. If not, further treatment is properly achieved before discharge. The solidified sediments are taken out from the press chamber and transported to the disposal site or used for reclaimed materials or fertilizer.

Experimentation

Dredged Sediments

The initial properties of sediments dredged from the bottom of a river are shown in Table 1. The sediments consist of mainly silt fraction and have a bulk density of 1.2 g/cm^3. The initial water content exceeds 200%. These properties are common for silty sediments in rivers and seas.

Agents for Solidification

To promote the solidification of sediments, several types of agents were examined respectively. These are slaked lime; polyaluminum chloride (PAC), to promote the aggregation of sediments; natural polymer (called "L-fresh"), to aggregate the sediments; and portland cement and resin (called "UKC-H") to strengthen the sediments. A small amount of the respective agent presented in Table 2 is added to the dredged sediments and mixed thoroughly in the mixing chamber by the rotation of vanes at 6 r/min. The mixing time was about 10 min. The mixing proportions are given by wet weight, as shown in Table 2. For the comparison, testing without adding any agent was also performed.

Extracting and Compression

In order to decrease the water content of the mixtures, the pore water was extracted in the press chamber, with a pressure of 600 to 630 kPa, through the filters. After 30 min, i.e., time required to fill up all the unit cells, the mixture was compressed with a maximum compressive pressure of 630 kPa during the extracting of water. The compression time was only 10 min. The sum of the time required to pour the mixture into the press chamber, compress it, and take it out of the chamber was only 55 min.

The cost for this treatment is estimated to be from $20 to $30 per 1 m^3.

TABLE 2—*Agents used for solidification.*

Slaked Lime	Portland Cement	UKC-H	PAC	L-Fresh
1.3 or 5%	1.3 or 5%	1.3 or 5%	0.5, 0.75 or 1.5%	0.1%

Results

Summary

The technique used can be evaluated based on the following items:

(a) decreasing rate of water content,
(b) solidification degree in terms of strength or hardness,
(c) ease of handling of solidified mixture or sediments, and
(d) water quality in terms of pH, biochemical oxygen demand (BOD), dissolved oxygen (DO), suspended solids (ss), etc.

Table 3 shows results in terms of the above items. The results are discussed below.

Decreasing Rate of Water Content

The water content of the mixture was decreased by extraction and compression. In this study, the change in water content of the sediments in the press chamber was determined by measuring the volume of the extracted water. Figure 2 shows the change in water content of the mixture with elapsed time in the press chamber. The initial breakpoint of the curves for the two types of agent, especially for polymer, may indicate that the high viscosity of the pore fluid due to the added polymer prevented the water from extraction. under a relatively low pressure at the initial stage. Therefore, a certain amount of time is needed to initiate the decrease in water content until the pressure applied becomes sufficient. Figure 2 shows that the decreasing rate of the water content varies with the type of agent added. The relatively rapid change in the water content at an elapsed time of about 30 min indicates the starting of compression operation. Therefore, all the curves break at an elapsed time of about 30 min. The lowest water content obtained at an elapsed time of 40 min, i.e., at the final state chosen in this operation, was 37.8% for the mixture with PAC, while the highest water content, 82%, was obtained for the sediments without adding any agent. As the polymer (L-

TABLE 3—*Experimental results obtained in the study.*

Agent	Without Agent	Slaked Lime	Portland Cement	UKC-H	PAC	L-Fresh
Adding Rate, %	—	3.0	3.0	3.0	0.75	0.1
Poured Volume, m^3	0.69	1.04	1.01	1.03	0.969	0.633
Soil						
Water Content, %	82	45	44.8	48.1	37.8	77.0
Unconfined Comp. Strength, kPa	—	400	450	1420	—	—
pH	6.3	11.2	11.8	11.4	7.14	7.2
Density, g/cm^3	1.39	1.73	1.7	1.73	1.81	1.54
Leachate						
Leachate pH	6.3	12.6	12.6	12.0	5.4	5.6
Neutralized pH	7.2	7.3	7.7	7.3	5.4[a]	7.3
ss (mg/L)	5.0	6.3	6.0	6.3	5.6	6.3
DO (mg/L)	8.5	6.3	7.5	6.3	5.6	6.3
BOD (mg/L)	15.7	16.4	16.2	16.4	16.2	16.4

[a] No treatment.

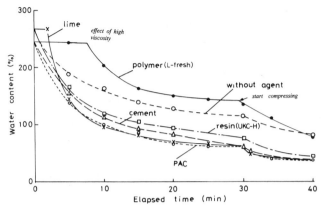

FIG. 2—*Change in water content during extraction and compression in press chamber.*

fresh) increases the viscosity of the pore water, the water can hardly be extracted, as shown in Fig. 2.

Unconfined Compressive Strength

In order to examine the degree of solidification after the compression, unconfined compression tests were performed on the specimens from the cake-like compressed mixture. The results are shown in Table 3. The highest unconfined compressive strength, 1.4 MPa, was obtained for the mixture with UKC-H. This agent has been developed to strengthen soft soil materials. The mixtures with portland cement or slaked lime show similar strength, approximately 400 kPa. The other mixtures and sediments without any agents were too soft for the test specimen to be prepared, except for the sediments mixed with 1.5% PAC. However, the unconfined compressive strength of the mixture with 1.5% PAC was only 140 kPa.

The results described above indicate that the unconfined compressive strength of the sediments is strongly influenced by the type of agent added to the sediments, but not necessarily the water content. For example, the sediments with 0.2% PAC, which have the lowest water content, were too soft for preparing the specimen. This is because the hardening of the mixture is primarily due to a chemical reaction or cementation process between the sediment particles and added agents.

Nature of Sediments After Treatment

The mixtures or sediments treated become plastic or solid as shown in Fig. 3. The solidified materials are disposed in the prepared site where a membrane is used to avoid the leaching from the materials as shown in Fig. 4. It is noted that the solidified materials with some agents will not be dispersed in water.

Water Quality

The "standard" for the water quality of river was established by law in Japan. The quality of industry and other discharge is regulated by the "standard for discharge" which depends on the site. Therefore, the water to be released must satisfy the level of standard for discharge for the site.

FIG. 3—*Solidified materials coming out from the presser.*

The pH values of the water extracted from the mixtures is dependent on the type of the added agent. When slaked lime, portland cement, or UKC-H was used as agents, the pH value of the water exceeds 12, as shown in Table 3. On the other hand, the agents such as PAC and L-fresh slightly decrease the pH value. Therefore, it is required that the water be neutralized using acid or alkaline, based on the standard for discharge. As shown in Table 3, the pH values after neutralization will range between 7.2 and 7.7, which will usually satisfy the standard for water quality.

The standard for river water quality permits a ss of 25 mg/L. Since a filter is used for extracting water, this level is usually satisfied. In fact, the obtained ss is relatively low and is in a range between 5.0 and 6.3 in any cases.

FIG. 4—*Disposal site of treated sediments.*

The DO values obtained are between 5.6 and 8.5 mg/L. These values usually satisfy the Japanese standard for discharge. However, the BOD does not satisfy the standard for discharge at this site. Therefore, in order to obtain a better quality, secondary treatment was needed.

Conclusion

The solidification technique of dredged sediments, for the cleanup of the bottoms of seas, rivers, lakes, etc., was developed and tested on river sediments. The results show that with some types of agents, the desired solidification of sediments was obtained. The quality of water from sediments was satisfactory in terms of dissolved oxygen, pH value, and suspended solids, but further treatment is required for biochemical oxygen demand.

Reference

Matsumoto, E. and Yokota, S., 1977, "Records of Pollution in Tokyo Bay Sediments," *Journal of Geochemistry,* Vol. 11, pp. 51–57 (in Japanese).

Restoration/Remediation

Stephen Garbaciak, Jr.[1] *and Jan A. Miller*[2]

Field Demonstrations of Sediment Treatment Technologies by the USEPA's Assessment and Remediation of Contaminated Sediments (ARCS) Program

REFERENCE: Garbaciak, S. Jr. and Miller, J. A., "**Field Demonstrations of Sediment Treatment Technologies by the USEPA's Assessment and Remediation of Contaminated Sediments (ARCS) Program,**" *Dredging, Remediation, and Containment of Contaminated Sediments, ASTM STP 1293,* K. R. Demars, G. N. Richardson, R. N. Yong, and R. C. Chaney, Eds., American Society for Testing and Materials, Philadelphia, 1995, pp. 145–154.

ABSTRACT: The Great Lakes National Program Office (GLNPO) of the U.S. Environmental Protection Agency (USEPA) has established the Assessment and Remediation of Contaminated Sediments (ARCS) Program in response to the Water Quality Act of 1987. The ARCS Program was charged with assessing and demonstrating remedial options for contaminated sediment problems in the Great Lakes. A set of technologies was identified, through a literature survey, for laboratory and field testing with sediment samples collected from five Great Lakes Areas of concern. Laboratory tests were conducted using nine processes, and pilot-scale (field-based) demonstrations of bioremediation, sediment washing, solvent extraction, and low-temperature thermal desorption were conducted at five sites. Analyses were performed on feed materials and all process residues showing polychlorinated biphenyl (PCB), polynuclear aromatic hydrocarbon (PAH), and heavy metals removal efficiencies ranging from >6% to 99%. This paper presents the results of the five pilot-scale demonstrations conducted by the ARCS Program.

KEYWORDS: sediment, Assessment and Remediation of Contaminated Sediments (ARCS) Program, treatment technology, pilot-scale demonstration, solvent extraction, sediment washing, thermal desorption, biodegradation

In natural water systems, bottom sediments serve as a sink for contaminants. The management of contaminated sediments has become an issue of critical importance to government agencies responsible for the management of waterways and the citizenry that live adjacent to such waterways. While there have been a number of remediation technologies developed for the treatment of soils, there is a general absence of data on how these technologies would perform with sediments.

The 1987 amendments to the Clean Water Act, Section 118(c)(3), authorized the USEPA GLNPO to conduct a five-year study and demonstration project relating to the control and removal of toxic pollutants from bottom sediments. In response to this authority, GLNPO organized and managed the Assessment and Remediation of Contaminated Sediments (ARCS) Program. The ARCS Program operated as a multi-agency endeavor from early 1988 through December 1993 (USEPA 1992).

---

[1] Environmental engineer, U.S. Environmental Protection Agency, Great Lakes National Program Office, Chicago, IL 60604-3590.
[2] Environmental engineer, U.S. Army Engineer Division, North Central, Chicago, IL 60606-3576.

A major aspect of the ARCS Program was the identification, selection, and testing of innovative technologies for the treatment of contaminated sediments from selected harbors and rivers on the Great Lakes. A literature search and screening process was performed, identifying 250 process options for the remediation of contaminated sediments from the Great Lakes. Of these process options, nine technologies were identified as having the greatest potential for successfully treating Great Lakes sediment, and were recommended for further study by ARCS (Averett et al. 1990).

The testing of the nine technologies was conducted through various mechanisms (consulting firms and government laboratories) at the bench- and pilot-scale level (Table 1). This paper will limit discussion to the four technologies that were demonstrated on-site, at a pilot scale, at the five ARCS Program demonstration sites. The technologies will be described, then the execution of each demonstration project, followed by a presentation of summary results from each of the demonstration projects, ending with a discussion of future technology demonstration needs.

Technologies Tested

The legislation authorizing the ARCS Program, and the subsequent Great Lakes Critical Programs Act of 1990, mandated that a pilot-scale sediment treatment technology demonstration be conducted on-site at each of the five ARCS demonstration sites. Following the laboratory testing of several promising technologies, four were selected for the on-site projects and matched with the contaminated sediments most likely to be successfully treated by the process. The sites and the technologies selected for demonstration are shown in Table 2.

Thermal desorption is the application, typically indirect, of heat to volatilize and remove the organic contaminants present in a solid matrix. The technology has established application in the treatment of hazardous wastes, sludges, and soils. A thermal desorption unit from Remediation Technologies, Inc. (RETEC) was tested at both the Buffalo River and

TABLE 1—*Technologies evaluated at bench and pilot-scale under the ARCS Program.*

	Ashtabula	Buffalo	Grand Calumet	Saginaw	Sheboygan
THERMAL					
EcoLogic					B-91
ReTec LTTS	B-91/P-92	P-91			
SoilTech ATP		B-91	B-91		B-91
Wet Air Oxidation			B-91		
PHYSICAL					
Bergmann Sediment Washing				P-91/92	
Bureau of Mines Particle Sep.	B-92	B-91	B-91	B-91	
In-situ Capping					P-90
Solidification/Stabilization		B-91/P-91	B-87		B-91
CHEMICAL/EXTRACTION					
B.E.S.T. Extraction		B-91	B-91/P-92	B-91	
KPEG Nucleophilic Substitution	B-90				B-91
BIOREMEDIATION					
EPA ERL-Athens	B-91	B-91		B-91	B-91/P-92

B-## indicates bench-scale evaluation conducted in 19##; P-## indicates pilot-scale demonstration conducted in 19##.

TABLE 2—*Sediment treatment technologies selected for pilot-scale demonstration under the ARCS Program.*

Demonstration Site	Technology
Ashtabula River	Thermal desorption
Buffalo River	Thermal desorption
Grand Calumet River	Solvent extraction
Saginaw River	Sediment washing
Sheboygan River	Bioremediation

Ashtabula River. The RETEC unit is based on twin hollow-augers using a molten eutectic material to heat sediments to temperatures between 150 and 260°C, volatilizing water and any organic contaminants present in the sediment matrix. This type of technology has been installed in full-scale applications, with a unit from the company SoilTech being employed at the Outboard Marine Company/Waukegan Harbor, Illinois, Superfund site for the remediation of 6000 m^3 of PCB-contaminated sediments. The technology condenses the volatilized contaminants and collects them as an oily residue of substantially less volume than the original sediment mass. Although no net destruction of contaminants is affected by this technology, the remaining volume of contaminated oil requiring further treatment is much smaller, thus potentially allowing the application of a costly destructive process in a more economical fashion.

Solvent extraction is the stripping and removal of organic contaminants from a solid or liquid matrix using the chemical and physical properties of a particular solvent or solvents. The Basic Extractive Sludge Treatment Process, or B.E.S.T. process, offered by the Resources Conservation Company, was used in the ARCS Program demonstration at the Grand Calumet River. The B.E.S.T. process uses triethylamine, a solvent with unique inverse miscibility properties, to separate organic contaminants and concentrate them as an oily residue. The B.E.S.T. process has had limited full-scale application as a waste treatment technology. As with thermal desorption, solvent extraction technologies do not destroy contaminants, instead concentrating them in a greatly reduced volume for further destructive treatment.

Sediment washing techniques such as hydrocyclones are used routinely in the mining and mineral processing industries to separate slurries into sets of different-sized particles. Contaminants tend to associate themselves with the finer-grained silt and clay particles, which represent only 20 to 30% of Saginaw River sediments. The sediment washing equipment produces separated fractions of sand, organic debris, and silts and clays. The organic contaminants, and some heavy metals, are concentrated in the silt and clay fraction. No contaminants are destroyed during the sediment washing process; however, the volume of contaminated material represented by the silts and clays is much smaller than the volume of the original contaminated sediment matrix. A sediment washing system developed by Bergmann USA was demonstrated on sediments from the Saginaw River.

Bioremediation, or the use of indigenous microorganisms to degrade and destroy organic contaminants, has received widespread public attention in the cleanup of oil spills in coastal areas. As a sediment treatment technology, the complete destruction of PCBs was attempted by cycling sediments between aerobic and anaerobic conditions. Such cycling was necessary since highly chlorinated PCBs are only degraded by anaerobic microorganisms, and then only to lower chlorinated compounds. Aerobic microorganisms can then take these compounds with three or less chlorine atoms and completely mineralize them. This technology was demonstrated on sediments from the Sheboygan River.

Methods

The pilot-scale demonstrations projects conducted under the ARCS Program began field operations in October 1991 and were completed in December 1992. The following section presents a description of the field operations at each of the five sites, in chronological order.

Buffalo River

The demonstration of the RETEC thermal desorption unit took place at the U.S. Army Corps of Engineers' Dike No. 4 Confined Disposal Facility (CDF) from 21–31 October 1991. Cold weather forced the demonstration project to be completed at RETEC's treatability laboratory in Acton, MA from 18–25 November 1991. An area of 1000 m^2 was cleared and leveled by RETEC, who then set up the two trailers that contained the thermal desorption unit and necessary support equipment (USACE Buffalo District 1993).

A clamshell dredge was used to collect 9 m^3 of contaminated sediment from an area of the Buffalo River known to contain elevated levels of PAHs. The sediments were screened of particles greater than 2 cm and placed into barrels for storage prior to processing. A total of 12 sediment batches was processed through the thermal desorption unit during the demonstration. Residence time, temperature, and initial moisture content were the process variables manipulated during the project.

Samples were collected of all process streams entering and exiting from the thermal desorption unit, including monitoring of the gas emissions for the formation of dioxins and dibenzofurans. The critical contaminants for the evaluation of the technology were PAHs, with PCBs and heavy metal contaminants being measured as well.

Saginaw River

The demonstration of the Bergmann USA sediment washing process took place on a barge-mounted plant anchored off of the U.S. Army Corps of Engineers' Saginaw Bay CDF from 31 October through 1 November 1991. As in the demonstration at the Buffalo River, cold weather forced the suspension of the project until the spring of 1992. The demonstration was resumed on 17 May and concluded on 29 May 1992. Following the ARCS Program demonstration, the USEPA Superfund Innovative Technology Evaluation (SITE) program continued operating the Bergmann USA equipment for an additional five days (USACE Detroit District 1994).

A clamshell dredge was used to collect 230 m^3 of contaminated sediments from an area of the Saginaw River known to contain moderate to low levels of PCBs. The sediments were transported by barge to the CDF, where they were stored in a cleared area which was lined with a geotextile, prior to processing.

The Bergmann USA sediment washing unit was comprised of a number of individual pieces of equipment that separated sediment particles based on size or density differences. Several different combinations of equipment were used during the demonstration, with the primary goal of the project to separate the sand particles from the silt and clay particles, making a cut around 75 μm in particle diameter. The main chemical contaminant monitored during the demonstration was PCBs, with the behavior of heavy metals considered as well.

Sheboygan River

At the Sheboygan River, the ARCS Program cooperated with a project that was in development through the Superfund program. Tecumseh Products, a potentially responsible party to the PCB contamination problem in the river, had constructed a sheetpile steel container, called a confined treatment facility (CTF), and filled it with 2100 m^3 of sediment from the river. The CTF was built with a piping system in the bottom to allow the delivery of oxygenated water and nutrient amendments to the sediments. The goal of the demonstration was to determine if PCBs in the Sheboygan River sediments could be either further dechlorinated anaerobically and then if they could be completely degraded aerobically (Blasland and Bouck and USEPA 1992).

Two experimental and two control cells were sectioned off in the CTF. Both control cells were maintained in anaerobic conditions by keeping them saturated with water. In the first experimental cell, aerated water was pumped through the sediments in an effort to stimulate the aerobic biodegradation of lower-chlorinated PCBs. In the second experimental cell, nutrient amendment was added to stimulate the further anaerobic dechlorination of the PCB molecules. The cell manipulations began in June 1992, and samples were collected from each CTF cell three times in 1992 and again in the spring of 1993.

Grand Calumet River

The B.E.S.T. process was demonstrated on sediments from the Grand Calumet River during July 1992, at a site provided by U.S. Steel Gary Works, along the river. This project was also a cooperative effort, between the ARCS Program and the Superfund Innovative Technology Evaluation (SITE) program. A hollow-pipe sampler was used to collect 1 m^3 of sediment from two sample locations in the Grand Calumet River. The material was homogenized and screened of all particles greater than 3 mm in diameter prior to processing in the B.E.S.T. unit (USACE Chicago District 1994).

The demonstration consisted of the processing of two sediment samples at three different operating conditions. One set of operating conditions was then repeated two additional times, for a total of ten batch runs of the unit. Samples were collected from all feed materials and process residuals to determine the efficiency of the unit in extracting PCBs and PAHs from the sediments.

Ashtabula River

The demonstration at the Ashtabula River was very similar to the project on the Buffalo River, with the same RETEC unit being used. The unit was set up on an unoccupied slip at Jack's Marine, a private marina on the Ashtabula River. A backhoe excavator was used to collect 9 m^3 of PCB-contaminated sediment from the river, and again the material was screened of particles greater than 2 cm and stored in barrels prior to processing (USACE Buffalo District 1994).

The RETEC unit was operated at the Ashtabula River during September 1992, processing 12 sediment samples. After the experience of the Buffalo demonstration, the monitoring plan for Ashtabula was modified to expand the sampling and analysis of the air emissions and activated carbon filters, to better determine the fate of certain heavy metals species. Additionally, the focus of the Ashtabula demonstration was on the ability of the RETEC unit to remove PCBs and other chlorinated hydrocarbons, as opposed to the focus on PAH removals at the Buffalo demonstration.

Results and Discussion

Extensive sampling and analyses were conducted at all five demonstration projects. This paper can give, at best, only a brief summary of the major highlights of each project. For each demonstration, the results related to the parameter or parameters that were the focus of that project are presented. The reader is encouraged to contact the first author for further information on the detailed results of each project, including requests for copies of the project reports.

Buffalo River Results

At this site, PAH concentrations were of primary concern. Seventeen individual PAH compounds were analyzed, with an average concentration in the feed sediments of 7.9 mg/kg and an average concentration in the treated solids produced by the RETEC process of 1.6 mg/kg. The relatively low contamination levels in the feed sediments, and the variation inherent in the analytical techniques for PAHs, make conclusions about the effectiveness of the thermal desorption process difficult. The data do show that the process was able to remove 78% of the PAHs from the sediments, on average. The results for PAHs and other parameters are presented in Table 3. Estimates developed for the use of this thermal desorption process for the remediation of Buffalo River sediments were $535 per cubic yard for a 10 000 yd^3 project and $352 per cubic yard for a 100 000 yd^3 project, excluding dredging costs and the cost of disposing of the treated solids residual. (NOTE: 1 yd^3 = 0.765 m^3.)

Saginaw River Results

For the Saginaw demonstration, the main performance parameters were the grain size distributions in the process residuals and the PCB concentrations in each of those streams. Results are summarized in Table 4. The feed material contained approximately 77% sand-sized particles greater than 75 μm in diameter. The coarse residual fraction that was produced by the Bergmann unit was comprised of 94% sand particles (>75 μm), showing that the unit was able to very effectively separate coarse from fine-grained particles.

The overall average concentration for PCBs in the feed sediments was 1.2 mg/kg, with a range of 0.74 to 3.2 mg/kg. The washed sand fraction produced by the process contained an average of 0.21 mg/kg PCBs, with a range of 0.14 to 0.38 mg/kg. This calculates to an average reduction in the PCB concentration from the raw sediments to the washed sand fraction of nearly 83%.

There were two other major residual streams generated by the Bergmann unit: particulate organics, which were a very small portion of the feed material, and the fine-grained fraction,

TABLE 3—*Summary of results from the Buffalo River thermal desorption demonstration.*

Parameter	Feed Material	Treated Solids	Percent Removal
Total PAHs	7.9	1.6	78
	(6.7 – 11.6)	(0.2 – 4.0)	(43 – 98)
Total PCBs	0.20	<0.09	>57
	(<0.02 – 0.35)	(<0.02 – 0.32)	(9 – >93)
Mercury	0.19	<0.06	>71
	(0.17 – 0.21)	(<0.0003 – 0.15)	(17 – >99.9)

All results are in mg/kg dry weight, presented as mean (range) for eight samples.

TABLE 4—*Summary of results from the Saginaw River sediment washing demonstration.*

Material	Sand[a]	Silt/Clay[b]	Total PCBs[c]	Heavy Metals[d]
Feed	77%	23%	1.2	170
Washed Sand	95%	5%	0.21	45.5
Fines	...	100%	4.6	743
Particulate Organics	94%	6%	3.9	344
Trommel Overs	59%	41%	NA	NA

NA = Not analyzed.
[a] Particles greater than 75 μm in diameter.
[b] Particles less than or equal to 75 μm in diameter.
[c] Concentrations in mg/kg dry weight, means of 10-14 samples.
[d] Sum of concentrations of copper, chromium, lead, nickel, and zinc, in mg/kg dry weight, means of 8-20 samples.

which was approximately 24% of the raw feed. The PCB concentrations in the particulate organics averaged 3.9 mg/kg, with a range of 1.64 to 9.37 mg/kg. The PCB concentrations in the fine-grained fraction averaged 4.6 mg/kg, much lower than expected from the concentration of the original mass of PCBs in a much smaller mass of sediment fine-grained particles. Analytical error may account for some of the discrepancy, while volatile losses may also be responsible for the rest. Estimates developed for the use of this sediment washing process for the remediation of Saginaw River sediments were $54 per cubic yard for a 10 000 yd^3 project and $24 per cubic yard for a 100 000 yd^3 project.

Sheboygan River Results

The data for the Sheboygan River bioremediation demonstration are still being analyzed at the time of the writing of this paper. Preliminary results show that the sediments originally placed into the CTF had an average PCB concentration of approximately 175 mg/kg, and that substantial anaerobic dechlorination had already taken place in the riverbed. Early results have also shown that insufficient oxygen was delivered to produce aerobic conditions in the sediments, thus leaving the prospects for demonstrating complete degradation of PCBs slim.

Grand Calumet River Results

PCBs and PAHs were again the main parameters of concern in the demonstration of the B.E.S.T. process. Two sediment samples were treated during the demonstration, sediment A and B. Sediment A contained, on average, 12.1 mg/kg PCBs and 548 mg/kg total PAHs; the treated solids residuals from the processing of sediment A contained, on average, 0.04 mg/kg PCBs and 22 mg/kg PAHs, for net removal efficiencies of 99.7% for PCBs and 96.0% for PAHS. Sediment B contained, on average, 425 mg/kg PCBs and 70 920 mg/kg total PAHs; the treated solids residuals from the processing of sediment B contained, on average, 1.8 mg/kg PCBs and 510 mg/kg PAHs, for net removal efficiencies of 99.5% for PCBs and 99.3 percent for PAHs. Results are summarized in Table 5. Additional analyses (not presented here) demonstrated that the PCBs and PAHs were captured by the solvent and oil separation equipment and concentrated in the oil fraction produced by the B.E.S.T. unit. Estimates developed for the use of this solvent extraction process for the remediation of Grand Calumet River sediments were $174 per cubic yard for a 25 000 yd^3 project and

TABLE 5—*Summary of results from the Grand Calumet River solvent extraction demonstration.*

	Sediment A	Sediment B
Total PCBs		
Feed Material	12.1	425
Treated Solids	0.04	1.8
Percent Removal	99.7	99.5
Total PAHs		
Feed Material	550	70 920
Treated Solids	22	510
Percent Removal	96	99.3

All concentrations in mg/kg, dry weight, mean of three samples.

$139 per cubic yard for a 100 000 yd^3 project, excluding dredging and solids residual disposal costs.

Ashtabula River Results

Thermal desorption was evaluated under the ARCS Program at both the Buffalo and Ashtabula Rivers, with different target compounds. As reported above, the process was shown to be effective for the PAH contaminants in the Buffalo River sediments. In the Ashtabula River sample, PCBs were the main compounds of concern, with a few additional chlorinated organic compounds of secondary interest. Summary results are presented in Table 6.

Despite lower than expected feed concentrations for the non-PCB-chlorinated organic contaminants, the thermal desorption unit was able to remove a high percentage of those compounds. Similar to the Buffalo River results, the process was also very effective in the removal of mercury, and the enhanced monitoring program put in place for this demonstration was able to show that the volatilized mercury was captured in the activated carbon stack, the result of both adsorption and physical filtration of mercury particulates. Costs for this process are expected to be similar to those developed for the application at the Buffalo River, since the technology is the same.

TABLE 6—*Summary of results from the Ashtabula River thermal desorption demonstration.*

Parameter	Feed Material	Treated Solids	Percent Removal
Total PCBs	4.0	0.28	>89
	(1.8 – 12)	(<0.20 – 0.58)	(68 – >98)
Hexachlorobutadiene	0.09	<0.02	>79
	(<0.02 – 0.31)		(>67 – >93)
Hexachlorobenzene	0.78	0.02	>94
	(0.16 – 3.3)	(<0.02 – 0.03)	(89 – >99)
Chlorobenzene	0.051	<0.0002	>97
	(0.003 – 0.200)		(>26 – >99.9)
1,2-dichloro-benzene	0.0043	<0.0002	>90
	(<0.0002 – 0.2300)		(>87 – >99.1)
Mercury	1.3	0.06	91
	(0.3 – 5.4)	(0.02 – 0.11)	(68 – 98)

All results are in mg/kg dry weight, presented as mean (range) for nine samples.

Future Research Directions

Contaminated sediments exhibit a number of characteristics such as high moisture content, small particle size, and relatively low contamination levels that make material handling and subsequent treatment difficult. While the technologies tested were effective, the treatment of sediments will, in most cases, be more costly than the treatment of soils. A need exists for the development of technologies that will be cost-effective in the remediation of contaminated sediments.

Further refinements of the sediment washing techniques demonstrated at the Saginaw River are also necessary prior to their widespread implementation as a sediment remediation technique. Specifically, quantification of the volatile losses of contaminants during the aggressively agitated unit operations that make up the sediment washing system are needed to ensure that air emissions are not of concern. The USEPA GLNPO is currently funding efforts by the U.S. Department of the Interior, Bureau of Mines Salt Lake City Research Center and the U.S. Army Engineer Waterways Experiment Station to investigate the volatile losses from sediment washing processes.

Disclaimer

Mention of trade names or commercial products does not constitute endorsement or recommendation for use by USEPA or the U.S. Army Corps of Engineers. This paper has not been subject to USEPA review procedures and does not reflect Agency opinion or policy.

Acknowledgments

The authors wish to give special thanks to the project managers at the U.S. Army Corps of Engineers who managed these demonstration projects for the ARCS Program: Thomas Kenna (Buffalo), James Galloway and Frank Snitz (Saginaw), Linda Diez and Jay Semmler (Grand Calumet), and David Conboy (Ashtabula). The authors also wish to recognize that portions of this paper have been presented or published previously (Garbaciak et al. 1993 and Timberlake and Garbaciak 1995).

References

Averett, D. E., Perry, B. D., Torrey, E. J., and Miller, J. A., 1990, *Review of Removal, Containment, and Treatment Technologies for Remediation of Contaminated Sediment in the Great Lakes*, U.S. Army Engineer Waterways Experiment Station, Vicksburg, MI.

Blasland and Bouck Engineers, P.C. and USEPA-Great Lakes National Program Office, 1992, *Sheboygan River and Harbor Biodegradation Pilot Study Work Plan*, Blasland and Bouck Engineers, P.C., Syracuse, NY.

Garbaciak, S. Jr., Fox, R. G., Tuchman, M. L., et al., 1993, "Laboratory and Field Demonstrations of Sediment Treatment Technologies by the United States Environmental Protection Agency's Assessment and Remediation of Contaminated Sediments (ARCS) Program," *Proceedings, CATS II Congress 1993: Characterisation and Treatment of Contaminated Dredged Material*, Antwerp, Belgium, Technologisch Instituut, Antwerp, Belgium, pp. 3.15–3.24.

Timberlake, D. and Garbaciak, S. Jr., 1995, "Bench-Scale Testing of Selected Remediation Alternatives for Contaminated Sediments," *The Journal of the Air and Waste Management Association*, Vol. 45, pp. 52–56.

U.S. Army Corps of Engineers Buffalo District, 1993, *Pilot-Scale Demonstration of Thermal Desorption for the Treatment of Buffalo River Sediments*, EPA 905-R93-005, USEPA-Great Lakes National Program Office, Chicago, IL.

USACE Buffalo District, 1994, *Pilot-Scale Demonstration of Thermal Desorption for the Treatment of Ashtabula River Sediments*, EPA 905-R94-021, USEPA-Great Lakes National Program Office, Chicago, IL.

USACE Chicago District, 1994, *Pilot-Scale Demonstration of Solvent Extraction for the Treatment of Grand Calumet River Sediments*, EPA 905-R94-003, USEPA-Great Lakes National Program Office, Chicago, IL.

USACE Detroit District, 1994, *Pilot-Scale Demonstration of Sediment Washing for the Treatment of Saginaw River Sediments,* EPA 905-R94-091, USEPA-Great Lakes National Program Office, Chicago, IL.

U.S. Environmental Protection Agency, 1992, *Assessment and Remediation of Contaminated Sediments (ARCS) Program 1992 Work Plan*, USEPA-Great Lakes National Program Office, Chicago, IL.

Matthew B. Hollifield,[1] *Jae K. Park,*[2] *William C. Boyle,*[2] *and Paul R. Fritschel*[2]

Factors Influencing the Development of a Biostimulant for the *In-Situ* Anaerobic Dechlorination of Polychlorinated Biphenyls in Fox River, Wisconsin Sediments

REFERENCE: Hollifield, M. B., Park, J. K., Boyle, W. C., and Fritschel, P. R., "**Factors Influencing the Development of a Biostimulant for the *In-Situ* Anaerobic Dechlorination of Polychlorinated Biphenyls in Fox River, Wisconsin Sediments,**" *Dredging, Remediation, and Containment of Contaminated Sediments, ASTM STP 1293,* K. R. Demars, G. N. Richardson, R. N. Yong, and R. C. Chaney, Eds., American Society for Testing and Materials, Philadelphia, 1995, pp. 155–169.

ABSTRACT: Polychlorinated biphenyl (PCB) contaminated sediments were collected from the Fox River, Wisconsin, and analyzed for the possible occurrence of reductive dechlorination. Evidence of *in-situ* dechlorination was observed. However, the extent of this *in-situ* dechlorination was less than that typically reported in the literature, suggesting that stimulation of further dechlorination was possible.

The use of nutrients and surfactants was explored for stimulating additional dechlorination. The nutrient amendment reported here was found to be inhibitory. Surfactants had varying effects, but none significantly improved dechlorination over control treatments. The most significant factors were observed to be the initial extent of dechlorination and PCB concentration. Additional dechlorination was most likely to be observed in sediments with higher PCB concentration and less initial dechlorination. All sediments converged on a common dechlorination level regardless of the initial state of the sediments.

KEYWORDS: polychlorinated biphenyl (PCB), bioremediation, dechlorination, reductive dehalogenation, biostimulant, nutrients, surfactants, Fox River

Sediments contaminated with polychlorinated biphenyls (PCBs) may pose a serious threat to ecosystems and human health. Since PCBs sorbed to sediments are a continuous source of pollution through desorption mechanisms in the Great Lakes and other bodies of water (Manchester 1993), decontamination of these sediments is desirable. Recent research has shown that anaerobes are capable of dechlorinating highly chlorinated PCB congeners (Quensen 1988). The resulting mixture of lightly chlorinated congeners is amenable to mineralization by aerobic organisms (Abramowicz 1990; Furukawa 1986). However, degradation does not occur in all environments; a strategy to stimulate this phenomenon in a wider variety of environments is desirable.

---

[1] Engineer, Radian Corporation, Milwaukee, WI 53214.
[2] Associate professor, professor, and associate instrument technician, respectively, Department of Civil and Environmental Engineering, University of Wisconsin–Madison, Madison, WI 53706.

The results of initial efforts to determine the appropriate formula for a biostimulant for use with indigenous microbial communities in *in-situ* bioremediation projects are presented below. The effect of a nutrient solution of mineral salts, trace metals, and vitamins has been explored. Additionally, the use of surfactants to increase the bioavailability of the PCBs is also reported. The ultimate goal of this work is to combine the biostimulant formula with a cost-effective delivery vehicle. The conceptual approach for the delivery of the biostimulant involves mixing the biostimulant with a biodegradable but slowly hydrolyzing material (retardate) to make a timed-release pellet. These pellet would be applied *in-situ* to remediate contaminated sediments.

Literature Review

Nomenclature and Background

Polychlorinated biphenyls are a family of chlorinated aromatic compounds. A PCB consists of a biphenyl molecule ($C_{12}H_{10}$) with varying degrees of chlorine substitution. The resulting chemical formula of a PCB is $C_{12}H_{(10-n)}Cl_n$, where n is the number of chlorines substituted on the biphenyl. Theoretically, there are 209 possible PCBs, commonly referred to as congeners. Throughout this text, congeners are described by the unique congener numbers proposed by Ballschmitter and Zell (1980) and adopted by IUPAC.

The PCB-contaminated sediments used in this research were obtained from the lower Fox River, WI. The lower Fox River is a 64 km reach of highly industrialized river basin beginning at the outlet of Lake Winnebago, near the cities of Neenah and Menasha, WI, and discharging into Green Bay at Green Bay, WI. Wastewater discharges from the deinking of recycled paper fibers have been identified as one of most significant sources of PCB discharges to the Fox River (Peterman et al. 1980). The primary form of PCB discharge was identified as Aroclor 1242, but identification of traces of Aroclor 1248 and 1254 were also reported (Wisconsin Department of Natural Resources 1978). Sediment contamination levels have been reported to range from <0.2 mg/kg to 100 mg/kg (Sullivan et al. 1983; Peterman et al. 1980; Wisconsin Department of Natural Resources 1978), and the highest sediment concentrations have been observed at the southern end of Little Lake Buttes des Morts. Recently, Manchester (1993) has reported that the Fox River is virtually the sole source of PCB contamination to Green Bay. Additionally, Green Bay has been identified as one of 42 areas of concern by the International Joint Commission. Therefore, the remediation of PCB contaminated sediments in the Fox River could significantly attenuate the PCB loading to Green Bay, and subsequently, Lake Michigan.

Anaerobic Dechlorination of PCBs

Anaerobic dechlorination of PCBs is a biologically mediated process where the PCB acts as an electron acceptor and is reduced through the removal of chlorine atoms. In general, anaerobic dechlorination changes more highly chlorinated congeners to more lightly chlorinated congeners. Anaerobic dechlorination does not result in the destruction of PCB molecules since the microorganisms only use the PCBs as electron acceptors, not electron donors.

The possibility of anaerobic alteration of PCBs was first proposed by Brown et al. (1984) following a review of PCB patterns in Hudson River sediments. Additional dechlorination patterns have subsequently been reported and discussed by Brown et al. (1987a, 1987b), Brown 1990, Quensen et al. (1990), Morris et al. (1992), Brown and Wagner (1990), and Abramowicz et al. (1989). Quensen et al. (1988) confirmed that this alteration could be biologically mediated anaerobic dechlorination in a series of laboratory experiments. Factors that have been reported to affect dechlorination and that are relevant to the results reported here are discussed below.

Nutrients—There is almost universal agreement that some form of nutrient amendment is beneficial to the dechlorination of PCBs. Almost all studies have been performed in a minimal media containing a complex mixture of buffer, mineral salts, and trace metals. However, Abramowicz et al. (1993) reported that the trace metals component of their media alone was just as effective as the entire media, indicating trace metals were the limiting nutrient in their Hudson River sediments. It is logical to assume that different components may be limiting in different sediment types. However, methods for the estimation of the bioavailability of sediment nutrients to microorganisms are lacking.

Surfactants—Abramowicz et al. (1993) observed sodium dodecyl benzene sulfonate (SDBS) to completely inhibit dechlorination in Hudson River sediments. Additionally, Triton X-100 and X-405 decreased the dechlorination rate to approximately one eighth of the rate observed in untreated controls. Contrary to the effects of the other surfactants, the high molecular weight surfactant Triton X-705 decreased the lag period prior to the onset of dechlorination by approximately one half, and slightly increased the extent of dechlorination. Abramowicz et al. (1993) also note that surfactants had positive effects in Woods Pond sediments, where dechlorination has not been as widespread as in Hudson River sediments.

Concentration Effects—Abramowicz et al. (1993) examined dechlorination rates over a range of initial PCB concentrations of 20 to 1500 ppm and observed that the greatest changes in total meta (m) para (p) chlorine levels per biphenyl occurred at 20 ppm. However, other researchers have noted difficulty in achieving dechlorination at these low levels (Quensen et al. 1988; Rhee et al. 1993a, 1993b). In contrast to observations relating the greatest removal of chlorine per biphenyl, Abramowicz et al. (1993) observed the most rapid specific activity (chlorine removed per unit time) above 750 ppm (including samples as high as 1500 ppm). At these higher levels, the rates were observed to become independent of concentration. At lower levels (0 to 250 ppm), the relationship between concentration and rate was nearly linear. Conversely, Quensen et al. (1988) observed an increase in dechlorinating activity with increasing concentration when dechlorination was expressed on a chlorine removed per biphenyl basis. Also, Rhee et al. (1993c) observed no dechlorinating activity at high levels (1000 and 1500 ppm) and reported that dechlorinating activity decreased with decreasing concentration below an upper activity limit of approximately 1000 ppm. Although the method of calculating the change in chlorine levels (change per biphenyl molecule versus chlorine removed per unit time) can affect the interpretation of results, the greatest dechlorination rates generally occur at higher PCB concentrations.

Other Factors—Assaf-Anid et al. (1992) report that vitamin B_{12} catalyzed the reductive dechlorination of a single congener in an abiotic system. However, Abramowicz et al. (1993) noted that the addition of vitamin B_{12} inhibited the dechlorination of a single congener in a biological system. Abramowicz et al. (1993) note that several researchers have linked vitamin B_{12} to the inactivation of certain enzymes.

Methodology

Sample Collection and Reactor Preparation

PCB-contaminated sediments were collected from the southern portion of Little Lake Butte des Morts, a portion of the lower Fox River, in the city of Neenah, WI, on 13 October 1991. Samples were collected with a metal core sampler, packed into 1-L mason jars sealed with Teflon® lined lids, transported under ice, and stored at 4°C upon receipt in the laboratory.

The laboratory reactors consisted of 150 mL serum bottles sealed with Teflon lined rubber septa and aluminum crimp caps. Anaerobic conditions were maintained during reactor setup and later sampling by flushing the serum bottles with a 60:40 (v:v) mixture of $N_2:CO_2$ rendered O_2-free by passage over a heated copper column. Each reactor contained approximately 75 g (wet weight) of Fox River sediments. Reactors receiving the nutrient amendment were filled to volume with a reduced anaerobic mineral medium (RAMM) to create a slurry. Reactors not receiving nutrients were filled to volume with deoxygenated site water. Controls (no amendments, but biologically active) and nutrient amended reactors were prepared in triplicate. Surfactant amended reactors were prepared in duplicate with the contents outlined below.

The nutrient solution, or RAMM, was composed of a buffer, macronutrients, trace nutrients, and vitamins. The buffer, macronutrient, and trace nutrient components were essentially the revised reduced anaerobic mineral medium of Shelton and Tiedje (1984). The vitamin solution was adapted from Wolin et al. (1963). The final RAMM contained the following: 272 mg/L KH_2PO_4, 348 mg/L K_2HPO_4, 535 mg/L NH_4Cl, 78.5 mg/L $CaCl_2 \cdot 2H_2O$, 101.5 mg/L $MgCl_2 \cdot 6H_2O$, 20 mg/L $FeCl_2 \cdot 4H_2O$, 500 µg/L $MnCl_2 \cdot 4H_2O$, 50 µg/L H_3BO_3, 50 µg/L $ZnCl_2$, 30 µg/L $CuCl_2$, 10 µg/L $Na_2MoO_4 \cdot 2H_2O$, 500 µg/L $CoCl_2 \cdot 6H_2O$, 50 µg/L $NiCl_2 \cdot 6H_2O$, 50 µg/L Na_2SeO_3, 20 µg/L biotin, 20 µg/L folic acid, 60 µg/L lipoic acid, 50 µg/L thiamin, 50 µg/L riboflavin, 50 µg/L nicotinic acid, 100 µg/L pyridoxal-HCl, 50 µg/L panthothenic acid, 50 µg/L cyanocobalamin (B_{12}), 50 µg/L p-aminobenzoic acid, 2.4 g/L $NaHCO_3$, 240 mg/L Na_2S as a reducing agent, 29 mg/L NaCl, 1 mL/L resazurin as a redox indicator, 164 mg/reactor of supplemental cyanocobalamin (vitamin B_{12}), and ASTM Type I deionized water.

Three surfactants were investigated, including sodium dodecyl sulfate (SDS), an anionic surfactant; Triton X-705, a high molecular weight nonionic surfactant; and Tween 20, also a high molecular weight nonionic surfactant. The surfactants were used at three concentrations in combination with the nutrient solution. The surfactant concentrations employed were 0.25, 0.50, and 1.00% for SDS (w:v) and Tween 20 (v:v) and 0.28, 0.60, and 1.00% for Triton X-705 (v:v). Reactors containing each surfactant at 1.00% (v:v for Triton X-705 and Tween 20 and w:v for SDS) and deoxygenated site water were also prepared.

Reactors were wrapped in aluminum foil to protect light-sensitive components of the RAMM and were stored at a constant temperature of 20°C. Reactors were incubated statically; no shaking, regular substrate amendments, or gas venting was performed.

Reactor Sampling and Extraction/Cleanup

Reactors were sampled at 8, 12, 18, 25, 38, and 52 weeks. During sampling, reactors were thoroughly mixed by shaking and anaerobic conditions were maintained as previously described. A sample volume of 4 to 5 mL of sediment-water slurry was collected with an autopipette fitted with a 2 mL disposable pasteur pipette. Samples were stored in tightly sealed 8 mL glass screw cap vials with Teflon-lined caps and transferred to a laboratory freezer.

All solvents used in the extraction and preparation of samples were pesticide grade reagents, and the sulfuric acid used in the cleanup was reagent grade. All steps requiring deionized water employed ASTM Type I water generated in the laboratory. Salts used in cleanup steps were Soxhlet extracted with a 60:40 (v:v) mixture of acetone/hexane prior to use. Metallic copper was cleaned with sulfuric acid, rinsed with two volumes of acetone, one volume of hexane, and dried on a rotary evaporation device. Pesticide grade silica gel was prepared by activation at or above 130°C and deactivation with 3% H_2O addition. Glassware was scrupulously cleaned in warm water containing a laboratory detergent, rinsed with deionized water, and ashed at 500°C for 6 to 7 h. Volumetric pipettes were soaked in dichromic acid, rinsed with deionized water, dried in a 110°C oven, and rinsed with four volumes of hexane and three volumes of iso-octane.

A 2 mL volume of sediment water slurry was spiked with 1 mL of a surrogate solution containing 3.60 µg/L of congener 014, 1.12 µg/L of congener 065, and 0.91 µg/L of congener 165. The resulting mixture was extracted with 40 mL of acetone for 1 h in an ultrasonic bath, followed by an additional hour of extraction with 30 mL of methylene chloride. Manchester (1993) has reported that this sonication technique resulted in the recovery of PCB mass that was statistically comparable to that of the Soxhlet method at the $p = 0.01$ level. Additionally, there were no differences in the congener distributions between samples extracted by sonication and those extracted by the Soxhlet technique.

The acetone/methylene chloride mixture was filtered into a separatory funnel containing 100 mL of deionized water and 11 mL of 2% NaCl. The water/acetone phase was separated from the methylene chloride phase by back-extraction with a total of three volumes of methylene chloride (the initial 30 mL volume from the ultrasonic extraction and two additional 20 mL volumes). Metallic copper was added, and the methylene chloride phase was reduced in volume on a rotary evaporator followed by transfer to a hexane phase. The sample was transferred to a Na_2SO_4/silica gel column containing 3 g of silica gel topped with 2.5 cm of Na_2SO_4 (1 cm inside column diameter). This column had previously been prepared by elution with 60 mL of methylene chloride and 20 mL of petroleum ether. The sample was dried on the sodium sulfate and the PCBs were separated from other compounds on the silica gel. The PCBs were eluted from the column with 60 mL of petroleum ether, and the resulting extract was extracted with 10 mL of concentrated sulfuric acid and then 20 mL of deionized water. This extract was eluted through a Na_2CO_3/Na_2SO_4 column (approximately 2.5 cm of each in a 1 cm inside diameter column) to remove traces of acid and to dry the sample. The sample was reduced in volume on a rotary evaporator and transferred to an iso-octane phase. The iso-octane phase was transferred to a volumetric flask and brought to a final volume of 5 mL. Extracts were prepared for gas chromatographic analysis by diluting 1:10 with iso-octane and spiking with 50 µL of an internal standard solution containing 2.40 µg/mL of congener 030 and 1.01 µg/mL of congener 204.

Gas Chromatographic Analysis

Gas chromatographic analyses were performed on a gas chromatograph equipped with an autosampler, electron capture detector, and a microcomputer-based workstation for identification, integration, and quantitation of analytical results. A 30 m DB-5 fused silica capillary column with a 0.25 mm inside diameter and 0.25 µm film thickness was used in all analyses. The carrier gas was helium (linear velocity of 40 cm/s) and O_2-free nitrogen was used as the makeup gas. Both the injector and detector temperatures were held constant and 300 and 350°C, respectively. The sample injection volume was 1.5 µL. The analysis program consisted of an initial hold of 1.00 min at an initial column temperature of 90°C. This was followed by an 0.80°C/min ramp to 170°C with no hold, a 1.0°C/min ramp from 170 to 245°C with no hold, and a final 20.0°C/min ramp from 245 to 300°C followed by a 30.00 min hold.

The calibration standard used for the calculation of relative response factors was composed of a solution containing Aroclors 1232, 1248, and 1262 at ratios of 25:18:18 (mass basis) at a total concentration of 915 ng/mL. The composition of this mixture has been determined by Mullin (1985). Additionally, the surrogates and internal standards were added to this calibration standard at concentrations of 72.0 ng/mL, 25.0 ng/mL, 22.4 ng/mL, 18.2 ng/mL, and 10.2 ng/mL for congeners 014, 030, 065, 166, and 204, respectively. Mullin (1985) identified 106 congeners or coeluting groups in this mixture (excluding surrogates and internal standards). A total of 95 congeners or coeluting groups was identified in this mixture due to the absence of five peaks and the inability to resolve six coeluting groups.

Quantitation and Data Reduction

Peaks eluting prior to and including 077 + 110 were quantitated relative to the first internal standard, congener 030, and the balance were quantitated relative to the second internal standard, congener 204. All analytical results were calculated using relative response factors derived from analysis of the calibration mixture described above. Calculation of surrogate spike recoveries were performed by dividing the analytical result for the surrogate by the theoretical concentration at 100% recovery. Once recoveries were calculated, the results for the individual congeners were corrected. Monochloro, dichloro, and trichlorobiphenyls were corrected by dividing by the recovery of congener 014; tetrachloro, pentachloro, and hexachlorobiphenyls were corrected by dividing by the recovery of congener 065; and heptachloro, octachloro, and nonachlorobiphenyls were corrected by dividing by the recovery of congener 166. Correcting results for recoveries was necessary because not all congeners were recovered uniformly. Without correcting for recovery, analytical results would not have been directly comparable with Aroclor product compositional data. After correcting results for extraction recovery, the molar percentage of each congener in the sample was calculated. When coeluting congener groups were composed of congeners from different homologs, the average molecular weight was used to calculate the moles in the congener group, and the resulting molar percentage was distributed equally among the homologs of each congener present in the group.

Results

PCB Composition of Fox River Sediments

A summary of the PCB composition of the original Fox River sediments prior to laboratory treatment is presented in Table 1. The composition of Aroclor 1242 is included for comparison (Manchester 1993). Information on the physical properties of the sediments, nutrients, and trace nutrients is presented in Tables 2 through 5. Examination of the average chlorines in the *m*, *p*, and ortho (*o*) positions as shown in Table 1 reveals important trends. First, the total average chlorine per biphenyl is lower in the Fox River sediments than that which is typical of Aroclor 1242. Secondly, the levels of *meta* chlorines in all three Fox River sediments are well below what is typical for Aroclor 1242. However, the levels of *ortho* and *para* chlorines are similar to those expected for Aroclor 1242, but slightly elevated. Thirdly, higher levels of *para* chlorines appear in conjunction with higher *meta* chlorine values.

TABLE 1—*PCB composition of Fox River sediments.*

Average Cl per Biphenyl	Sample One	Sample Two	Sample Three	Aroclor 1242
meta	0.70	0.65	0.68	0.96
para	0.90	0.83	0.85	0.82
ortho	1.46	1.44	1.41	1.40
Total	3.06	2.92	2.94	3.18
Concentration Dry weight basis	67 mg/kg	89 mg/kg	72 mg/kg	N/A

TABLE 2—*Available micronutrients and total organic matter.*

Sample	Estimated CEC, meq/100 g	Avail. P, ppm	Avail. K, ppm	Avail. Mn, ppm	Avail. Zn, ppm	Avail. SO_4-S, ppm	Total Organic Matter, %
1	22	100	55	3	65	118	11.7
2	23	150	45	2	42	80	15.0
3	21	200	40	2	6	92	12.4

Finally, the *ortho* chlorine level does not appear to be correlated with trends in the *meta* and *para* chlorine levels. Many researchers have reported the *meta* position to be the most active dechlorination site with some dechlorination also occurring at the *para* position. Also, previous researchers have noted that the PCB contamination in the Fox River is characterized by a congener pattern very similar to Aroclor 1242 with traces of more heavily chlorinated products (Wisconsin Department of Natural Resources 1978; Sullivan and Delfino 1982). Therefore, the marked decrease in *meta* chlorine values, slightly increased *ortho* and *para* chlorine values, lack of relationship between the *meta* and *para* chlorine trends and the *ortho* chlorine trends are consistent with the anaerobic dechlorination of a mixture of Aroclor 1242 with traces of more heavily chlorinated products in the Fox River. These observations support the potential for dechlorination in Fox River sediments.

Two additional points are important to note. First, the PCB composition of the Fox River samples is not uniform. Examination of Table 1 shows that the chlorine distribution of samples two and three compares well, but sample one shows a homolog distribution characterized by less enrichment of lightly chlorinated congeners. Additionally, the total concentration of PCBs in the samples varies. Second, other researchers (Brown et al. 1987a, 1987b) have reported more extensive transformation of mono-*ortho* congeners than are observed in these Fox River sediments. The levels of terminal monochloro mon-*ortho* and dichloro mono-*ortho* congeners are not as high as would be expected based on reductions in the more chlorinated congeners. Therefore, potential for additional dechlorination should be possible.

Experimental Treatments

Although experimental reactors were sampled six times, only two time points were completely extracted and analyzed. The means of the average chlorine values of the 38-week and 52-week samples were compared with a student's *t*-test for each treatment. The *t*-test showed that the results for all treatments at the two time points were not statistically different at the $p = 0.05$ level. However, anomalous chlorine levels were observed in two treatments. The 1.00% SDS treatment without RAMM addition and the 1.00% Triton X-705 treatment with RAMM addition displayed average *meta* chlorine levels higher than those observed in

TABLE 3—*Exchangeable calcium and magnesium, total nitrogen, ammonia, nitrate, and chloride.*

Sample	Exchangeable Ca, ppm	Exchangeable Mg, ppm	Total N, %	NH_4-N, ppm	NO_3-N, ppm	Cl^-, ppm
1	3900	360	0.54	109.0	11.5	136
2	4050	360	0.57	127.0	41.5	133
3	3800	250	0.40	116.5	16.5	172

TABLE 4—*Analysis results of metals by ICP (ppm).*

Sample	Cd	Cr	Cu	Mo	Ni	Zn	Li	Co	Mn	As	Pb
1	4.9	108	151	9.2	22	1089	16	11	315	<28	504
2	1.8	105	106	9.6	20	349	15	12	256	<28	463
3	2.2	128	105	10.0	19	434	14	11	203	<28	760

the original sediments. This rise in *meta* chlorines is not explained by known anaerobic transformation processes, but could be due to a local variation in the PCB composition in the sediments used in these reactors. The results for these two treatments are not presented here. The average chlorine values for the other treatments are summarized in Table 6.

The results in Table 6 show clear differences between the *meta* chlorine and total chlorine values of the control and nutrient amended treatments. The confidence intervals for the *para* chlorine values overlap and the means and confidence intervals for the *ortho* chlorine values show that they are essentially the same. Comparison of the *meta* chlorine value for the controls with the *meta* chlorine value for Fox River sample one (Table 1) also shows the reduction in *meta* chlorines in the control reactors. A similar comparison between the nutrient amended reactors and Fox River sample three (Table 1) shows the lack of dechlorination in the nutrient amended reactor.

The results of the surfactant amendments are less clear. The results tabulated in Table 6 show that there is little variation among the SDS reactors. Furthermore, the difference between the SDS reactors and the original sediments is slight. The Triton reactors are less consistent, with some reactors, such as the 1.00% amendment, showing evidence for dechlorination and others, such as the 0.68% + RAMM amendment showing little evidence of further dechlorination. In contrast, the Tween reactors are uniformly comparable to the controls, although the change between the original sediments and the Tween reactors is less dramatic than the difference between the controls and the original sediments.

The data contained in Table 6 indicate that across all experimental treatments there was little change in the average *ortho* chlorine values. The *para* chlorine values all occur in the range (0.79, 0.83). In contrast, the sediments used to create these reactors were initially found to contain *para* chlorine values in the range (0.83, 0.90). Additionally, despite the differences observed in the total dechlorination occurring under the various treatments, all treatments have resulted in similar *para* chlorine values. The greatest variability can be observed in the *meta* chlorine values which occur in the range (0.60, 0.71) in the experimental samples. In contrast, the original sediment levels occurred in the range (0.65, 0.70) and Aroclor 1242 contains approximately 0.96 *meta* chlorines. These results indicate that the *meta* dechlorinating activity of the Fox River culture is the most significant.

Since the original Fox River sediments varied slightly in composition, the percent decrease in average chlorine values relative to the original sediments and Aroclor are presented in Table 7. These results show that the laboratory efforts had varying degrees of success. In

TABLE 5—*Analysis results of micronutrients by ICP (ppm).*

Sample	P	K	Ca	Mg	S	B	Fe	Cu	Al	Na
1	1311	1207	68 330	16 250	7924	12	14 300	150	11 960	366
2	1463	1092	63 860	16 500	8050	12	13 660	105	10 820	380
3	2207	742	74 380	8 950	4837	8	8 100	104	12 290	321

TABLE 6—*Summary of average chlorine levels.*[a]

Treatment	meta Cl ± 95% C.I.	para Cl ± 95% C.I.	ortho Cl ± 95% C.I.	Total Cl ± 95% C.I.
Control	0.62 ± 0.01	0.80 ± 0.01	1.43 ± 0.02	2.85 ± 0.04
Nutrient	0.71 ± 0.03	0.83 ± 0.02	1.42 ± 0.01	2.96 ± 0.05
SDS (0.25% + Nutrient)	0.66 ± 0.02	0.81 ± 0.02	1.40 ± 0.02	2.87 ± 0.02
SDS (0.50% + Nutrient)	0.68 ± 0.01	0.81 ± 0.02	1.40 ± 0.04	2.89 ± 0.04
SDS (1.00% Nutrient)	0.68 ± 0.01	0.80 ± 0.03	1.41 ± 0.02	2.89 ± 0.04
Triton (1.00%)	0.61 ± 0.01	0.80 ± 0.01	1.41 ± 0.02	2.81 ± 0.03
Triton (0.28% + Nutrient)	0.61 ± 0.03	0.80 ± 0.02	1.43 ± 0.03	2.85 ± 0.07
Triton (0.60% + Nutrient)	0.67 ± 0.02	0.82 ± 0.02	1.40 ± 0.04	2.89 ± 0.06
Tween (1.00%)	0.61 ± 0.03	0.80 ± 0.01	1.42 ± 0.04	2.84 ± 0.08
Tween (0.25% + Nutrient)	0.60 ± 0.02	0.80 ± 0.01	1.40 ± 0.04	2.80 ± 0.05
Tween (0.50% + Nutrient)	0.61 ± 0.06	0.79 ± 0.02	1.41 ± 0.01	2.81 ± 0.07
Tween (1.00% + Nutrient)	0.62 ± 0.04	0.81 ± 0.02	1.42 ± 0.02	2.85 ± 0.07

[a] Grand average of 38 and 52 week sample sets.

the case of the nutrient amendment, no additional dechlorination appears to have occurred in the laboratory. However, for most of the treatments, the laboratory treatments resulted in 30% to 100% increases in the extent of dechlorination. Decreases in average chlorine values in reactors receiving nutrient amendments and surfactants suggest that the surfactants offset the inhibitory effects of the nutrient formulation. The comparison of the experimental treatments with Aroclor 1242 also shows the convergence of the various treatments on a common degree of dechlorination; that is, regardless of the initial average chlorine values, no treatment was able to result in dechlorination better than that of the controls.

A regression analysis was performed to determine the significance of the initial extent of dechlorination and PCB concentration on the final extent of dechlorination at the end of the experiment. The independent variables employed in this analysis were initial PCB concen-

TABLE 7—*Average changes in chlorine levels.*[a]

	Percent Decrease Relative to—			
	Original Sediments		Aroclor 1242	
Treatment	Total Cl	meta + para Cl	Total Cl	meta + para Cl
Control	6.86	11.2	10.4	20.2
Nutrient	−0.68	−0.65	6.92	13.5
SDS (0.25% + Nutrient)	6.21	8.12	9.75	17.4
SDS (0.50% + Nutrient)	3.67	4.79	9.12	16.3
SDS (1.00% + Nutrient)	5.56	7.50	9.12	16.8
Triton (1.00%)	3.77	4.73	11.6	20.8
Triton (0.28% + Nutrient)	3.18	4.73	10.4	20.8
Triton (0.60% + Nutrient)	3.67	4.79	9.12	16.3
Tween (1.00%)	2.74	4.73	10.7	20.8
Tween (0.25% + Nutrient)	4.11	5.40	11.9	21.3
Tween (0.50% + Nutrient)	4.10	6.98	11.6	21.3
Tween (1.00% + Nutrient)	2.73	4.98	10.4	19.7

[a] Grand average of 38 and 52 week sample sets.

tration, the initial average *meta* + *para* chlorines per biphenyl, and the level of nutrient, SDS, Triton, or Tween present in the reactor. The actual sediment PCB concentration and average *meta* + *para* chlorines were used as inputs into the regression equation. For all other variables, dummy values were used. A zero was used when the treatment was not present, 1 was used for the lowest concentration, 2 for the medium concentration, and 3 for the highest concentration. The dependent variable was the change in the *meta* + *para* chlorine levels between the treatment results and the appropriate initial sediment values. *Meta* + *para* chlorine values were used because examination of the data discussed above did not show activity at the *ortho* position. The regression results are presented in Table 8.

The regression analysis shows that the initial *meta* + *para* chlorine level has the most significant effect on the success of dechlorination. The coefficient was negative, indicating that higher initial *meta* + *para* chlorine levels resulted in greater decreases in *meta* + *para* chlorines. Since the most successful dechlorination for any treatment was about the same regardless of the initial extent of dechlorination in the original sediments, there is a high likelihood that there is a lower limit beyond which the process will not proceed. The initial PCB concentration in the sediments was also observed to have a significant effect on the dechlorination process. This coefficient was negative, indicating that higher initial PCB concentrations resulted in greater decreases in *meta* + *para* chlorines. The coefficients for the effects of the nutrient, SDS, Triton, and Tween treatments were not significant at the 95% confidence level; the trends previously discussed should be considered qualitative.

Discussion

Treatment Effects

Nutrient Amendment—There is general agreement in the literature that the use of minimal media has a positive effect on the anaerobic dechlorination of PCBs. However, the nutrient amendment used in this experiment was inhibitory when compared to control reactors receiving no nutrient amendment. A variety of nutrient amendments have been described in reports of the anaerobic dechlorination of PCBs, the most common of which is the RAMM formulation of Shelton and Tiedje (1984). This formulation has been used without modification in the research of Quensen et al. (1988, 1990); with minor modifications by Van Dort and Bedard (1991) and Ye et al. (1992) and with more extensive modification by Morris et al. (1992) and Abramowicz et al. (1993). The formulations of Morris et al. (1992) and Abramowicz et al. (1993) both employed the vitamins of Wolin et al. (1963), as in this experiment. Other nutrient amendments include the formulation of Balch et al. (1979), that has been used by Rhee et al. (1993a, 1993b, 1993c, 1993d) with minor modifications, the

TABLE 8—*Regression analysis of dechlorination results.*[a]

	Coefficient	Standard Error	Lower 95%	Upper 95%
Intercept	4.9729	0.6325	3.6982	6.2476
meta + *para* Chlorines	−2.7755	0.3361	−3.4528	−2.0982
Concentration	−0.0105	0.0018	−0.0140	−0.0069
Nutrient	0.0034	0.0123	−0.0214	0.0283
SDS	0.0111	0.0077	−0.0043	0.0266
Triton	0.0018	0.0079	−0.0141	0.0178
Tween	−0.0018	0.0072	−0.0163	0.0126

[a] Multiple $r = 0.82$, $r^2 = 0.67$, adjusted $r^2 = 0.62$, standard error = 0.04, observations = 51.

media of Healy and Young (1979), used by Alder et al. (1993), and an adaptation of the media of Owen et al. (1979) as reported by Nies and Vogel (1990). None of these researchers reported inhibition due to the use of minimal media, and all had more successful dechlorination in at least some of their treatments than was observed in the experiments contained in this report.

A review of the media described by the above researchers shows that the concentrations of the mineral salts and trace metals vary considerably. Since all these media have produced favorable dechlorination results, it would appear that the anaerobic communities responsible for dechlorination were relatively insensitive to the concentration of these compounds—their presence may have been required for optimal growth, but no inhibition was observed over concentration ranges as large as two to three orders of magnitude for some of the compounds. The most notable difference between the media used in this experiment and the media of other researchers is the concentration of vitamin B_{12}. Virtually all the media employ vitamin B_{12} at a concentration of 1 μg/L. However, the experimental media reported here contained vitamin B_{12} at 50 μg/L, and each reactor was supplemented with an additional 0.44 mol of vitamin B_{12} per mol of PCBs present ($\approx$ 164 mg/reactor). Assaf-Anid et al. (1992) showed that a concentration of 0.44 mol of vitamin B_{12} per mol of PCB catalyzed the reductive dechlorination of a PCB congener in an abiotic system. However, Abramowicz et al. (1993) have reported that vitamin B_{12} can be inhibitory to dechlorination at concentrations as low as 100 μg/L. The total concentration used in this experiment was over three orders of magnitude higher that the levels studied by Abramowicz et al. (1993). If this is true, the high concentrations employed here could prove toxic and this would explain the observed inhibition. Although Assaf-Anid et al. (1992) showed that vitamin B_{12} could catalyze the reductive dechlorination of PCBs under abiotic conditions, their work provided no way of assessing the effect of this level of vitamin B_{12} on microorganisms.

Surfactants—The results of surfactant amendments to the experimental reactors are complex. It was previously discussed that SDS may have slightly attenuated the inhibitory effects of the nutrient solution. The SDS doses investigated here ranged from 0.625 g per 250 mL to 2.50 g per 250 mL. All of these doses would exceed the critical micelle concentration (CMC) (2385 mg/L) in a purely aqueous solution. Although the surfactant was in the presence of a sediment, it is not likely that significant sorption interactions occurred since SDS is anionic. Rather, SDS is expected to be repulsed from most natural surfaces. Since the formation of micelles is highly likely, the observed attenuation may have been due to slight solubilization of PCBs in the palisade layer of the surfactant micelles, resulting in increased bioavailability and biological activity. However, the attenuation may have also been due to interactions between the surfactant and inhibitory components of the nutrient amendment. If the surfactant decreased the bioavailability of certain inhibitory compounds, biological activity may have increased. The possibility of micelle formation at all concentrations may also account for the apparent insensitivity of SDS to the concentration at which it was tested.

In the presentation of the results, it was noted that Triton X-705 appeared to be sensitive to the concentration at which it was used. The cause of this effect is difficult to ascertain. Triton X-705 is a very large surfactant molecule that does not form stable micelles due to its size (Wall 1993). Additionally, significant sorption interactions could be anticipated to occur between Triton X-705 and the sediments contained in the experimental reactors. The differences in Triton X-705 effects could be due to the presence or absence of micelles or toxicity of Triton X-705 to the microorganisms.

Abramowicz et al. (1993) previously noted that Triton X-705 at 600 mg/L had a positive effect on the rate and extent of dechlorination and the lag period that preceded it. Abramowicz et al. (1993) performed their experiments with 500 ppm of a mixture of Aroclors 1242/1254/1260 in reactors containing a 2:3 (v:v) ratio of wet sediment to liquid media.

In this research, the doses of Triton X-705 employed ranged from approximately 3000 to 10 800 mg/L. The observed differences could be due to toxicity of Triton X-705 at high concentrations.

In general, Tween 20 amendments appear to have had the most positive effect on dechlorinating activity. Inhibition due to the nutrient amendment was not observed at a variety of concentrations ranging from 2700 to nearly 11 000 mg/L, and the final extent of dechlorination was as good as, if not modestly better than, that of the controls. These concentrations are probably in excess of the CMC (60 mg/L), even when sorption to the sediments is considered. Laha and Luthy (1992) have previously noted that Tween surfactants are considered relatively nontoxic. The lack of inhibition by the nutrient mixture may have been due to increased solubilization of PCBs or trapping of nutrient components in a biologically unavailable phase by the surfactant. Since Tween 20 has such a low CMC (60 mg/L), it is possible that the same effects could be observed at much lower concentrations if the observed trends are micellular phenomena.

Since dechlorination studies are time-consuming and tedious, simpler tests to explore the behavior of surfactants in the presence of Fox River sediments may increase the chances of success in later dechlorination studies using surfactant amendments. Determination of the concentrations required for micelle formation in the presence of the sediments would allow experiments to be conducted that assess differences in the effects caused by the presence of surfactant micelles or monomers. Additionally, performing simpler biological methane potential tests at various surfactant concentrations might provide clues as to what concentrations, if any, are inhibitory to microorganisms. Finally, physical screening of surfactants could help determine those that are likely to meet with success in the field. The size of micelles, ability of the surfactant to solubilize PCBs, and the occurrence of liquid crystal formation and/or the formation of coacervate phases are all important. Determination of these characteristics would contribute to the design of an experiment that would yield more conclusive information on the mechanism of surfactant action on PCB dechlorination.

Other Significant Factors—The experimental results show that the initial level of dechlorination had a very significant effect on the success of further dechlorination in the laboratory. Samples that had been dechlorinated more extensively in the field were not dechlorinated well in the lab, and samples that had not undergone extensive field dechlorination were more successfully dechlorinated in the lab. These results suggest that they may be some lower limit beyond which dechlorination will not proceed in Fox River sediments.

Concentration effects were also observed. However, the concentrations employed in this research spanned a small range. A better assessment of the effect of PCB concentration on dechlorination would require the collection of environmental samples encompassing a wider range. Additionally, determination of PCB porewater concentrations might provide a more realistic correlation between PCB concentration and dechlorinating activity.

The full implications of PCB concentration and the extent of natural dechlorination on efforts to stimulate further dechlorination of PCBs are not known. A more thorough review of the levels and congener patterns found in Fox River sediments could help assess the chances of bioremediating PCBs in the Fox River under anaerobic conditions. Sampling of various locations to determine a congener specific distribution may be required. However, analytical data from various researchers who have characterized the occurrence of PCBs in the Fox River may be available for such comparisons. A comprehensive evaluation of the state of PCB congener distributions in Fox River sediments is beyond the scope of this work.

Unstudied Factors—Two factors that may have a significant effect on the dechlorination of PCBs in Fox River sediments were not studied. These two factors are the role of cosubstrates and the occurrence of cocontaminants. Cosubstrates have been shown to affect the dechlorination of PCBs in some matrices, but a strong correlation with specific sediment properties has not been provided. The collection of sediments from a variety of locations could allow better comparisons to be developed on the effect of cosubstrates in relation to the background methane potential of the sediment and the organic carbon content. An investigation of this type might help define when cosubstrates are appropriate and when their use is unlikely to have beneficial effects.

Cocontaminants could influence dechlorination, particularly if they were other halogenated compounds. Since a variety of halogenated compounds can be reductively dehalogenated (Mohn and Tiedje 1992), the presence of cocontaminants could cause plateaus in the dechlorination of PCBs. These gaps in PCB dechlorination could result during periods when other compounds were being dehalogenated because they were more thermodynamically favorable. Since PCB dechlorination appears to occur in a stepwise fashion, it is plausible that other compounds could dominate portions of the overall dechlorinating activity at certain time points. During this initial research, a detailed screening of Fox River sediments for other contaminants was not performed. However, without a vigorous screening of Fox River sediments for other compounds that are potential electron acceptors, it is difficult to determine if this may be occurring. This topic is of particular concern since a large, unidentified peak was present in PCB chromatograms from the Fox River sediments. This peak eluted slightly after congener 044.

Conclusions

The results of analysis of Fox River sediments are consistent with *in-situ* dechlorination of PCBs. The extent of this dechlorination ranged from 3.77 to 8.18% of the total chlorine and 10.1 to 16.9% of the *m* and *p* chlorines (all percentages relative to the composition of Aroclor 1242). This extent of dechlorination is significantly less than that encountered in many other locations. The *in-situ* dechlorination appears to have occurred primarily at the *meta* and *para* positions, with a preference for the *meta* position noted.

Attempts to further dechlorinate Fox River sediments in the laboratory met with limited success. The range of additional dechlorination in the laboratory ranged from -0.68 to 6.86% on a total chlorine basis and -0.65 to 11.2% on a *meta* and *para* chlorine basis. The most significant factor influencing dechlorination was the initial extent of dechlorination *in situ*. Furthermore, all samples displaying dechlorination in the laboratory tended to converge on a common chlorine distribution (removal of $\approx 10\%$ of the total chlorine and $\approx 20\%$ of the *meta* and *para* chlorines, relative to Aroclor 1242). The concentration of PCBs in the sediments was also observed to have an effect. Those sediments with higher PCB concentrations were observed to undergo more successful dechlorination.

The nutrient amendment employed in this research was inhibitory to dechlorination, and comparison of the media of this research with the media of other researchers suggests the high level of vitamin B_{12} in the media is the cause of the inhibition. The three surfactants investigated had varying effects. The anionic surfactant sodium dodecyl sulfate appeared to slightly attenuate the inhibition caused by the nutrient amendment. The nonionic surfactant Triton X-705 displayed erratic effects, and the performance of this surfactant may be related to its inability to form stable micelles in solution. The nonionic surfactant Tween 20 appeared to attenuate the inhibition of the nutrient amendment at all concentrations at which it was used.

Factors not studied in this research may also influence dechlorination of PCBs in Fox River sediments. These unstudied factors include the presence of cocontaminants (both organic and inorganic), the organic composition of the sediment as it relates to the adsorption of PCBs to the sediment matrix and the availability of substrates for bacterial growth, and the surface chemistry of the sediments as it relates to interactions with surfactants and nutrients. Further research to identify cocontaminants that may interfere with PCB dechlorination is recommended. In addition, physical characterization of the sediment matrix at this and other sites is desirable so that the effect of physical properties on the dechlorination of PCBs can be assessed.

References

Abramowicz, D. A., Brennan, M. J., and Van Dort, H. M., 1989, "Microbial Dechlorination of PCBs: I. Aroclor Mixtures," in *Research and Development Program for the Destruction of PCBs, Eighth Progress Report,* H. L. Finkbeiner and S. B. Hamilton, Eds., General Electric Corporate Research and Development Center, Schenectady, NY, pp. 49–60.

Abramowicz, D. A., 1990, "Aerobic and Anaerobic Biodegradation of PCBs: A Review," *CRC Critical Reviews in Biotechnology,* Vol. 10, pp. 241–251.

Abramowicz, D. A., Brennan, M. J., Van Dort, H. M., and Gallagher, E. L., 1993, "Factors Influencing the Rate of Polychlorinated Biphenyl Dechlorination in Hudson River Sediments," *Environmental Science and Technology,* Vol. 27, pp. 1125–1131.

Alder, A. C., Häggblom, M. M., Oppenheimer, S. R., and Young, L. Y., 1993, "Reductive Dechlorination of Polychlorinated Biphenyls in Anaerobic Sediments," *Environmental Science and Technology,* Vol. 27, pp. 530–538.

Assaf-Anid, N., Nies, L., and Vogel, T. M., 1992, "Reductive Dechlorination of a Polychlorinated Biphenyl Congener and Hexachlorobenzene by Vitamin B_{12}," *Applied and Environmental Microbiology,* Vol. 58, pp. 1057–1060.

Balch, W. E., Fox, G. E., Magrum, L. J., et al., 1979, "Methanogens: Reevaluation of a Unique Biological Group," *Microbiological Reviews,* Vol. 43, pp. 260–296.

Ballschmiter, K. and Zell, M., 1980, "Analysis of Polychlorinated Biphenyls (PCB) by Glass Capillary Gas Chromatography," *Fresenius Z. Anal. Chem.,* Vol. 302, pp. 20–31.

Brown, J. F., Wagner, R. E., Bedard, D. L., et al., 1984, "PCB Transformations in Upper Hudson Sediments," *Northeastern Environmental Science,* Vol. 3, pp. 166–178.

Brown, J. F., Bedard, D. L., Brennan, M. J., et al., 1987a, "Polychlorinated Biphenyl Dechlorination in Aquatic Sediments," *Science,* Vol. 236, pp. 709–712.

Brown, J. F., Wagner, R. E., Feng, H., et al., 1987b, "Environmental Dechlorination of PCBs," *Environmental Toxicology and Chemistry,* Vol. 6, pp. 579–593.

Brown, J. F., 1990, "Differentiation of Anaerobic Microbial Dechlorination Processes," in *Research and Development Program for the Destruction of PCBs, Ninth Progress Report,* H. L. Finkbeiner and S. B. Hamilton, Eds., General Electric Corporate Research and Development Center, Schenectady, NY, pp. 87–90.

Brown, J. F. and Wagner, R. E., 1990, "PCB Movement, Dechlorination, and Detoxication in the Acushnet Estuary," *Environmental Toxicology and Chemistry,* Vol. 9, pp. 1215–1233.

Furukawa, K., 1986, "Modification of PCBs by Bacteria and Other Microorganisms," in *PCBs and the Environment,* Volume II, J. W. Waid, Ed., CRC Press, Boca Raton, FL, pp. 89–100.

Laha, S. and Luthy, R. G., 1992, "Effects of Nonionic Surfactants on the Solubilization and Mineralization of Phenanthrene in Soil—Water Systems," *Biotechnology and Bioengineering,* Vol. 40, pp. 1367–1380.

Manchester, J., 1993, "The Role of Porewater in the Remobilization of Sediment-Bound Polychlorinated Biphenyl Congeners," Ph.D. dissertation, University of Wisconsin–Madison, Madison, WI.

Mohn, W. W. and Tiedje, J. M., 1992, "Microbial Reductive Dehalogenation," *Microbiological Reviews,* Vol. 56, pp. 482–507.

Morris, P. J., Mohn, W. W., Quensen, J. F., et al., 1992, "Establishment of a Polychlorinated Biphenyl-Degrading Enrichment Culture with Predominantly *meta* Dechlorination," *Applied and Environmental Microbiology,* Vol. 58, pp. 3088–3094.

Mullin, M. D., 1985, PCB Workshop, U.S. EPA Large Lakes Research Station, Grosse Ile, MI.

Nies, L. and Vogel, T. M., 1990, "Effects of Organic Substrates on Dechlorination of Aroclor 1242 in Anaerobic Sediments," *Applied and Environmental Microbiology,* Vol. 56, pp. 2612–2617.

Owen, W. F., Stuckey, D. C., Healy, J. B., et al., 1979, "Bioassay for Monitoring Biochemical Methane Potential and Anaerobic Toxicity," *Water Research,* Vol. 13, pp. 485–492.

Peterman, P. H., Delfino, J. J., Dube, D. J., et al., 1980, "Chloro-Organic Compounds in the Lower Fox River, Wisconsin," in *Hydrocarbons and Halogenated Hydrocarbons in the Aquatic Environment,* B. K. Afghan and D. MacKay, Eds., Plenum, New York, pp. 145–160.

Quensen, J. F., Tiedje, J. M. and Boyd, S. A., 1988, "Reductive Dechlorination of Polychlorinated Biphenyls by Anaerobic Microorganisms from Sediments," *Science,* Vol. 242, pp. 752–754.

Quensen, J. F., Tiedje, J. M., and Boyd, S. A., 1990, "Dechlorination of Four Commercial Polychlorinated Biphenyl Mixtures (Aroclors) by Anaerobic Microorganisms from Sediments," *Applied and Environmental Microbiology,* Vol. 56, pp. 2360–2369.

Rhee, G.-Y., Sokol, R. C., Bush, B., and Bethoney, C. M., 1993a, "Long-Term Study of the Anaerobic Dechlorination of Aroclor 1254 with and without Biphenyl Enrichment," *Environmental Science and Technology,* Vol. 27, pp. 714–719.

Rhee, G.-Y., Sokol, R. C., Bethoney, C. M., and Bush, B., 1935b, "Dechlorination of Polychlorinated Biphenyls by Hudson River Sediment Organisms: Specificity to the Chlorination Pattern of Congeners," *Environmental Science and Technology,* Vol. 27, pp. 1190–1192.

Rhee, G.-Y., Bush, B., Bethoney, C. M., et al., 1993c, "Reductive Dechlorination of Aroclor 1242 in Anaerobic Sediments: Pattern, Rate, and Concentration Dependence," *Environmental Toxicity and Chemistry,* Vol. 12, pp. 1025–1032.

Rhee, G.-Y., Bush, B., Bethoney, C. M., et al., 1993d, "Anaerobic Dechlorination of Aroclor 1242 as Affected by Some Environmental Factors," *Environmental Toxicology and Chemistry,* Vol. 12, pp. 1033–1039.

Shelton, D. R. and Tiedje, J. M., 1984, "General Method for Determining Anaerobic Biodegradation Potential," *Applied and Environmental Microbiology,* Vol. 47, pp. 850–857.

Sullivan, J. R. and Delfino, J. J., 1982, *A Select Inventory of Chemicals Used in Wisconsin's Lower Fox River Basin,* University of Wisconsin Sea Grant Institute, Madison, WI, WIS-SG-82-238.

Sullivan, J. R., Delfino, J. J., Buelow, C. R., and Sheffy, T. B., 1983, "Polychlorinated Biphenyls in the Fish and Sediment of the Lower Fox River, Wisconsin," *Bulletin of Environmental Contamination and Toxicology,* Vol. 30, pp. 58-64.

Van Dort, H. M. and Bedard, D. L., 1990, "Reductive *ortho* and *meta* Dechlorination of a Polychlorinated Biphenyl Congener by Anaerobic Microorganisms," *Applied and Environmental Microbiology,* Vol. 57, pp. 1576–1578.

Wall, G., 1993, Sigma Chemical Company Technical Services, St. Louis, MO, Personal communication, with author.

Wisconsin Department of Natural Resources, 1978, "Investigation of Chlorinated and Nonchlorinated Compounds in the Lower Fox River Watershed," *EPA Report No. EPA 950/3-78-004.*

Wolin, E. A., Wolin, M. J., and Wolfe, R. S., 1963, "Formation of Methane by Bacterial Extracts," *Journal of Biological Chemistry,* Vol. 238, pp. 2882–2886.

Ye, D., Quensen, J. F., Tiedje, J. M., and Boyd, S. A., 1992, "Anaerobic Dechlorination of Polychlorobiphenyls (Aroclor 1242) by Pasteurized and Ethanol-Treated Microorganisms from Sediments," *Applied and Environmental Microbiology,* Vol. 58, pp. 1110–1114.

Hao Zhang,[1] William Davison,[1] and Geoffrey W. Grime[2]

New *In-Situ* Procedures for Measuring Trace Metals in Pore Waters

REFERENCE: Zhang, H., Davison, W., and Grime, G. W., "**New *In-Situ* Procedures for Measuring Trace Metals in Pore Waters**," *Dredging, Remediation, and Containment of Contaminated Sediments, ASTM STP 1293*, K. R. Demars, G. N. Richardson, R. N. Yong, and R. C. Chaney, Eds., American Society for Testing and Materials, Philadelphia, 1995, pp. 170–181.

ABSTRACT: The most mobile and biological and chemically active fractions of trace metals in sediments are the dissolved components present in pore waters. Measuring metals in pore waters is complicated by the requirement for anoxic handling procedures. Due to the dynamic nature of sediment, steep concentration gradients extending over as little as 1 mm may develop at the sediment-water interface. New procedures for measuring metals in pore waters using polyacrylamide gels as *in-situ* probes are described. The gel can be used to establish a diffusive equilibration in a thin-film (DET). Because the film is typically less than 1 mm thick, equilibration is achieved within five minutes and insertion of the gel assembly causes minimal disturbance of sediment. An alternative procedure is to use a diffusive gradient in a thin-film (DGT), whereby a monolayer of chelating resin is incorporated at one side of the gel. Such a technique provides a kinetic measurement of labile species in solution. If the supply of metal from solid phase sediment to pore waters is fast enough, DGT provides a quantitative estimate of labile metal concentration. Alternatively, it measures directly the rate of supply of metal from solid phase to pore waters. As both DET and DGT are simple procedures capable of submillimetre spatial resolution, they provide previously unobtainable information on trace metal concentrations and fluxes. Furthermore, DGT has the potential to be used as a long-term monitor, providing mean concentrations of metals in sediment pore waters over periods of days, weeks, or even months.

KEYWORDS: *in-situ* measurement, polyacrylamide gel, trace metals, pore waters, diffusive equilibration in thin-film (DET), diffusive gradient in thin-film (DGT), ion-exchange resin

Large chemical gradients at the sediment-water interface are ultimately due to the interception of settling particulate material that reacts in a self-created chemical and microbial micro-environment (Santschi et al. 1990). The labile organic fraction decomposes, ultimately to yield inorganic nitrogen, phosphorous and trace elements. This oxidation of organic material is accompanied by the reduction of electron acceptors such as O_2, NO_3^-, Mn(IV), Fe(III) and SO_4^{2-}. The conversion of solid phase iron and manganese to their reduced, soluble forms of Mn(II) and Fe(II) may liberate other trace components adsorbed or associated with oxides. Thus the process of organic decomposition fuels a wide range of interrelated chemical transformations and generates sharp chemical gradients in the immediate vicinity of the interface. As the supply of material occurs as a series of episodes rather than a steady state, the

---

[1] Research associate and professor, respectively, Division of Environmental Science, Lancaster University, Lancaster, LA1 4YQ, U.K.
[2] Research lecturer, Scanning Proton Microprobe Unit, Nuclear Physics Laboratory, University of Oxford, Oxford, OX1 3RH, U.K.

chemical concentrations and gradients can be expected to change with time. This dynamic interfacial exchange is most pronounced in productive systems, but the same processes operate even in impoverished situations.

Many studies have shown that the pore waters of subsurface marine sediments differ significantly in composition from the overlying sea water (Hartmann and Muller 1982; Klinkhammer et al. 1982; Burdige and Gieskes 1983; Gieskes 1983; Sawlan and Murray 1983). Such differences are produced mainly by slow reactions between the solids of sediments and the trapped interstitial solutions. The chemical gradients between the pore fluids and the overlying ocean water that result from these reactions produce diffusive fluxes of constituents across the seawater-sediment interface. Even if the gradients are very small, these fluxes may be of great importance in geochemical mass balances.

To understand the fate of chemical components in marine, estuarine, and lacustrine systems, it is necessary to know the flux of material across the sediment-water interface. Two general approaches to measuring fluxes have been employed: direct measurements of exchange and studies of pore water composition. Direct procedures include inference from a mass balance of components measured in the overlying water column, calculation from chemical gradients measured over large distances (usually metres) in the overlying waters, measurement of concentration changes in the waters overlying a sediment core, and measurement of concentration changes in benthic chambers which enclose an area of sediment and its overlying water. All these procedures have problems, in particular, they may reflect only the average flux over a period of time, and they may perturb the system such that the process investigated no longer reflects accurately the undisturbed, natural state. Although such measurements can be used to infer chemical gradients, they do not look at them directly and therefore their use for investigating detailed mechanisms occurring at the interface is limited.

Profiles of pore water composition have been used to estimate fluxes across the interface (Sayles 1979; Reimers and Smith 1986). There are several problems inherent in the use of pore water profiles to determine fluxes. To estimate adequately the flux and to consider the interdependence of individual chemical reactions, it is necessary to measure the chemical concentrations directly using a scale which, at the very least, clearly defines the shape of the gradient in the vicinity of the interface. For several years now oxygen gradients have been measured directly using microelectrodes (Jorgensen and Revsbech 1985; Gunderson and Jorgensen 1990) capable of a spatial resolution better than 100 µm. The gradients at the sediment water interface have usually been found to extend over distances of a few millimetres, but in extreme cases they are complete within a millimetre. Therefore, to begin to appreciate the mechanisms that generate such gradients of O_2 and other elements, chemical profiles must be measured with at least 1 mm resolution and in some situations at intervals of 100 µm.

Microelectrodes have also been used for measuring nitrate (Sweets and de Beer 1989), sulphide (Revsbech and Jorgensen 1986), and ammonia (de Beer and van den Heuvel 1988) in pore waters. However, most pore water measurements have been made after extracting a sediment core or by performing an *in-situ* equilibration or reaction.

Core Extraction

Core extraction has been the method most widely used (Carignan et al. 1985; Bender et al. 1987; Davison et al. 1982). It enables the measurements of pore water as well as sediment from one sample and produces reliable information on the relationship between solids and pore waters. Although spatial resolution has usually been restricted to a 1 cm interval, extraction to a resolution of a few millimetres is possible (Bender et al. 1989). The main disadvantage is that the extraction of pore water does not take place under *in-situ* conditions. Pore water has usually been extracted from sediment cores by squeezing (Reeburgh 1976; see also ASTM D 4542, Test Method for Pore-Water Extraction and Determination of the Soluble Salt Content of Soils by Refractometer) and centrifugation followed by filtration (Elderfield et al. 1981). It has been shown, however, that temperature increases during recovery alter the concentration of many components in the interstitial solutions. Components of the carbonate system may also be affected by large pressure changes during recovery (Fanning and Pilson 1971). It was noted that in some pore water samples that were high in calcium, magnesium and carbonates, precipitation of carbonate minerals occurred during centrifugation and filtration due to degassing of dissolved CO_2 and the accompanying pH increase. These changes will undoubtedly affect trace element speciation. Centrifugation and filtration methods are also prone to artifacts resulting from sample oxidation (Bray et al. 1973; Lyons et al. 1979). A variation of the squeezing technique where the sample passes through the overlying sediment layer is capable of near-millimetre spatial resolution, but concentrations may be modified by reaction with the sediment (Bender et al. 1989). Sometimes, instead of squeezing and centrifugation, pore water is collected with a syringe fitted with a plastic needle pushed through a sealed opening in the core tube wall (Davison 1982). The sediment slurry is subsequently forced by hand through a filter assembly. Although handling is rapid, minimizing oxidation effects, only a few millilitres of pore water is recovered and the resolution may be poor due to water flowing from different layers into the syringe needle. The measurement of trace metal concentrations, which typically do not exceed 10^{-7} mol/L in pore water samples, is prone to contamination in all the procedures. Moreover, because filtration is used, it is likely that colloidal components are present in the operationally defined dissolved fraction.

In-situ **Direct-Suction Sampling**

Because of the problems of core extraction, *in-situ* extraction devices have been developed that work by suction through filter-covered ports (Sayles et al. 1976). The separation of pore water and sediment then takes place under reliable *in-situ* conditions. However, such samples have generally failed to produce the precise concentrations and distributions needed to predict benthic fluxes. The main reason is that *in-situ* samplers disturb the surface sediment during emplacement. Moreover, the sampled solutions represent unknown fluid pathways withdrawn from known depths, rather than specific well defined depth intervals. The sediment volume from which interstitial water is withdrawn is not readily demarcated, so that the minimum distance between two neighboring openings is large (2.5 to 10 cm) and concentration profiles are consequently coarse.

In-situ **Diffusion Sampling**

In the most frequently used variation of this technique, a series of compartments in a plastic assembly is covered with a dialysis membrane. The assembly is inserted into the sediment and equilibrium allowed to establish between the pore waters and internal solution (usually deoxygenated distilled water). This procedure overcomes most of the problems encountered in coring and suction techniques and provides good estimates of pore water concentrations (Mayer 1976; Hesslein 1976; Carignan et al. 1985). To ensure complete equilibrium, deployment for a period of weeks is necessary. Construction difficulties and limitations concerning usable sample volumes usually restrict resolution to centimetre intervals. There is also a possibility of solid or colloidal phases such as FeS forming within the dialysis cells, leading to overestimation of solution components.

The technique of diffusive equilibration in a thin film (DET) can be used to make measurements at submillimetre intervals (Davison et al. 1991, 1994). It relies on a similar equilibration principle to the dialysis technique, but rather than confining the solution to compartments, it uses a thin film of polyacrylamide gel to provide the medium for solution equilibration. Diffusive equilibrium can be established rapidly (within minutes) in a thin film of gel inserted in the sediment. The DET technique has been used successfully in measuring iron, manganese, nitrate, and sulphate concentrations in lacustrine pore water at submillimetre resolution (Davison et al. 1991, 1994; Krom et al. 1994).

This paper reports some of the strengths and weaknesses of the DET procedure and provides some examples of its further application. The development of a new procedure known as diffusion gradient in thin-films (DGT) is described. It involves *in-situ* chemical reactions and is capable of measuring trace metals at millimetre resolution.

Experimental

Gel Preparation

The formation of polyacrylamide gel begins with two monomers: (1) acrylamide, which is a small organic molecule that terminates in an aminocarbonyl (-$CONH_2$) group, and (2) bisacrylamide (cross-linker), which consists of two acrylamide units that are linked through their aminocarbonyl groups. The monomers are dissolved in water. Ammonium persulphate and tetramethyl ethylene diamine (TEMED) are then added; their reaction produces an unpaired electron that initiates polymerization. The result is a tangled web of polyacrylamide that can support a large proportion of water (>90% wet volume). The properties of the gel, including its hydration, pore size, and elasticity, depend on the concentration and proportion of acrylamide and cross-linker.

A polyacrylamide gel composed of 15% by volume acrylamide and 0.3% by volume cross-linker (trade name AcrylAide) was used for this work. After mixing an appropriate amount of acrylamide and cross-linker solution, about 0.6% by volume initiator (ammonium persulphate (10%)) and about 0.2% by volume catalyst (TEMED) were added. The well-mixed solution was immediately cast between two glass plates separated by plastic spacers and maintained at about 40°C for about 30 to 45 min until the gel was completely set.

Gels used for the DGT technique, in which a layer of chelating resin (Chelex-100, Bio-Rad) is incorporated at one side of the gel, were prepared by the same procedure, with an appropriate amount of resin added in the gel solution mixture.

All gels were hydrated in high-purity deionized water (MilliQ-Millipore Ltd) for at least 24 h before use. The volume of gel increases by a factor of 3.2 and the water content of the gel is about 95% after full hydration. Gels were always stored in MilliQ water to maintain their dimensional stability.

Probe Preparation and Field Deployment

A probe designed for inserting the gel into the sediment comprises two perspex plates (15 by 5 by 0.1 cm), one of which has a 10 by 1 cm window (Fig. 1). The gels were placed between the two plates which were held together by plastic clips along their long edges. Large particles in sediments may damage the gel surface during insertion and authigenic deposition of iron, manganese, and other elements may occur on the gel surface after inser-

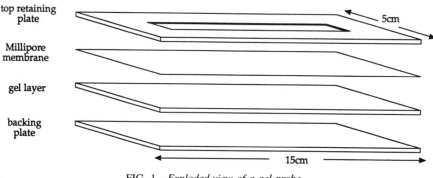

FIG. 1—*Exploded view of a gel probe.*

tion (Belzile et al. 1989). To prevent these problems, a 0.45 μm cellulose nitrate membrane was placed between the gel and window plate. The probes were immersed in MQ water and vigorously deoxygenated with nitrogen for at least 24 h. They were sealed in anoxic containers during their transport to the field.

Gel probes were deployed in the sediments of the anoxic hypolimnion at a 14 m depth in Esthwaite Water in the English Lake District. Probes were inserted directly into lake sediment by divers and were also inserted into freshly collected Jenkin Cores. The position of the interface was read from a scale marked on the probes inserted in cores. For *in-situ* deployment, the interface was judged by the visible discontinuity of particulate material adhering to the probe. The exposure of gels to air during insertion into cores was restricted to a few seconds during the insertion procedure. The probes were allowed to equilibrate in the sediment for 24 h prior to retrieval. Diffusional transport can be expected to restore original chemical conditions and the structure of the pore water gradients over a distance of more than a centimetre in 24 h, thus minimizing the effect of sediment disturbance.

Gel Sample Treatment

After removal from the sediment, DET gel probes were immediately immersed in 10 mmol L^{-1} NaOH solution to fix iron and manganese by oxidation and hydrolysis, thus preventing further diffusion within the gel. The time from initiating retrieval to immersion in NaOH solution was about 10 s. DGT gel probes were placed in clean polyethylene bags directly after retrieval without any treatment as trace metals are automatically fixed by the resin.

Some of the gels were dried onto a 0.45 μm cellulose nitrate membrane using a gel drier so that they could be measured by PIXE (see next subsection). Other gels were sliced into 1 to 2.5 mm strips and each strip was equilibrated with a known volume of 2 mol L^{-1} HNO_3 for at least 6 h. The concentration of zinc in the equilibrated solution was analyzed by AAS.

Analytical Methods

Concentrations of iron and manganese in the dried gels were analyzed by Proton-Induced X-ray Emission (PIXE) using a scanning proton microprobe (Johanson and Campbell 1988). Theoretically, it is possible that a spatial resolution of 1 μm can be achieved by PIXE technique (Grime et al. 1991), but for these measurements 100 by 100 μm rasters were used. Due to problems with the internal calibrant which is necessary to account for shrinkage of the gel during drying, the PIXE results reported here are not calibrated. Absolute concentrations cannot therefore be compared between gels, but relative changes in concentration within a gel are accurate.

Sliced gels were analyzed by graphite furnace atomic absorption spectroscopy (AAS) for the concentrations of trace metals (zinc, iron, and manganese) in each strip of gel. A known volume of acid (2 mol L^{-1} HNO_3) was added to a known volume of gel, enabling accurate calculation of the concentrations of trace metals in the gel and hence the pore water.

Principles of DET Technique

The principle of the DET technique is based on Fick's second law of diffusion

$$dC/dt = D\,(d^2C/dx^2) \qquad (1)$$

where C is concentration, t is time, x is distance, and D is the diffusion coefficient. When an assembly is inserted into the sediment, a diffusional equilibrium between the surrounding solution and the aqueous phase of the gel will be reached after a certain time. The time taken to establish this equilibrium depends on the rate of diffusion of ions into the gel and the gel thickness. Experiments that followed the diffusion of $MnCl_2$ through the gel (Davison et al. 1994) showed that complete equilibration was achieved in a 0.4 mm thick gel within 6 min. This equilibration time was in good agreement with the theoretical prediction from the numerical solution of the diffusion equation (Eq 1), indicating that the diffusion of ions in the polyacrylamide gel is effectively the same as in water.

Since NH and CO groups are the basic units of a polyacrylamide gel, there is the possibility of metal ions reacting with the gel. However, when gels were exposed to solutions containing a wide range of concentrations of metal ions at pH 5 and pH 7, the concentrations of metal ions in the gel were the same as in solution, indicating that there was no reaction with the gel.

At neutral pH, Fe(II) diffusing in from the pore waters will be oxidized to oxyhydroxides if O_2 is present in the gel when it is inserted into the sediment. This precipitated iron within the gel leads to an overestimate of the pore water concentration of iron. The oxidation of iron may also affect other trace metals which can be adsorbed onto oxyhydroxides. To prevent this oxidation effect, gels are deoxygenated in nitrogen-purged MQ water for 24 h before insertion. However, oxygen may enter the gel when it is transferred from the deoxygenated solution to the sediment. Experiments have shown that the error caused by exposure to air for up to 15 s is negligible.

The diffusion of ions within the gel during the time between gel retrieval and insertion in NaOH solution allows some relaxation of concentration gradients in the gel. Calculations of diffusional re-equilibration have shown that to maintain the original spatial structure of pore water concentrations with high fidelity to a resolution of 1 mm, the delay prior to fixing must be restricted to 20 s.

Further investigations of the spatial resolution of DET techniques have shown that the thickness of the gel also effectively limits the achievable resolution, which is consistent with the diffusion process occurring in three-dimensions within the gel. As the thickness of the gels used throughout this work is 0.4 mm, a reliable spatial resolution of 1 mm can be easily achieved.

Principles of DGT Technique

The diffusion gradient in thin-films (DGT) technique has been developed and applied to *in-situ* speciation measurements of trace metals in seawater (Davison and Zhang 1994). It offers the possibility of making measurements with *in-situ* preconcentration and *in-situ* fixing. The spatial limitations of DET associated with fixing have already been mentioned. DET is also unsuitable for measuring trace metals as the small volume of individual gel slices (typically 10 μL) places constraints on detection limits.

The DGT technique incorporates an ion-exchange resin layer separated from solution by an ion-permeable, polyacrylamide gel layer. Metal ions are rapidly bound by the resin, creating a steep concentration gradient in the gel layer. If this concentration gradient remains constant with time, the amount of trace metal accumulating in the resin layer can be related directly to the concentration in the pore waters and the exposure time. Local buffering of pore water concentrations by interaction with the solid phase can maintain a constant concentration gradient. This technique relies on measurement of a flux through the gel, rather than equilibration as in DET.

It is because mass transport through the gel is diffusion-controlled and well-defined that it is possible to quantify concentration. The measurement is based on Fick's first law of diffusion which can be expressed by the following equation:

$$F = DC_b/\Delta g \qquad (2)$$

where F is the flux of trace metal ions diffusing from pore water to resin layer, C_b is the concentration of trace metal in bulk pore waters, Δg is the thickness of the gel layer, and D is the diffusion coefficient of a metal ion in the gel. Equation 2 can be rearranged and written as

$$C_b = C_r \, \Delta g \, \Delta r / Dt \qquad (3)$$

where C_r is the concentration of trace metal accumulated in the resin layer after a given time, t, and Δr is the thickness of resin layer.

Assuming a typical value of D of 10^{-5} cm^2 s^{-1}, for a 24 h immersion in sediment, a gel layer thickness of 0.5 mm and a resin layer thickness of 0.1 mm, the concentration in the resin layer will be 1728 times greater than the concentration in sediment pore waters. This *in-situ* preconcentration not only enables the measurement of ultra trace concentrations of metals in pore waters, but also overcomes the contamination problems that beset conventional techniques. The concentration measured using DGT is the time averaged mean concentration in the solution. DGT is actually a kinetic speciation technique which is expected to measure those species that can readily dissociate during diffusion through the gel and that have a stability constant smaller than that characterizing the binding of the metal to resin. Where pore water concentrations are not well buffered, DGT measures directly the combined flux due to diffusive supply through the pore waters and local dissociation from the solid phase.

Trace metals in the resin layer are spatially fixed. No relaxation of concentration gradients recorded by the assembly has been found even after six months of storage subsequent to sampling. The ultimate spatial resolution of the DGT technique depends on the size of resin particles and the thickness of the gel layer. The ion-exchange resin Chelex-100 (75 to 150 μm spheres) was used for this work, with a gel layer thickness of 0.4 mm. Therefore, the achievable resolution is probably about 0.4 mm. Better resolution should be possible using a thinner diffusion layer, smaller resin beads, and PIXE as the analytical tool.

Applications

On the 25 May and 8 July 1993, both DET and DGT techniques were used to make measurement at sediment-water interface in Esthwaite Water, a productive lake in the English

Lake District. The gel probes were inserted directly into the lake sediment by divers as well as into freshly retrieved sediment. Three gel samples obtained from separate sediment cores in May 1993 were dried on filter membranes and the distributions of iron and manganese were measured by PIXE (Fig. 2). Owing to the absence of an internal standard, the PIXE counts could not be converted to concentration. Within a given gel sample, however, the counts are quantitatively related to one another. A gel sample from May was also sliced and analyzed by AAS (Fig. 3). In this sliced gel there are well defined peaks, less than 1 cm wide, of iron and manganese at the interface. There is also an indication of minor concentration maxima at a depth of 5 to 6 cm. All the gels measured by PIXE clearly show the maximum concentrations of iron and manganese close to the sediment interface, but the

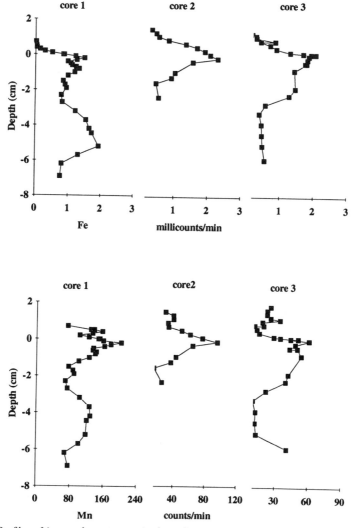

FIG. 2—*Profiles of iron and manganese in the sediment pore water in Esthwaite Water on 25 May 1993. Gels were inserted in three separate cores and analyzed by PIXE.*

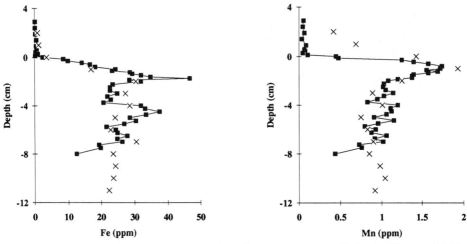

FIG. 3—*Profiles of iron and manganese in the sediment pore water in Esthwaite Water on 25 May 1993 sampled by Jenkin cores followed by the DET technique (■) and conventional syringe technique (×). The gel was sliced and analyzed by AAS.*

detailed structure varies between the three gels. In one case there is a well defined secondary maximum for both iron and manganese at a depth of 4 to 5 cm. These detailed concentration profiles demonstrate the variability that exists in pore water gradients measured in different sediment cores collected from the same sampling site. The structure in the pore water chemistry which is apparent at depth, away from the interface, may be associated with microniches of high local activity. Because we were unable to convert PIXE counts to concentration, it is not possible to assess the variability in the concentration gradients associated with the surface maximum. Such gradients form the basis of flux estimates from the interface to the overlying water. Conventional sampling at larger intervals is likely to underestimate the surface concentration gradient and hence the flux from sediment to water.

Samples of pore water were extracted by syringe from a separate core collected in May. The results, plotted on the same graph as the gel sliced data (Fig. 3) confirm the magnitude of the concentrations measured by the gel procedure. Although the Mn peak was well defined, there was no surface Fe peak, illustrating how coarse interval sampling may fail to define the interfacial chemistry.

When gel assemblies were inserted directly into the sediment by divers, in July, the profile shapes were similar to those obtained by insertion in cores in May (Fig. 4).

The concentrations and distribution of Zn in pore waters in July, measured by the DGT technique, are shown in Fig. 5. A sharper peak than that observed for Fe and Mn is apparent near the sediment-water interface. Unpublished measurements of other trace metals also show a sharp interfacial peak, indicating that this single point is not an artifact. Evidently, better spatial resolution than 1 mm is required to define precisely the concentration of Zn at the interface. The concentrations of Zn measured by DGT were similar to those measured by extracting pore waters.

Measurements in solution by DGT must be made in the presence of convection to minimize the effect of the diffusive boundary layer (Davison and Zhang 1994). There is no such convection in pore water where only molecular diffusion can operate. The good agreement with the alternative technique therefore implies that there is a rapid (minutes) supply of Zn

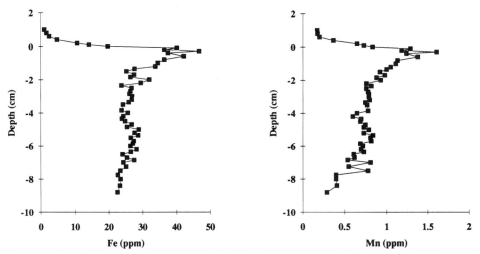

FIG. 4—*Profiles of iron and manganese in the sediment pore water in Esthwaite Water on 8 July 1993, sampled in-situ by the DET technique with slicing and determination by AAS.*

to the solution phase within a few 10's of micrometres of the filter surface. Rapid supply of dissolved Zn from Zn adsorbed to particles is the most likely explanation.

The DGT technique provides an integrated record of metal concentration during its deployment. It could, therefore, in principle be deployed for periods of days, weeks, or months to act as a long-term monitor of trace metal concentrations in sediments and the overlying water.

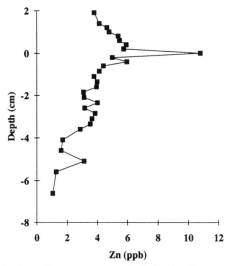

FIG. 5—*Profile of zinc in the sediment pore water in Esthwaite Water on 8 July 1993, measured by the DGT technique with slicing and determination by AAS.*

Conclusions

The new techniques of DET and DGT can be used to measure concentrations in pore water at millimetre intervals. Measurements of Fe and Mn by DET and of Zn by DGT show fine structures on a millimetre scale. They illustrate that sampling intervals of 1 cm, most commonly achieved by previous sampling methods, such as pore water extraction and use of dialysis cells, fail to define concentration gradients. Even 1 mm resolution is insufficient to establish accurately the concentration gradients of Zn at the sediment water interface and, hence, its fluxes to the overlying water and underlying sediments. In its present form, the spatial resolution attained by DET is limited to about 1 mm by the time it takes to fix the ions and prevent them from undergoing further diffusional re-equilibration. In DGT, the ions are automatically fixed *in-situ*. Using a beam technique, such as PIXE, it may be possible to achieve a spatial resolution of 100 to 200 μm.

DGT provides a direct measurement of an *in-situ* flux. It is only when pore water concentrations are well buffered, by interaction between solution and solid phase, as is the case for Zn, that the mean flux measured during the time of deployment may be interpreted as a concentration. DGT therefore provides a means of assessing *in-situ* the interaction between pore waters and solids. It effectively measures the available metal in the sediments, and defines a time scale for this availability, determined by the gel layer thickness, of typically 5 min. Transport across the gel membrane has parallels with transport across a biological membrane and so this *in-situ* measurement may be a good surrogate for bioavailability. DGT devices may therefore be suitable probes for assessing the likely toxicity of sediments. They have the great advantage of being deployed *in-situ* and therefore minimally disturbing the natural balance of sediment chemistry.

References

Belzile, N., De Vitre, R. R., and Tessier, A., 1989, "*In-situ* Collection of Diagenetic Iron and Manganese Oxyhydroxides from Natural Sediments," *Nature*, Vol. 340, pp. 376–377.

Bender, M., Jahnke, R., Weiss, R., et al., 1989, "Organic Carbon Oxidation and Benthic Nitrogen and Silica Dynamics in San Clemente Basin, A Continental Borderland Site," *Geochimica et Cosmochimica Acta*, Vol. 53, pp. 685–697.

Bender, M., Martin, W., and Hess, J., 1987, "A Whole-Core Squeezer for Interfacial Pore Water Sampling," *Limnology and Oceanography*, Vol. 32, pp. 1214–1225.

Bray, J. T., Bricker, O. P., and Troup, B. N., 1973, "Phosphate in Interstitial Water of Anoxic Sediments: Oxidation Effects during Sampling Procedures," *Science*, Vol. 180, pp. 1362–1364.

Burdige, D. J. and Gieskes, J. M., 1983, "A Pore Water/Solid Phase Diagenetic Model for Manganese in Marine Sediments," *American Journal of Science*, Vol. 283, pp. 29–47.

Carignan, R., Rapin, F., and Tessier, A., 1985, "Sediment Porewater Sampling for Metal Analysis: A Comparison of Techniques," *Geochimica et Cosmochimica Acta*, Vol. 49, pp. 2493–2497.

Davison, W., Grime, G. W., and Morgan, J. A. W., 1991, "Distribution of Dissolved Iron in Sediment Pore Waters at Submillimetre Resolution," *Nature*, Vol. 352, pp. 323–325.

Davison, W., Woof, C., and Turner, D. R., 1982, "Handling and Measurement Techniques for Anoxic Interstitial Water," *Nature*, Vol. 295, pp. 582–583.

Davison, W. and Zhang, H., 1994, "*In-situ* Speciation Measurements of Trace Components in Natural Waters Using Thin-Film Gels," *Nature*, Vol. 367, pp. 546–548.

Davison, W., Zhang, H., and Grime, G. W., 1994, "Performance Characteristics of Gel Probes Used for Measuring Pore Waters," *Environmental Science and Technology*, Vol. 28, pp. 1623–1632.

de Beer, D. and van den Heuvel, J. C., 1988, "Response of Ammonium-Selective Microelectrodes Based on the Neutral Carrier Nonactin," *Talanta*, Vol. 35, No. 9, pp. 728–730.

Elderfield, H., McCaffrey, R. J., Luedtke, N., et al., 1981, "Chemical Diagenesis in Narragansett Bay Sediments," *American Journal of Sciences*, pp. 1021–1055.

Fanning, K. A. and Pilson, M. E. Q., 1971, "Interstitial Silica and pH in Marine Sediments: Some Effects of Sampling Procedures," *Science*, Vol. 173, pp. 1228–1231.

Gieskes, J. M., 1983, "The Chemistry of Interstitial Waters of Deep Sea Sediments: Interpretations of Deep-Sea Drilling Data," in *Chemical Oceanography*, Vol. 8, J. P. Riley and R. Chester, Eds., Academic Press, London, pp. 221–269.

Grime, G. W., Dawson, M., Marsh, M., et al., 1991, "The Oxford Submicron Nuclear Microscopy Facility," *Nuclear Instruments and Methods in Physics Research B54*, Elsevier Science Publishers, B.V. (North-Holland), pp. 52–63.

Gunderson, J. K. and Jorgensen, B. B., 1990, "Microstructure of Diffusive Boundary Layers and the Oxygen Uptake of the Sea Floor," *Nature*, Vol. 345, pp. 604–607.

Hartmann, M. and Muller, P. J., 1982, "Trace Metals in Interstitial Waters from Central Pacific Ocean Sediments," *The Dynamic Environment of the Ocean Floor*, K. A. Fanning and F. Manheim, Eds., Lexington Books, Lexington, MA.

Hesslein, R. H., 1976, "An *In-situ* Sampler for Close Interval Pore Water Studies," *Limnology and Oceanography*, Vol. 21, pp. 912–914.

Johanson, S. A. E. and Campbell, J. L., 1988, *PIXE—A Novel Technique for Elemental Analysis*, John Wiley, New York.

Jorgensen, B. B. and Revsbech, N. P., 1985, "Diffusive Boundary Layers and the Oxygen Uptake of Sediments and Detritus," *Limnology and Oceanography*, Vol. 30, pp. 111–122.

Klinkhammer, G., Heggie, D. T., and Graham, D. W., 1982, "Metal Digenesis in Oxic Marine Sediments," *The Earth and Planetary Science Letters*, Vol. 61, pp. 211–219.

Krom, M. D., Davison, P., Zhang, H., and Davison, W., 1994, "High Resolution Pore Water Sampling Using a Gel Sampler: An Innovative Technique," *Limnology and Oceanography*, Vol. 39, pp. 1967–1972.

Lyons, W. B., Gaudette, H. E., and Smith, G. M., 1979, "Pore Water Sampling in Anoxic Carbonate Sediments: Oxidation Artifacts," *Nature*, Vol. 277, pp. 48–49.

Mayer, L. M., 1976, "Chemical Water Sampling in Lakes and Sediments with Dialysis Bags," *Limnology and Oceanography*, Vol. 21, pp. 909–912.

Reeburgh W. S., 1976, "An Improved Interstitial Water Sampler," *Limnology and Oceanography*, Vol. 12, pp. 163–165.

Reimers, C. E. and Smith, K. L. Jr., 1986, "Reconciling Measured and Predicted Fluxes of Oxygen Across the Deep Sea Sediment-Water Interface," *Limnology and Oceanography*, Vol. 31, pp. 305–318.

Revsbech, N. P. and Jorgensen, B. B., 1986, "Microelectrodes: Their Use in Microbial Ecology," *Advances in Microbial Ecology*, Vol. 9, K. C. Marshall, Ed., Plenum, New York, pp. 293–352.

Santschi, P., Hohener, P., Benoit, G., and Brink, M. B., 1990, "Chemical Processes at the Sediment-Water Interface," *Marine Chemistry*, Vol. 30, pp. 269–315.

Sawlan, J. J. and Murray, J. W., 1983, "Trace Metal Remobilization in the Interstitial Waters of Red Clay and Hemipelagic Marine Sediments," *The Earth and Planetary Science Letters*, Vol. 64, pp. 213–230.

Sayles, F. L., 1979, "The Composition and Diagenesis of Interstitial Solutions: 1. Fluxes Across the Seawater-Sediment Interface in the Atlantic Ocean," *Geochimica et Cosmochimica Acta*, Vol. 43, pp. 527–545.

Sayles, F. L., Mangelsdorf, P. C. Jr., Wilson, T. R. S., and Hume, D. N., 1976, "A Sampler for the *In-situ* Collection of Marine Sedimentary Pore Waters," *Deep-Sea Research*, Vol. 23, pp. 259–264.

Sweets, J. R. A. and de Beer, D., 1989, "Microelectrode Measurements of Nitrate Gradients in the Littoral and Profundal Sediments of a Meso-Eutrophic Lake (Lake Vechten, The Netherlands)," *Applied and Environmental Microbiology*, Vol. 55, pp. 754–757.

Kiyoto Yoshinaga[1]

Mercury-Contaminated Sludge Treatment by Dredging in Minamata Bay

REFERENCE: Yoshinaga, K., **Mercury-Contaminated Sludge Treatment by Dredging in Minamata Bay,"** *Dredging, Remediation, and Containment of Contaminated Sediments, ASTM STP 1293,* K. R. Demars, G. N. Richardson, R. N. Yong, and R. C. Chaney, Eds., American Society for Testing and Materials, Philadelphia, 1995, pp. 182–191.

ABSTRACT: To eradicate Minamata Disease, caused by the discharge of sewage containing methyl mercury and its accumulation in fish and shellfish through the food cycle, a large-scale sediment disposal project was conducted with special care taken to prevent new pollution resulting from the project itself. The basic approach to sediment disposal was to construct a highly watertight revetment to reclaim the inner area of the bay and then confine sediment dredged from the remaining contaminated area in the reclamation area through surface treatment. Before sediment disposal, boundary nets were installed to enclose the work area to prevent the mixing of contaminated and noncontaminated fish. Dredging work was successfully carried out by using four cutterless suction dredgers, newly developed in advance for minimizing resuspension of sediments. Dredged material was discharged into the reclamation area, filled up to sea level, and covered with a sandproof membrane, lightweight volcanic ash earth, and mountain soil.

KEYWORDS: mercury-contaminated sludge, dredging, sediment disposal

This paper presents a summary of the project to dispose of mercury-contaminated sediment in Minamata Bay as a representative public pollution control project. In Minamata City, located at the southernmost end of Kumamoto Prefecture in southwestern Japan as shown in Fig. 1, sewage from a chemical factory caused Minamata disease. Methyl mercury contained in the sewage accumulated in fish and shellfish in Minamata Bay, causing toxic central nervous system disease in those who ate such fishery products in large quantities over a long period. Over 2000 people have been designated as Minamata disease patients so far, of whom roughly 900 people have died, including people who have died of old age.

To solve this environmental problem at its roots, it was planned that polluted fish in the bay would be made inaccessible to consumers and that mercury sediment deposited over an area of 2 090 000 m² in the bay would be disposed to restore the bay's clean environment. A plan in line with the approach to sediment disposal having been finalized by the Kumamoto Prefectural Government, disposal work commenced in 1977. The project was commissioned by the Kumamoto Prefectural Government and carried out by the Japanese Ministry of Transport. As a result of active efforts by a number of people who participated in the work, sediment dredging work finished in December 1987. Then, on 26 February 1988, Kumamoto Prefecture's Minamata Bay Public Pollution Control Project Supervisory Committee confirmed, on the basis of sediment sampling and analysis result, that all sediment with mercury

---

[1] Director, Planning Division, First District Port, Construction Bureau, Ministry of Transport, Niigata, Japan.

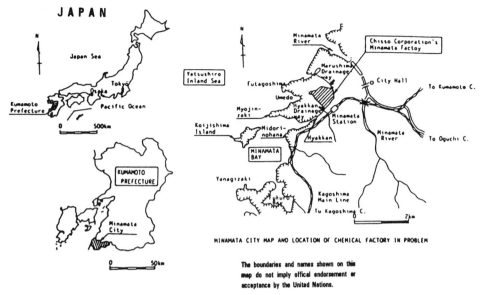

FIG. 1—*Location of Minamata city and city map.*

concentration exceeding the maximum limit had been removed. During the whole period of work, special care was taken to complete the work as early as possible and to prevent new pollution due to the project itself (Hirose 1990).

Basic Plan

Developments Towards Implementation

In March 1974, the Director General of the Environmental Agency, the Minister for Transport, and the Governor of Kumamoto Prefecture reached a basic agreement on the implementation of Minamata Bay public pollution control work. Subsequently, the relevant governmental agencies and experts conducted detailed studies, based on which the Kumamoto Prefectural Government in June 1975 developed the Minamata Bay Sludge Deposit Disposal Plan and the Basic Plan for Supervision of Minamata Bay Sludge Deposit Disposal. In accordance with the Disposal Plan, the Kumamoto Prefectural Government commenced in October 1977, prior to beginning the major project, installing boundary nets to block fish movement in and out of the work area.

Basic Plan for Mercury Sediment Disposal

Disposal of sediment with 25 ppm or higher total mercury concentration was planned. This value of mercury concentration for sediment disposal was estimated on the basis of the Provisional Standards for Removal of Mercury Contaminated Sludge, established in August 1973 by the Environmental Agency, and finalized through the deliberation of the Kumamoto Prefectural Anti-Pollution Measures Council. Accordingly, disposal of roughly 1 510 000 m^3 of sediment was necessary from an area of about 2 090 000 m^2 in Minamata Bay. Contamination was serious in terms of both mercury concentration and sediment thickness

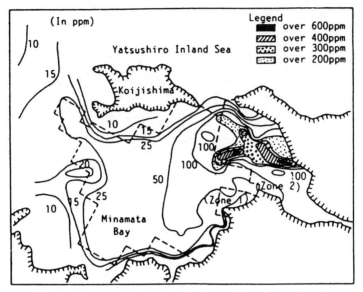

FIG. 2—*Mercury concentration distribution before disposal work.*

in the bay's inner area, into which sewage flowed out from Chisso's Minamata factory, and gradually lessened toward the outside of the bay as shown in Fig. 2.

It was planned that the most severely polluted inner area, 580 000 m² in size, would be enclosed with a highly watertight revetment (Fig. 3), reclaimed with sludge dredged from the remaining work area, and ultimately covered with good quality mountain soil.

Existing port facilities in Minamata Bay, all in the newly planned reclamation area, were to be totally phased out of service with the progress of work. To replace them, zone 1 (for

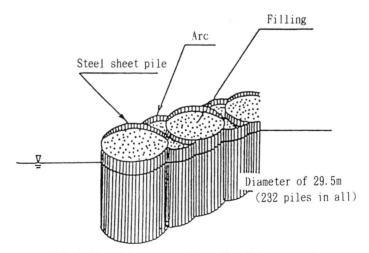

FIG. 3—*Watertight revetment (sheet pile cellular structure).*

constructing the so-called Midori wharf in advance through the preceding reclamation) was planned outside the main reclamation area (called zone 2), shown in Fig. 4. In this zone, all stages of work, from revetment construction to dredging and reclamation, were to be conducted, but on a small scale compared to zone 2; therefore, this zone was also regarded as a pilot zone for checking the safety of the work method to be adopted.

To prevent polluted fish from mixing with nonpolluted fish, double fish netting was to be installed to enclose the work area during work and for an appropriate length of time thereafter if necessary, except on the navigation route, where an acoustic fish way controller was used to keep fish off. After completion of sediment disposal, all fish in the work area were to be caught.

Prevention of New Pollution

A number of measures was taken to prevent renewed pollution due to the project's implementation.

First, to stabilize water in the dredging area and in front of the reclamation area, a cofferdam was temporarily constructed between Koijishima Island and the end of Myojinzaki prior to work. This cofferdam stagnated the intrabay current, thus accelerating suspended sludge sedimentation and preventing it from spreading out of the bay.

Second, until discharged sludge was completely confined with mountain soil, the reclamation area was kept under water to prevent direct sludge exposure to sunlight and air, which, as was pointed out, might methylate inorganic mercury in sludge.

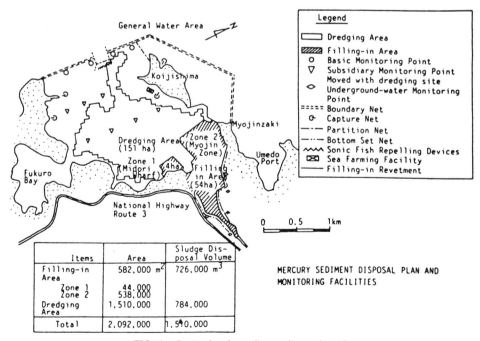

FIG. 4—*Basic plan for sediment disposal work.*

Third, to anticipate and avoid a new outbreak of pollution, the Kumamoto Prefectural Government established a system for monitoring the influence of mercury sediment dredging on water quality, fishery products, and plankton.

Besides, prompt feedback of monitoring results of water quality to the project was thus ensured. It was also decided that these monitoring results be promptly announced officially to residents and referred to the supervisory committee, organized by experts, resident representatives, etc., and that if such results were found to deviate from the supervisory standards, work should be suspended, or other relevant measures taken.

Mercury Sediment Dredging

Basic Concept

Important points kept in mind in sediment dredging, were: minimization of the work period, elimination of resident anxiety as early as possible, and control of muddiness to prevent a new outbreak of pollution. On the other hand, it should be noted that: (a) most of the mercury contained in the sediment in Minamata Bay is inorganic and does not easily dissolve into water and (b) methylation of inorganic mercury in dredging poses no problem if aerobic conditions can be avoided.

Special care was also taken in order to minimize extra dredging so as to permit the accommodation of all dredged sediment in the reclamation area as planned and to achieve completion of the project as early as possible. Mercury was contained basically only in the upper thin layer of bottom sludge in the bay. Whereas the reclamation area, planned in the inner part of the bay, had a fairly thick layer of contaminated sediment compared to other parts, the dredging area required only roughly 50 cm deep dredging on average. The thickness of the layer to be removed by dredging, however, was accurately and carefully identified to ensure the elimination of all polluted sediment, as well as to minimize extra dredging so as to save time and money. From this point of view, dredging work was actually done along the shelf-shaped profile drawn in advance, based on the mercury concentration distribution and the specifications of each suction dredger.

Since bottom sediment disturbance due to dredging could cause mercury to dissolve, a dredging method was carefully chosen so as to minimize muddiness. A conventional suction dredger cuts into the sea bottom at the suction mouth and then pumps up sediment to be removed, with the sediment disturbance by the cutter causing muddiness. A suction dredger without a cutter was developed especially for sediment dredging in Minamata Bay. Since the cutterless suction dredger had already been used with great success in sludge dredging, though on a small scale, in Tokuyama Bay and Yokkaichi Port in Japan, the use of this type of suction dredger was decided upon with a slight modification for the present work.

Implementation and Execution Control

Four cutterless suction dredgers were used in Minamata Bay because only four of all the existing cutterless suction dredgers were found to be applicable to the present work with a slight modification. They were different from each other mainly in terms of dimensions. The work area was divided into four portions for convenience. Then, the above-mentioned four dredgers were assigned to each portion in consideration of their own properties.

All the suction heads of these dredgers, including the mouth, were modified in advance. Major modifications were as follows:

(a) The suction mouth shape was modified to suit Minamata Bay's sediment properties after experiments with similar soft mud.
(b) Provisions were made for monitoring muddiness near the suction mouth by TV camera.

For monitoring purposes, two kinds of equipment were adopted and continuous measurement of turbidity and direct monitoring with an underwater TV camera were undertaken. For turbidity measurement, continuous type turbidimeters were attached to the suction heads, with readings displayed in the dredger operator room. For visual checking of muddiness, an underwater TV camera was mounted on the suction head, together with sight vanes positioned at constant intervals. Muddiness was checked on the TV monitor in the operation room on the basis of the view of the vanes. In addition, we confirmed in advance through the model experiment that the vane visibility distance was correlated with turbidity. These two systems enabled the dredger operator in the operation room to check muddiness around the suction head and thereby take necessary measures such as slowing down the dredging speed.

Procedures undertaken to minimize extra dredging were as follows: (1) Prior to dredging, submarine topography and thickness of sediment to be removed were accurately measured in the dredging area; (2) Submarine topography after ideal sediment removal without any extra dredging was expressed with contour lines; (3) Since equal thickness dredging was not technically feasible, a certain area was to be dredged to a uniform level; therefore, the dredging area was divided, in 30 cm units of thickness to be removed on the basis of the above-mentioned contour lines, into segments in which dredging to a uniform water depth would eliminate at least all subject sediment; (4) To ensure high precision, enhance efficiency and monitor dredging conditions, dredgers were furnished with various sensors, whose data were processed via microcomputer for easy operator perception and displayed on the monitor screen in the operation room. Major data obtained through the above operation management system, based on the measurements of various sensors loaded on each dredger and the basic principle to obtain these data, were as follows:

(a) Dredging Thickness
The suction mouth was on the side of the suction head, which swung to the left and right while pumping up sediment. Therefore, echo sounders were installed on both left and right sides of the head to obtain information on dredging thickness from the difference between the readings of the two sounders. Since the inclined head gave a distorted measurement, a clinometer was also attached to the head to correct the resulting thickness data.
(b) Submarine Topography Before and After Dredging
The dredger moved forward while swinging its suction head and dredging. Therefore, sounders identical to the above-mentioned ones were also installed on both the front and rear sides of the head to survey submarine topography before and after dredging. Errors due to the head inclination were corrected in the same manner as described in dredging thickness.
(c) Suction Head Position, Swing Speed, and Swing Direction
In addition to electric positioning equipment, the suction head was provided with a gyrocompass to measure the head position, swing speed, and swing direction.
(d) Dredged Volume and Mud Concentration
To measure these data, a current meter and a density meter were installed in the

dredger discharge pipe, which were also useful for managing the effluent treatment system.

The resulting measurements, together with other necessary data for execution control, were processed via microcomputer on board and displayed in real time on the monitor screen in the operation room.

The discharge pipe was also improved to prevent sludge exposure to air and sunlight, which promote methylation of inorganic mercury in sludge. In addition to this improvement, the discharge outlet in the reclamation area was submerged in water, with discharged sludge kept underwater during work.

Effluent Treatment

A Prime Minister's Office ordinance stipulates that effluent (which is to be discharged back to sea) from the land reclaimed with harmful earth/sand must not contain mercury exceeding 0.005 ppm. Since mercury analysis takes considerable time, turbidity (which was correlated with mercury concentration and can be measured in real time) was also monitored to obtain the prompt information needed for efficient operation of the effluent system. Through experiments, we adopted 40 ppm as the maximum limit of turbidity for effluent discharge.

The following points were kept in mind in designing effluent treatment:

(1) To increase natural sedimentation in the reclamation area, effluent flow distance was maximized by commencing reclamation from the innermost portion and by providing partition cofferdams in the area;
(2) Precipitation in the reclamation area was estimated on the basis of past data and then the result was incorporated into the effluent treatment plan.
(3) Before discharge, effluent was to pass through the effluent treatment facilities, which are composed primarily of a mixing tank, flocculant mixer, flocculation tank, settling tank and a rapid filter.
(4) Since effluent treatment, unlike dredging, can be controlled even at night, 24 hour treatment was designed.

The sludge discharged into the reclamation area had to be kept underwater at all times to prevent direct exposure to air and sunlight. The water depth was checked every morning to ensure at least 50 cm as a general rule.

Since rainwater inflow from the existing land into the reclamation area lowers effluent treatment efficiency, a side ditch was installed between the land and the reclamation area to divert rainwater.

With these points kept in mind, effluent was treated in the following manner:

(1) After natural sedimentation in the reclamation area, supernatant water was first subjected to water quality testing to determine the need for flocculant addition and, if needed, the appropriate amount.
(2) Water was sent to the mixing tank, where a flocculant was added if necessary, then to the flocculation tank, and then to the settling tank, where flocs settled.
(3) The resultant effluent flowed into the monitoring tank for water quality testing.
(4) If the discharge criterion was exceeded, effluent was sent back to the rapid filter for thorough purification.

It was planned that if muddiness exceeded the supervisory standard, effluent discharge would be suspended. However, muddiness has remained far below the standard so far, so no hindrance occurred.

Sludge Confinement

Sludge discharged into the reclamation area had to be covered with mountain soil and other good quality earth/sand for confinement. Since the sludge had a high proportion of fine particles as well as an extremely high water content and low bearing capacity, earth covering work would be inefficient and almost impossible if conducted over the sludge as it was. To increase bearing capacity sufficiently for subsequent earth covering work, the sludge surface was subjected to the following two processes:

(1) The entire reclamation area was covered with a sandproof membrane to disperse the concentrated load to be imposed;
(2) Sirasu (volcanic ash earth/soil of low specific gravity, locally produced), which generates strength with light weight, was sprinkled in the condition of slurry over the membrane to the thickness of 80 cm to improve the bearing capacity.

These two processes, always carried out underwater (50 cm or more deep) to prevent the sludge's exposure to air, imparted the required bearing capacity to the sludge surface, making smooth and efficient earth covering work possible. After water in the reclamation area drained as effluent, good quality mountain soil was placed over the Sirasu layer and leveled for complete confinement.

Supervision

Monitoring Plan and Results

A thorough monitoring plan was developed to prevent a new outbreak of pollution due to sediment disturbance and dispersion with work or to exudation of harmful substances from the reclamation area. In response to local residents' anxiety regarding safety, continuous monitoring was planned for water quality and fishery product contamination. Regarding water quality, basic monitoring points were provided along the work area boundary to prevent the project from having any effect beyond that area. Subsidiary monitoring points were also provided between the basic monitoring points and the current working points to predict water quality at the basic monitoring and for expeditious assessment of the work. Also, to assess fishery products contamination, periodic sampling and analyses were conducted, together with fish culture experiments in Minamata Bay and plankton investigation, to ensure that work implementation would not pollute the aquatic life in the bay. The locations of the water quality monitoring points and the sea farming facility for fishery product monitoring are shown in Fig. 4.

These monitoring practices were carried out under the auspices of the Kumamoto Prefectural Government. To confirm work safety from an objective standpoint, the Minamata Bay Public Pollution Control Project Supervisory Committee was established in December 1976 as an advisory organization to the Prefectural Government. The Committee consists of eight experts, four staff members of administrative agencies concerned, three members of the Prefectural Assembly, and ten resident representatives. Committee meetings are open to the public as a general rule.

Water quality testing in accordance with the monitoring plan revealed that turbidity, hydrogen ion exponent, chemical oxygen demand, and dissolved oxygen showed variations probably attributable to natural environmental changes (rainfall, red tide, etc.) but that mercury concentration and all other items remained below determination limits or did not exceed supervisory standards. Work safety was thus confirmed.

None of the fish sampled outside the work area exceeded the supervisory standard for total mercury concentration. Continuous investigation into total mercury concentration changes in fishery products in the work area revealed no noteworthy changes suggestive of influence by the project.

Confirmation of Sediment Dredging Completion

It was determined in advance that sediment dredging would be adjudged completed if the average mercury concentrations of four grid points of a mesh was below 25 ppm in each of the meshes established (200 m interval meshes as a general rule) to determine the dredging area and the investigation points of mercury concentration.

In December 1987, just after the completion of dredging work, the Kumamoto Prefectural Government sampled and analyzed sediment in the dredging area and its peripheral area to determine mercury concentrations in individual meshes shown in Fig. 5. According to Fig. 5, the highest concentration was found to be 8.75 ppm. In addition, the comparison between the results in Figs. 2 and 5 shows the great improvement of the bay environment after the dredging work. This finding was reported on 26 February 1988 at the 60th Supervisory

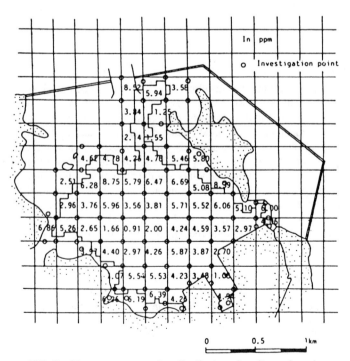

FIG. 5—*Mercury concentration distribution after disposal work.*

FIG. 6—*View of completed reclaimed land in Minamata Bay.*

Committee meeting, where completion of Minamata Bay mercury sediment dredging was formally confined.

Conclusion

Sediment dredging work, which was the main part of the whole work planned, was successfully completed in December 1987. Completion of the work without the occurrence of any outbreak of pollution relieved residents' anxiety. We were obliged to develop and improve the working method and equipment.

After about 10 years' effort to remediate the mercury sediment that caused Minamata Disease, the Minamata Bay Public Pollution Control Project, with financial aid from the Japanese government, was officially completed in March 1990 with the final act of earth covering. Figure 6 shows the view of the completed reclaimed land, confining Minamata Bay's mercury sediment.

In Minamata Port, the so-called "Marine Town Project" (a series of the waterfront development) has been planned and actively started for local revitalization. We hope that this revitalization program will be aided by the effective usage of the sludge reclaimed land.

Reference

Hirose, M., 1990, "Mercury Sediment Disposal Work in Minamata Bay," *Civil Engineering in Japan*, Vol. 13, No. 421.

Containment and Isolation

G. N. Richardson,[1] D. M. Petrovski,[2] R. C. Chaney,[3] and K. R. Demars[4]

State of the Art: CDF Contaminant Pathway Control

REFERENCE: Richardson, G. N., Petrovski, D. M., Chaney, R. C., and Demars, K. R., **"State of the Art: CDF Contaminant Pathway Control,"** *Dredging, Remediation, and Containment of Contaminated Sediments, ASTM STP 1293*, K. R. Demars, G. N. Richardson, R. N. Yong, and R. C. Chaney, Eds., American Society for Testing and Materials, Philadelphia, 1995, pp. 195–219.

ABSTRACT: Contaminants from sediments contained within a confined disposal facility (CDF) can be discharged to the environment via six potential pathways. These pathways include three waterborne pathways, the direct uptake of the contaminants by plants or animals, and airborne emission of contaminants. Conventional CDF design focuses on retention of sediment particles with perimeter dikes. Depending upon the nature of the site, the contaminants of concern, method of dredging, physical properties of the dredged material, operational aspects, and many other factors, including sociopolitical factors, supplemental environmental design criteria may be required for the CDF. This paper reviews design alternatives to control contaminant loss from the CDF basin through the six identified contaminant pathways. These alternatives include the use of both additional design components and operational constraints. The need for a specific pathway control measure is shown to depend on both site and sediment specific evaluation criteria.

KEYWORDS: dredged material, contaminant, pathway, containment, confined disposal facility (CDF)

Throughout the Great Lakes, about 4 million cubic yards (5.2 million cubic metres) of sediments are dredged annually to maintain navigation in channels and harbors for commercial, military, and recreational users, and as part of environmental remediation projects (EPA 1990). Within the United States, about 400 million cubic yards (520 million cubic metres) of sediments are dredged annually (COE 1987). Sediment is the material that settles to the bottom of a body of water and includes soil particles consisting of clays, silts, and sands; organic matter; shells; and residuals from industrial discharges, which can include organics and heavy metals. Many of the waterways are adjacent to urban and industrial areas and the sediments in these areas are often contaminated from various adjacent sources. Such contamination is typically historical and predates modern regulatory controls. A portion of the sediments are so highly contaminated with anthropogenic substances that they require remedial action by EPA. About one half of the total sediments dredged in the Great Lakes [approximately 2 million cubic yards (2.6 million cubic metres)] are sufficiently contaminated to be problematic sediments and require placement in a confined disposal facility

---

[1] Principal, G. N. Richardson & Associates, Raleigh, NC 27603.
[2] Geologist, U.S. Environmental Protection Agency, Region 5, CERCLA Division, Chicago, IL 60604.
[3] Professor, Humbolt State University, Arcata, CA 95521.
[4] Professor, University of Connecticut, Storrs, CT 06269.

(CDF). These contaminated sediments require special consideration during dredging and disposal operations because of the potentially adverse impact on water quality and local organisms. Sound planning and design of dredging operations and disposal facilities are necessary to protect the environment and yet keep these activities economically viable. The remaining half of the sediments are generally classified by EPA as clean and suitable for unconfined, open water disposal. This paper focuses on CDF contaminant pathway control in freshwater environments. In the following sections the issues of the regulation of dredging activity, contaminant pathway control, CDF basin design recommendations, and closure practices will be discussed.

Regulation of Dredging Activity

The regulatory requirements for the disposal of dredged material are determined by both the type and level of the contaminants associated with the dredged material, as well as the extent to which the contaminants could potentially be released from the sediments to proximal air, ground water, or surface water. Although release routes of concern commonly include emissions to the atmosphere and can include more exotic pathways such as the discharge of free phase organics, the ubiquity of water in all aspects of sediment dredging and disposal have focused consideration on contaminant release routes to water. This is reflected in Table 1, which depicts several pertinent relationships between the level of sediment contamination, the degree of contaminant partitioning to the water associated with the sediments, three conceptual categories under which sediment disposal can occur, and the significant disposal regulations. As depicted on Table 1, these three approaches to sediment disposal are labeled "beneficial use or open water disposal," "solids retention," and "hydraulic isolation."

TABLE 1—*CDF design criteria based on contaminant level and pathway.*

Uncontaminated Sediments				Highly Contaminated Sediments
	CWA 404 Guidance	Partitioning in Excess of State WQS	Significant Partitioning	
Minor Sediment Contamination Minimal Partitioning			PROBLEMATIC SEDIMENTS	
	Minor Partitioning	Moderate Partitioning	Extensive Partitioning	RCRA/TSCA Sediments
Beneficial Use or Open Water Disposal CWA 401 Certification for Dredged Material Discharge	Solids Retention Approach Increasing Degree of Sediment Isolation Filter Intake CDFs CWA 401 Certification for Diffuse Discharge	Hybrid Approach Hydraulic Isolation During Sediment Disposal Only CWA 401 Certification for Point or Diffuse Discharge	Hydraulic Isolation Approach Increasing Degree of Hydraulic Isolation CWA 401 Certification for Point Discharge	Hydraulic Isolation CDF MTG Design CWA 402 Permit for a Point Discharge Conformance with RCRA-C Minimum Technology Guidance
←------ 87 COE CDF DESIGN MANUAL ------→				
←------------ SHORELINE AND IN-LAKE CDF DESIGNS ------------→				
←------------ UPLAND CDF DESIGNS ------------→				
←------------ REGION 5 CDF GUIDANCE DESIGNS ------------→				
←------ CWA SECTION 404 AND NEPA, EPA/COE FRAMEWORK ------→				
←------------ INCREASING DEGREE OF SEDIMENT CONTAMINATION ------------→				

The Clean Water Act (CWA) governs the discharge of dredged material into "waters of the United States." If the level of sediment contamination is sufficiently low so that the uncontained release of the sediments into the environment would not have an unacceptably adverse environmental impact and would not result in an exceedance of the applicable State Water Quality Standards (WQS), the sediments could be disposed of at an approved open water disposal site or placed in the environment in an unrestricted manner. An example of such unrestricted or unconfined placement would be the use of dredged material for beach nourishment or as fill. As shown in Table 1, compliance with the State WQS for the disposal of dredged material via open water or beneficial use could entail the issuance of a Section 401 Certification under the CWA. Dredged materials that cannot meet the CWA standards for open water disposal or beneficial use, e.g., problematic dredged materials, must be segregated from the environment to some extent. Sediments that cannot be released to the environment in an unrestrictive manner are labeled problematic dredged material in Table 1. The disposal of problematic dredged materials is the focus of this paper.

The U.S. Army Corps of Engineers (COE) uses CDFs to contain contaminated sediments that cannot be released without control to the environment. As shown in Fig. 1, CDFs can be located at both upland and in-lake sites depending on the level of isolation that the sediments under consideration warrant. CDF designs can be grouped based on their extent of isolation: (1) CDFs that physically isolate the sediment solids from the adjacent environment (solids retention) and (2) CDFs that hydraulically isolate the sediments and any derived effluent from the adjacent environment (hydraulic isolation) (see Table 1).

Dredged materials typically contain large amounts of water. Depending upon the method used to excavate the materials, dredged materials typically have solids to water ratios of 5 to 50% by weight. The disposal of large quantities of material with a high percentage of both solids and water presents both technical and regulatory challenges unique to dredged materials. Generally, the disposal of wastes that have a high percentage of water is regulated by the CWA, while the disposal of wastes that are not liquids, e.g., solids, is regulated under the Resource Conservation and Recovery Act (RCRA). Given the large quantities of water and solids, CDFs commonly incorporate aspects of both regulatory programs into their designs.

The basic framework for federal water pollution control regulation was established by the Federal Water Pollution Control Act (FWPCA) established in 1972. In 1977, FWPCA was renamed the Clean Water Act (CWA) and amended to provide regulatory control of toxic water pollutants. Section 404 of the 1977 amendments to the CWA provide the source of the current regulatory control over dredging and disposal activity. The CWA is currently undergoing reauthorization. The need for a given federally sponsored dredging activity must be evaluated under the National Environmental Policy Act (NEPA) based on a review of alternative solutions. Note that NEPA does not establish performance objectives, but simply requires the evaluation process.

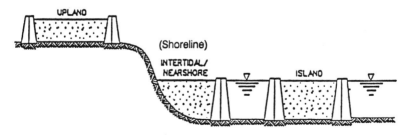

FIG. 1—*Upland, shoreline, and island CDF.*

Contaminated Dredged Sediments

Dredged materials are considered contaminated when the ambient or leachable concentration of metals or organic compounds exceed federal RCRA or Toxic Substance Control Act (TSCA) regulatory limits or when contamination exists in high enough concentrations and are sufficiently available to affect human or ecosystem health, or both. RCRA and TSCA regulations provide for the disposal of contaminated materials in landfill cells that hydraulically isolate the waste from proximal ground and surface waters. Wastes placed within such landfill cells must not be a liquid as defined by the paint filter test (EPA Method 9095). RCRA and TSCA land disposal regulations are therefore directed at wastes that are very high in solids and have little or no free liquids.

Confined Disposal of Problematic Dredged Materials

Problematic sediments are defined as those that contain contaminants that have the potential to adversely impact human health or the proximal ecosystem but that are not regulated by RCRA-C or TSCA, but can be controlled by Section 404 of the CWA or the Rivers and Harbors Act (RHA). For example, sediments can become highly contaminated by point and nonpoint source contamination related to preregulatory industrial activities. Such contamination may not trigger RCRA-C or TSCA regulatory control but can lead to the degradation of navigable water and as such can be regulated by the combination of National Environmental Policy Act (NEPA) Environmental Impact Statement (EIS), CWA Section 404, and state authority.

The placement of problematic sediments within a CDF may pose design problems not adequately addressed by current TSCA and RCRA-C Minimum Technology Requirements (MTR). However, containment of contaminants remains a high priority in federal regulations. Recently promulgated RCRA Subtitle D (RCRA-D) regulations have extended containment requirements in common municipal solid wastes. Under these regulations, a composite (soil plus geomembrane) liner is required unless the designer can demonstrate that contaminants in the upper most aquifer will remain below the Maximum Contaminant Level (MCLs) at a specified point of compliance adjacent to the landfill. Similarly, CDFs for problematic sediments may require some form of lining system unless it can be demonstrated that WQS will not be exceeded in the ground water or surface water adjacent to the facility.

Contaminant Pathway Control

Contaminants from sediments contained within a CDF can be discharged to the environment via six potential pathways. These pathways are shown in Fig. 2 and include three waterborne pathways, two pathways related to the direct uptake of the contaminant by plants or animals, and an airborne pathway.

Waterborne Contaminants

The control of waterborne contaminants must consider both the contaminants dissolved in the effluent and the solid contaminant fraction associated by sorption or ion exchange with the total suspended solids (TSS) within the effluent (Thackston and Palermo 1990). Given sufficient retention time in a containment area, noncolloidal suspended solids will settle out of the effluent and be retained. This design practice of "solids retention" is the basis for the COE design process but may be limited by contaminant concentration and partitioning as previously shown in Table 1. The approach presented in this paper considers the impact of contamination associated with both the TSS and the aqueous phase.

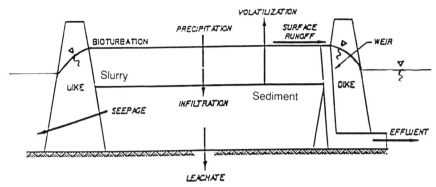

FIG. 2—*Contaminant pathways for in-lake CDF.*

For noncolloidal dredged materials, the suspended solids concentration in effluent water is influenced by the pond surface area and residence time for the effluent within the CDF. The conventional COE design procedures for determining the surface area or the residence time required for sedimentation to meet effluent suspended solids requirements are summarized in Figs. 3 and 4, respectively. Both design procedures assume that the CDF dikes are impermeable to the effluent and that the dredged material was hydraulically dredged. Contaminated dredged materials generally have a high percentage of colloidal material that remains in suspension due to physiochemical factors.

The procedure for calculating the minimum required surface area, shown in Fig. 3, recognizes that sediments fall out of suspension quicker in ponds having larger surface areas and shallow depths. In deep ponds, the falling sediment particles begin to collide, accumulate, and create a condition of compression settling, e.g., consolidation under self weight. Compression settling is very slow and therefore results in an increase in the required retention time. The minimum required surface area for the pond is calculated in two ways: (1) based on the storage volume required for the dredged material once it settles out of suspension and (2) based on the surface area for flocculent sedimentation. The larger of the two areas is then assumed to be the minimum required surface area for the pond within the CDF.

Having solved for the minimum pond area the COE procedure then calculates the minimum residence time for the supernatant within the pond as shown in Fig. 4. A percent solids removal versus time relationship is first established based on the pond depth and a range of assumed times. Knowing the suspended solids requirement for the effluent, the residence time (T_d) based on pond depth is established. The actual design residence time (T) is obtained by modifying T_d based on the actual length and width of the pond.

Effluent Flow through Weirs and Filters—Typical effluent filter systems include pervious dikes and downflow/upflow weirs or cartridges. The pervious dikes are designed to filter out the suspended solids by selecting the filter media particle size distribution using conventional geotechnical filter criteria. Clogging of the pervious dike is minimized by using stratified or baffled dike sections that provide a control over the maximum flow gradients that develop within the dike section. The larger the potential flow gradient, the smaller the potential for clogging of the filter.

Pathway controls, filter weirs, and cartridges for contamination related to TSS in effluent leaving the CDF through a weir are presented in Fig. 5. Current COE weir design procedures focus on the removal of suspended solids from the effluent. Four design alternative courses of action for the control of effluent have been presented by Krizek et al. (1976). Pervious dikes are recommended by the COE to filter effluents with concentrations of TSS up to 0.5

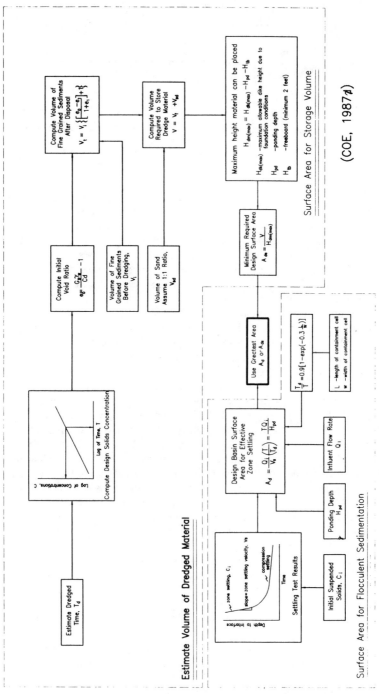

FIG. 3—*COE design procedure for determining the surface area required to meet effluent suspended solids requirement.*

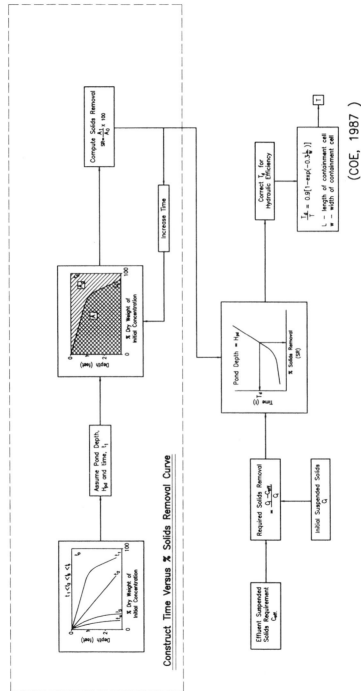

FIG. 4—*COE design procedure for determining residence time (T) required to meet effluent suspended solids requirement.*

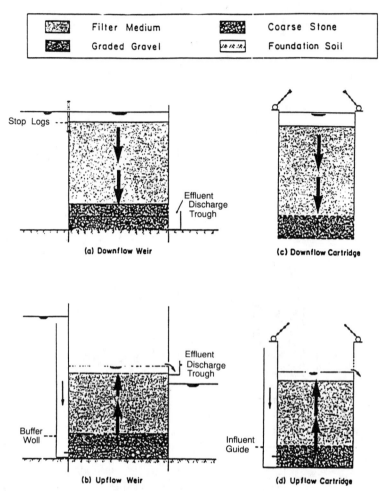

FIG. 5—*Filter weirs and cartridges.*

g/L. Such dike filters constitute a low-maintenance filter that is characterized by long effective lifetimes. For cases where the effluents are expected to have TSS concentrations up to 1 to 2 g/L, the sandfill weir offers an attractive alternative. Sandfill weirs designed without backwash capabilities require maintenance to replace clogged filter media at periods significantly shorter than pervious dike lifetimes. Although the type of effluent to be treated with the sandfill weir is similar to that for pervious dikes, its mode of operation is much more flexible. Granular media cartridges can be used with waters having TSS up to 10 g/L; however, maintenance requirements are expected to be excessive at loads higher than a few grams per litre.

The TSS concentration of an effluent can be significantly influenced by the length of the weir (sharp crested, rectangular, or shaft type) and the depth of the pond as shown in Fig. 6. Waterborne suspended solids and the associated contaminants that cannot be removed by basin or weir design may render the effluents from disposal areas unacceptable for discharge to open waters, and it may be necessary to employ a filter system or chemical methods (Schroeder 1983) to clarify disposal area supernatant.

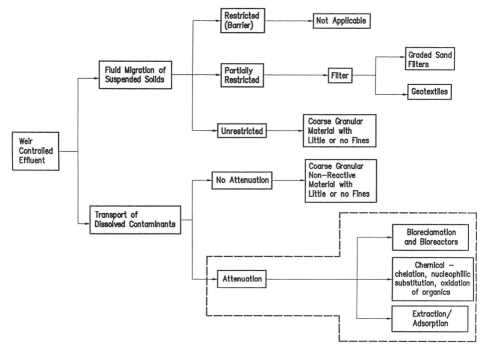

FIG. 6—*Control of contaminant pathway through weir.*

Dike Seepage—Movement of contaminated pond water through the dike structure will occur unless a mechanism is provided to either restrict the actual water migration or attenuate the waterborne contaminants. The flow through a dike can be restricted either by controlling the hydraulic gradient that exists across the dike or by reducing the permeability of the dike. Dye tracer studies performed in Great Lakes CDFs (Pranger and Schroeder 1986) showed that zones of significantly higher dike seepage existed at several CDFs that had dikes constructed using both sheet pile cutoff walls within a rip-rap dike and cores of crushed limestone gravels and sands. During the studies by Pranger it was also observed that significant decreases in outflow occurred in areas where deltas of previously dredged material were placed against the dike within the CDF.

Design considerations to limit the release of contaminants through dike seepage are shown in Fig. 7. The design of dike seepage control systems must reflect the nature of the subgrade upon which the dike is built. Vertical barrier systems must intercept and key into an underlying low-permeability soil layer (i.e., aquitard) that prevents contaminated ground water from flowing beneath the barrier. Obviously the economy of a vertical barrier system is significantly influenced by the depth of penetration required to intercept such a confining layer. Lacking a natural aquitard, a low-permeability liner may need to be incorporated in the basin design.

Mechanisms to totally restrict water migration through the dike involve constructing an impermeable barrier in or on the dike consisting of either a low-permeability soil, a geomembrane, or a geosynthetic clay liner (GCL). The low-permeability soil barrier can be constructed using either a compacted clay line (CCL) or GCL designed into the CDF dikes or by an intentional operations placement of clean, fine-grained sediments against the surface of the dike structures. Table 2 reviews the placement of alternative barriers to restrict the

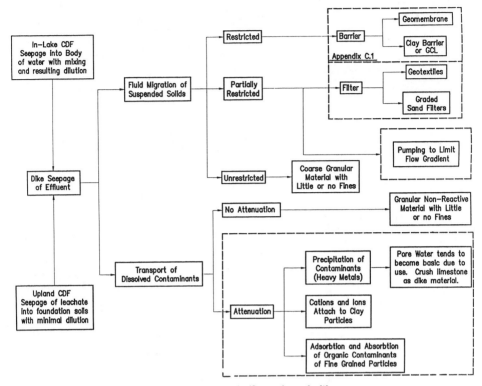

FIG. 7—*Seepage of effluent through dike.*

flow through perimeter dikes. While geomembrane and clay barriers will be highly restrictive, the operationally placed barrier may be only partially restrictive depending upon the sediment particle size and placement method.

In summary, adequate attenuation of water borne contaminants in the seepage can require the removal of both suspended solids and dissolved contaminants from the effluent. The suspended solids can be allowed to settle out of the effluent or can be physically filtered as described above. Removal of soluble contaminants from the effluent may require changing the effluent water chemistry.

Foundation Seepage of Leachates—The migration of contaminants into the foundation soils involves two different mechanisms, as shown in Fig. 8. These two mechanisms are: (1) advective transport of suspended particles or dissolved contaminants, and (2) diffusion. Laboratory and field investigations have shown that foundation leachate problems can exist even when the discharged effluent quality is acceptable (Chen et al. 1978). Foundation leachate problems were observed in facilities where the supernatant and effluent had very low contaminant concentrations.

Advective transport of contaminants into the foundation soils is caused by the flow of leachate under Darcy's Law. The rate of flow is controlled by the hydraulic gradient and the permeability of the soils. Shoreline CDFs are commonly located in ground water discharge zones (e.g., wetlands) that have natural flow gradients towards the CDF. Such natural inward gradients minimize leachate migration into foundation soils. Under these conditions, leachate migration into foundation soils is driven by hydraulic heads produced by the varying depth of effluent within the CDF, surface water, (e.g., precipitation) infiltration, and excess pore

TABLE 2—*Dike barrier system applications in CDFs.*

	Barrier System	Barrier Location*	COE Usage	In-Lake CDF	Shoreline CDF	Upland CDF
"Impermeable" Barriers	Compacted Clay Liner (CCL)	C		N/A	N/A	●
	Geomembrane Liner (GML)	C	X	○	○	●
	Geosynthetic Clay Liner (GCL)	C		N/A	○	●
	Geomembrane Cut-off Wall	A		●	●	●
	Bentonite Slurry Cut-off Wall	A	X	●	●	●
	Fabric Form w/ Grout	C		●	●	●
"Low" Permeability Barriers	Clean 'Fine' Sediments	C		●	●	N/A
	Clogged Geotextile	C	X	○	●	N/A
	Graded Soil Filter	B	X	○	●	N/A
	Fabric Form w/Sand	C		●	●	N/A

N/A Not Applicable   ● Good Application   ○ Maybe Difficult to Construct

* Barrier Locations

pressures generated during placement of dredged sediments. The extent or significance of surface water infiltration is dependent upon whether a final cover is in place or interim conditions exist. During interim periods where the dredged material is exposed, the advective mobility of contaminants may be greater due to the chemical actions of acid rains or oxidation related breakdown of contaminants.

The actual flow of contaminants into the foundation soils is controlled by the type of liner material in the CDF. The types of liner material are the following: (1) restricted/no flow barrier, (2) partially restrictive barrier, and (3) unrestrictive barrier. The fully restricted or no flow barrier is typically the result of a CDF foundation liner of compacted clays or a geomembrane. In contrast, a partially restricted barrier can be a function of site stratigraphy, site hydrogeology, or the presence of a filter media at the bottom of the CDF. The filter media can be constructed of a clean dredged fine sediment layer either by itself or in combination with a geosynthetic filter fabric. An unrestricted barrier is typically comprised of clean granular foundation soils.

The advective transport of soluble contaminants in the leachate can potentially undergo attenuation depending on three different processes. These processes are (1) attachment of anions/cations to clay minerals, (2) adsorption/absorption of organic contaminants on humic materials, and (3) biodegradation of the contaminant.

Diffusion transport of dissolved contaminants is driven by concentration gradients and is evaluated using Fick's First Law. This contaminant transport can occur in the absence of

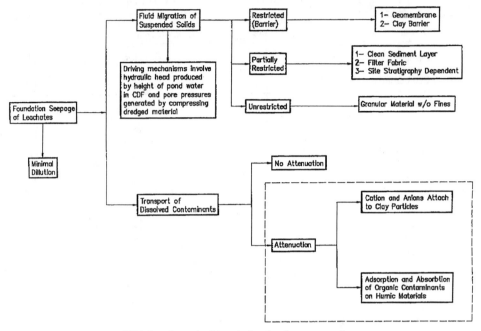

FIG. 8—*Control of foundation leachate generation.*

water movement. Fortunately, the concentration of contaminants within the pore water is typically low and does not lead to significant diffusion rates.

Runoff Contaminant Control—Precipitation falling on a CDF can contribute to contaminant migration by increasing infiltration and surface water runoff as shown in Fig. 9. The acidic nature of precipitation in the northern United States may lower the pH of the exposed dredged material and supernatant and place more metal contaminants into solution. The potential for contaminant transport due to runoff is influenced by whether the dredged material remains exposed to air during interim operations such that geochemical changes in the sediments are possible, and whether a final cover has been placed over the dredged material.

During interim operations, the surface water runoff over air exposed dredged materials can cause significant erosion losses if no vegetation or large slopes (<5%) exist. Contaminant losses from runoff during the interim operations period can be reduced by providing surface vegetation, limiting slopes to less than 5%, and ensuring that runoff does not over top perimeter dikes. It should be noted that interim operations may last several decades so that the interim vegetation of exposed dredged materials may be very cost-effective. During this same period, consolidation settlement of the dredged material may result in ponds forming within the CDF. Such ponds provide the potential for plant and animal uptake.

Once the final cover is placed over the dredged material, the significance of surface water runoff as a contaminant transport mechanism significantly diminishes. The presence of a thick vegetative layer improves the resistance to soil erosion and the removal of suspended solids and soluble nutrients such as ammonia, nitrogen, and soluble phosphorous (Chen 1978). Such runoff will have no contact with contaminants if a barrier layer is incorporated in the final cover and therefore would not be a potential contaminant transport mechanism.

The primary control of the surface water runoff contaminant pathway is conventional erosion control practices such as limiting slopes and surface vegetation. Fortunately

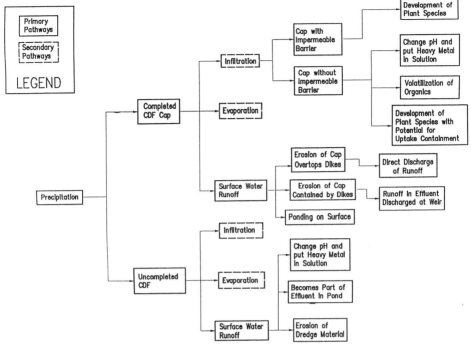

FIG. 9—*Surface water mechanisms related to precipitation of CDF.*

vegetation/revegetation of CDFs commonly occurs naturally and with vigor. A typical erosion control program will include limiting slopes to <5%, vegetation of cover surfaces with native grasses, and temporary containment of all runoff in sedimentation basins.

Contaminant Pathway Control for Proximal Plant and Animal Communities

Pathway controls to protect plant and animal communities immediately adjacent to the CDF can be identified using a general food web flow relationship as presented in Fig. 10. Such food web flow diagrams show the general movement of nutrients and contaminants within and between the plant and animal community. This movement can be "up" the food chain, e.g., fish in ponds within the CDF, or "down" the food chain, e.g., excretion or death. Such movement of contaminants by biological mechanisms is in addition to the movements caused by the nonbiological flow shown in Fig. 11. Interactions between various metals and other contaminants influence the toxicity and bioavailability to organisms. There are three basic relationships identified for systems containing concentrations of two or more toxic metals (Kelly 1988):

1. Additive—the combined toxic effect of the metals equals the sum of the individual toxicities.
2. Synergism—the combined toxic effect exceeds the sum of individual toxicities.
3. Antagonism—the combined toxic effect is less than individual toxicity.

Contaminant pathways to plant and animal communities must be considered during both the operational and postclosure time periods. The uptake of contaminants by plants and animals

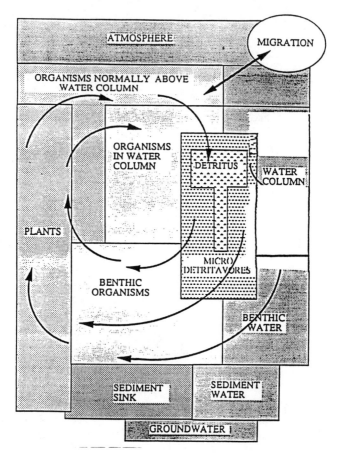

FIG. 10—*Food web drawing showing general flow of nutrients and metals.*

can be estimated by the use of simple mathematical relationships using empirical constants (animal bioconcentration factor (BCF), plant uptake factor). In particular, ponds that form on the dredged materials during the decades-long operation of the CDF may offer the most significant opportunities for plant and animal uptake. The ability to limit these contaminant transport mechanisms prior to placement of the final cover should be evaluated.

Airborne Emissions Control

The control of airborne contaminant emission from a CDF includes limiting the direct volatilization of contaminants into the air and the wind-related loss of soil particles that are contaminated. The direct volatilization of contaminants is a concern during placement of the dredged materials in the CDF and when the dredged materials rise above the supernatant and are exposed to the atmosphere. Thus the volatilization of contaminants can be limited by keeping the dredged materials below the elevation of the supernatant. Alternatively, the problematic dredged materials can be covered with clean dredged materials to limit volatilization.

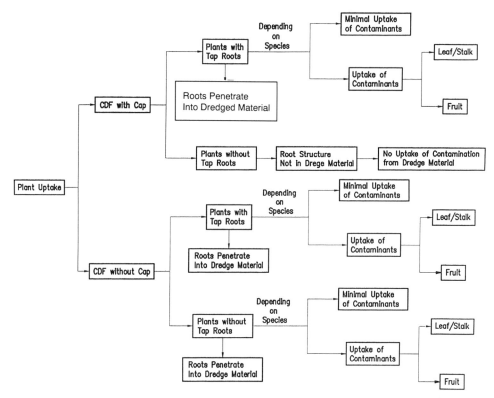

FIG. 11—*Mechanisms involved in plant uptake of contaminants in capped and uncapped CDFs.*

The wind-induced erosion of the dredged material is controlled by limiting the exposure of contaminated dredged material. If historic surface wind velocities and the grain-size distribution of the dredged material, e.g., the soil has a large percentage of soil particles smaller than 10 μm, indicates significant wind erosion, then the use of a clean granular soil cover over the air-exposed dredged materials would be warranted.

CDF Basin Design Recommendations

Conventional CDF design focuses on retention of sediment solids within the facility. Depending upon the nature of the site, the contaminants of concern, method of dredging, physical properties of the dredged material, operational aspects, and many other factors, including sociopolitical factors, supplemental environmental design criteria may be required for the CDF. This section presents design alternatives to control contaminant loss from the CDF basin through (1) effluent discharge through the dikes, and (2) leachate drained from the sediments that discharge into the ground water. These alternatives include the use of both additional designed components and operational constraints.

The basic containment basin of a CDF is formed by the perimeter dikes and the subgrade of the site. Water can potentially leave the basin as a nonpoint source by either seepage through the perimeter dikes or by leaching into the underlying subgrade. The control of either pathway is therefore dependent upon limiting hydraulic gradients and/or the design of a barrier to limit advective transport of contaminants, or design of a filter to attenuate the flow of the TSS itself. Hydraulic gradients may be significantly influenced by the type of CDF, e.g., in lake CDFs typically have very low gradients as compared to high gradients common to upland CDFs.

Effluent Discharge through the Dikes

Water carried by the dredged sediments must be removed from the CDF to provide space for additional sediments and to develop a stable base for construction of the final cover over the dredged material. Effluent can leave the CDF by seeping through perimeter filter dikes or through a weir point discharge system. The latter is particularly attractive if the effluent must be processed to remove or attenuate contaminants. Monitoring of effluent release through conventional CDF dikes indicates that point discharges from porous zones in the dikes occur rather than uniform seepage along the entire dike structure (Schroeder 1984).

The flow of water beneath the dike can be controlled using an impermeable basin liner or a dike vertical barrier that penetrates into a lower natural barrier layer. Impermeable dike barrier liner systems include the following: (1) compacted clay liner (CCL), (2) geomembrane liner (GML), (3) geosynthetic clay liner (GCL), (4) geomembrane cut-off wall (GCW), (5) bentonite slurry cutoff wall, (6) fabric form with grout, (7) clean fine sediments, (8) clogged geotextile, (9) graded soil filter, and (10) fabric forms with sand.

Integration of barrier systems into CDF dike sections must (1) not impair the stability of the dike, (2) allow construction of the barrier using conventional construction technology, and (3) key into a lower low permeable layer to minimize effluent discharge beneath the dike. A summary of applicable dike barrier systems is given in Fig. 12.

Leachate Discharge to the Ground Water

The possible flow of contaminated waters into the environment through the bottom of the CDF is site-dependent. Shoreline CDFs may be located in areas of ground water discharge such that ground water flows into the CDF and limits outward leachate movement into the environment. For CDFs located where effluent from contaminated sediments can move into the underlying soils, a barrier is required to seal the bottom. The basic barrier system previously discussed in the subsection, *Effluent Discharge through the Dikes*, can serve this function, with the exception of those appropriate only for vertical cutoff wall type systems. These barrier systems include CCLs, GMLs, GCLs, clogged geotextiles, and graded soil filters.

Construction of a bottom liner system in shoreline or in-lake CDFs would appear to significantly favor the use of a clean fine-grained slurry placed into the CDF initially to intentionally clog the native underlying soils. Extensive geotechnical data for dredged materials exist that clearly show they are capable of achieving field permeabilities low enough to equal the other natural material barriers. The actual permeability achieved is influenced by the plasticity of the dredged material and the loading that it experiences. This slurry could be hydraulically dredged silty sediments selected to seal the CDF and not specifically selected to meet a given dredging need. Operationally, such a sealing layer would only require an initial dredging operation contract that is specific about the type of dredging to be performed (hydraulic), the type of sediments to be moved (fine-grained), and a period of time for placement and limited consolidation of the sediments. On-going placement of sediments within the CDF increases the vertical effective stress acting on the sealing layer. This increase in vertical stress further consolidates the sealing sediments and reduces their permeability.

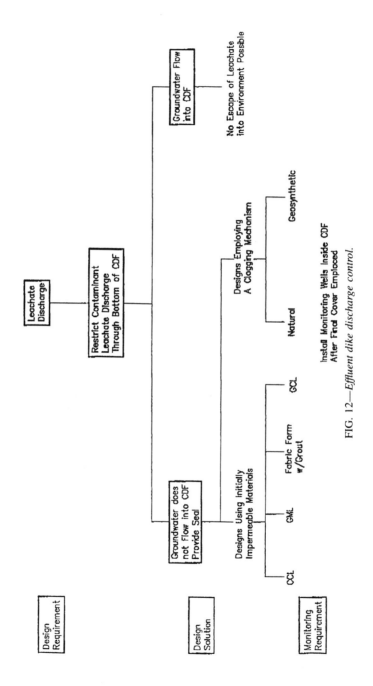

FIG. 12—*Effluent dike discharge control.*

Impoundment Basin Design Guide

The need for a low-permeable dike or ground water barrier can be evaluated using Fig. 13. A dike barrier layer is required for all upland CDFs to prevent discharge of effluent to the surrounding lands and underlying ground water. The economics of dike construction will favor those barriers commonly placed in nonaqueous site conditions. Such barrier layers include CCL, GCL, and GML systems. A dike barrier layer would generally be required in shoreline or in-lake CDFs if the WQS of the dike discharge exceeds applicable WQS beyond the zone of dilution. The zone of dilution should be estimated assuming random point discharge through the dike unless the designed dike section is specifically designed to prevent this occurrence.

A low-permeability barrier is required between the dredged material and ground water if both significant mobile contamination exists within the dredged material and significant potential for contamination of the ground water exists. For example, a CDF sited on low-permeability ($<1 \times 10^{-7}$ cm/s) glacial tills would not require a ground water barrier even if the problematic dredged sediments were close to regulatory limits of contamination. This is applicable to all CDF types and ground water discharge/recharge zones.

Closure Practices and Design Recommendations

The role played by the interim and final covers of a CDF is heavily dependent upon the nature of the contaminant and the type of CDF environment. For example, heavy metals

CDF Characteristic*	Dike Barrier Alternatives**	Groundwater Barrier Alternatives**
Minimum Cont. in DM Minimum Potential Cont. of GW	UP–CCL,GCL,GML SL or IL – None Required	None Required
Minimum Cont. in DM Maximum Potential Cont. of GW	UP–CCL,GCL,GML SL or IL – None Required	UP–CCL,GCL,GML SL or IL – Layer of Clean, Fine Grained DM
Minimum Cont. in DM Existing Cont. of GW	UP–CCL,GCL,GML SL or IL – None Required	None Required
Maximum Cont. in DM Minimum Potential Cont. of GW	UP–CCL,GCL,GML SL or IL – WQS OK – None Required WQS Exceeded – See Table 6.1	None Required
Maximum Cont. in DM Maximum Potential Cont. of GW	UP–CCL,GCL,GML SL or IL – WQS OK – None Required WQS Exceeded – See Table 6.1	UP–CCL,GCL,GML SL or IL – Layer of Clean, Fine Grained DM
Maximum Cont. in DM Existing Cont. of GW	UP–CCL,GCL,GML SL or IL – WQS OK – None Required WQS Exceeded – See Table 6.1	None Required

Cont. = Contaminated
DM = Dredged Material
GW = Ground Water
UP = Upland CDF
SL = Shore CDF
IL = Inlake CDF
CCL = Compacted Clay Liner
GCL = Geosynthetic Clay Liner
GML = Geomembrane Liner
WQS = Water Quality Standards

FIG. 13—*Impoundment basin design guide.*

contaminants potentially present in CDF sediments can be oxidized in a wetland environment common within an operational CDF. This precipitates the metals out of the water column to the sediments. Conversely, acidic rains common to the midwest can leach metals from the sediments and mobilize them. Specific design objectives for the cover must therefore consider the partitioning of the contaminants present, the impact of surface water generated infiltration and runoff, and the long-term environment at the CDF.

The cover system selected for a CDF can influence all of the contaminant pathways that have been evaluated in this document, with the exception of the effluent discharge during filling operations. As such, the design of a cover system for a CDF must focus on minimizing the impact of such pathways when required. It is assumed in this paper that final CDF covers will be constructed above adjacent water surfaces, e.g., no submarine final covers are considered. A listing of cover related pathway control systems is presented in Fig. 14.

Cover systems must also be designed to ensure low maintenance, be easily monitored, and be economical to construct. Final covers are placed once the CDF is full and the dredged material is stable enough that future settlements will not damage the cover. Interim covers may be required when the facility is either inactive for a prolonged period of time or when the dredged material is unstable and future subsidence could impair the function of a final cover. This section reviews the existing cover criteria used by COE and EPA for CDFs or waste containment systems and concludes with a review of alternate cover systems that use geosynthetic components or sediment disposal strategies, or both.

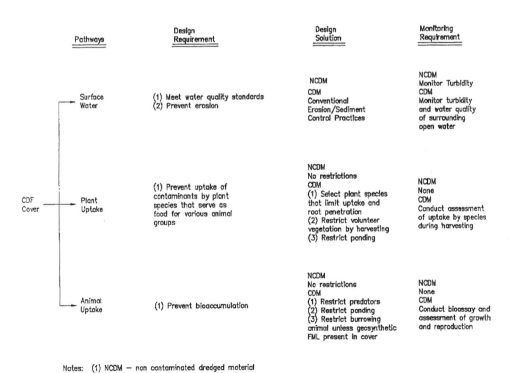

FIG. 14—*Cover related contaminant pathway control.*

COE Closure Objectives

The type of closure selected for a particular CDF is often dependent upon the needs and desires of the local sponsor. Local sponsors are typically a city, county, or state governmental agency. The local sponsor is required to provide all lands, easements, and rights of way to the COE for the CDF. Under the Diked Disposal Program (PL 91-611, Section 123) for constructing CDFs on the Great Lakes, a local government sponsor is required. It should be noted however, that local sponsors are not required for disposal of dredged materials from the Great Lakes connecting channels in the State of Michigan. The local sponsor may have planned or implemented productive and beneficial uses for CDFs. These uses include the development of recreational areas, new or expanded marinas, wildlife refuges, etc. The planned development must be compatible with the structural integrity of the facility and the types of sediments contained. These lands cannot be transferred from the local sponsor without the approval of the COE. In recent years, the COE has phased out the construction of CDFs under the authority of PL 91-611. Future maintenance dredging will, however, still require confined disposal. Future CDFs will be constructed under the operation and maintenance (O & M) authorities of individual navigational projects.

EPA Closure Objectives

The COE has played a significant technical role in the development of the current EPA hydraulic isolation covers incorporated in RCRA and TSCA. The COE involvement has ranged from development of design guidance documents (e.g., EPA 1979) to development and continued support of the HELP (Hydrologic Evaluation of Landfill Performance) computer model (Schroeder 1987, 1988). The HELP program is a water-balance model used to evaluate the effectiveness of hydraulic barriers. It has played a major role in the development of the EPA closure program. This COE technical assistance to EPA will aid in establishing applicable closure technologies for CDFs.

RCRA Subtitle C Hazardous Waste Landfills—The development of the RCRA Minimum Technology Guidance (MTG) cover resulted from the need to meet the requirements of RCRA 40 CFR 264.310. Here RCRA specifies that the final cover be designed and constructed to accomplish the following:

(a) promote long-term minimization of liquids infiltrating through the cover,
(b) function with minimal maintenance,
(c) promote surface water drainage to minimize erosion or abrasion of the cover,
(d) accommodate settling and subsidence so that the cover's integrity is maintained, and
(e) have a permeability less than or equal to the permeability of any bottom liner system or natural subsoil present.

COE prepared a technical guidance manual for EPA to assist in the implementation of the RCRA closure requirements (EPA 1985).

Under RCRA, final covers must as a minimum consist of a vegetated or erosion resistant top layer, a middle drainage layer to prevent mounding of surface water infiltration, and a composite barrier layer consisting of a geomembrane over a 2 ft (0.6 m) compacted clay liner (CCL). The vegetated layer minimizes erosion and can be replaced with hardened erosion control layers such as rip-rap or asphalt paving. The middle drainage layer prevents mounding of surface water infiltration and in that way limits the hydraulic head acting on the barrier system. The barrier layer limits infiltration of surface waters to the waste and limits the movement of gas from the waste through the cover.

Supplemental layers for the RCRA cover were proposed by COE to provide for gas collection from the waste and for biotic barriers to limit intrusion into the cover by burrowing animals. MTG for closure issued by EPA in 1989 has been consistent with COE closure guidance.

RCRA Subtitle D Nonhazardous Waste Landfill Covers—EPA closure criteria for RCRA-D nonhazardous waste landfill covers were established on 9 October 1991. The promulgated regulations require that RCRA-D covers have a permeability less than or equal to the liner system or natural soils beneath the waste. A minimum cover under RCRA-D for an existing unlined landfill consists of a 6-in. (15.2 cm) layer of top soil with vegetation and an 18-in. (45.7 cm) soil infiltration layer having a permeability less than 1×10^{-5} cm/s. For covers over landfill that incorporated a composite liner beneath the waste, EPA has recently (Federal Register, 26 June 1992) interpreted the closure criteria to allow an increase in the permeability and decrease in the thickness of the soil component of the cover composite barrier layer. Thus a landfill having a liner that consists of a geomembrane over 2 ft of 1×10^{-7} cm/s CCL would only need a cover barrier layer consisting of a geomembrane over 18-in. of 1×10^{-5} cm/s CCL.

The RCRA-D cover for unlined landfills represents EPA's perspective on what a minimum cover must contain. In general, the cover system components will be more rigorous than the minimal. For example, the 6-in. (15.2 cm) vegetated soil cover will commonly be substantially thicker to provide sufficient water storage so that the vegetation will not die during periods of extended drought. The thickness of the vegetated soil cover will be determined by the root depth of the selected cover vegetation and long-term soil moisture predictions made using the HELP model.

Additional Regulatory Closure Criteria—The RCRA covers described in the previous two sections are required for new landfill facilities receiving newly generated waste. Covers over older wastes being remediated *in-situ* under the Comprehensive Environmental Response, Compensation, and Liability Act (CERCLA) program show considerably more flexibility. This same flexibility would appear appropriate for CDF closures since sediment contamination is typically both at very low concentrations and due to historic events.

CERCLA closures vary significantly depending upon the potential exposure of the waste to either ground water or humans (EPA 1991). This cover design versus exposure concept is illustrated in Fig. 15. The required CERCLA cover ranges from the RCRA-C MTG cover for sites having high contaminant concentrations and high exposure of ground water and humans to a no action alternative for covers having low contaminant levels and little potential exposure to ground water or humans. Such a risk-based cover design is consistent with the pathway control alternatives discussed in this paper.

Design of Closure Components

As shown in Table 3, the existing closure of CDFs is commonly achieved by seeding grass on or allowing volunteer vegetation to germinate in the last lift of dredge material placed in the CDF. Dredged materials are typically rich in phosphorous, nitrogen, and potash, which promote rapid growth of vegetation. Only a limited number of presently closed CDFs incorporate a clay barrier layer in the cover. As an example, the Michigan City CDF cover has both a clay layer and a surface vegetated soil layer. The cover system must provide the following:

(a) an effective permeability that is low enough that the rate of infiltration through the cover is less than or equal to the rate of leakage of leachate through the liner,
(b) sufficient flexibility that long-term settlements of the waste will not damage the cover, and
(c) a design life that exceeds the projected life of the potentially mobile contaminants.

216 CONTAMINATED SEDIMENTS

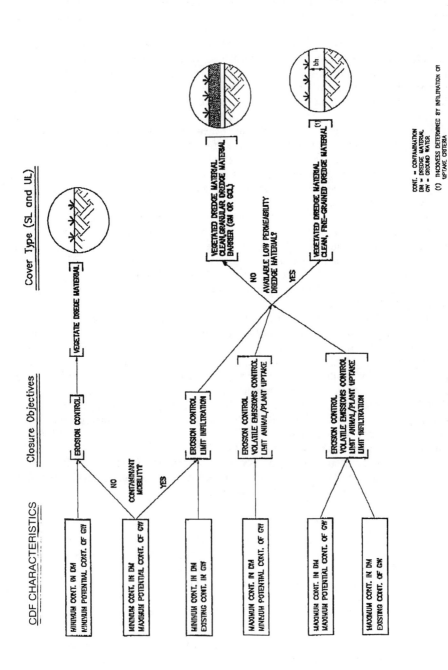

FIG. 15—*CDF cover selection guide.*

TABLE 3—*Summary of closed CDF's in the Great Lakes region.*

Facilities	Years of Operation	Capacity, yd³	Local Sponsor	Cap Design	Ultimate Use
Harbor Island (Grand Haven)	1974–1985	310 000	State of Michigan	none	public use
Harsen's Island	1975–?	100 000	State of Michigan	restored to former soil level and revegetated	upland nesting habitat for waterfowl
Kawkawlin River	n/a	n/a	Bay County, MI	none	n/a
Kidney Island (Green Bay)	1979–1986	1 200 000	City of Green Bay	none	wildlife habitat
Monroe (Edison)	–1984	n/a	n/a—private site	none	Detroit Edison
Port Sanilac Village	1979–1983	143 300	Village of Port Sanilac	none	municipal landfill
Verplank	1974–n/a	134 000	Verplank Coal and Dock Company	none	not available
Whirlpool (St. Joseph Harbor)	1978–1990	25 000	State of Michigan	none	n/a
Windmill Island	1978–1988	160 000	State of Michigan	none	park facility
Cleveland Dike #12	1974–1979	2 760 000	Cleveland-Cuyahoga County Port Authority	none	waterfront development
Small Boat Harbor, Buffalo	1968–1972	1 500 000	Niagara Frontier Transportation Authority	6 ft (1.8 m) of soil	wildlife area/parking
Toledo (Grassy Island)	1967–1978	5 000 000	Toledo-Lucas Port Authority	none	wildlife/recycle as top soil
Michigan City	1978–1987	50 000	City of Michigan City	clay cap with top soil cover	park land
Bayport (Green Bay)	1965–1979	650 000	City of Green Bay	City of Green Bay plans to cap site	industrial development/marine terminal facility
Clinton River	1978–1990 (98% filled)	370 000	State of Michigan	clay	public access site and MDNR field station
Frankfort Harbor	1982–1990	30 000	State of Michigan	site to be seeded when completed	cherry orchard
Grassy Island (Detroit River)	1960–1984	4 320 000	none, U.S. Fish and Wildlife Service owns land	none	wildlife area

This section presents basic cover design concepts that may be required for the problematic sediments associated with CDFs.

Barrier Layer Design—The primary function of a cover is to limit the infiltration of precipitation and surface water into the waste. This barrier function can be achieved by incorporating a low-permeability barrier within the cover system or by designing the cover system as a water-balance system. EPA has favored the use of barrier systems in the cover while current cover development by the Department of Energy (DOE) favors the use of a water-balance cover. The DOE covers are generally designed to provide a minimum 1000-year service life.

Low-permeability barrier layers are typically constructed using either a single barrier layer or a composite barrier system. The single barrier layer may be a CCL, a geomembrane (GM), or a GCL. The composite barrier is a geomembrane over a CCL or GCL. The flow of water through a well constructed CCL is controlled by the permeability of the clay and the hydraulic gradient acting on the clay (Darcy's Law). A reduction in infiltration through a CCL requires either (1) reducing the permeability of the clay with admixtures or (2) reducing the hydraulic gradient by limiting the head of water standing on the barrier or increasing the barrier thickness. CCLs must be protected from freezing to maintain their low permeability (Zimmie and LaPlante 1990). In Region 5, this may require a minimum of 4 ft (1.2 m) of soil cover over the CCL.

Flow through a GM can be by diffusion of the liquid through the GM or flow through holes or defects. For most GMs, the rate of diffusion is very low [<1 gal/acre/day (1.5 L/hectare/day)] such that the flow through holes in the GM will dominate the leakage. The flow through a given penetration of a GM is significantly impacted by the permeability of the soil layer immediately below the GM (Giroud 1992). A composite barrier system places a CCL or GCL immediately beneath the GM to limit the flow rate through a penetration of the GM. The need for the CCL or GCL will depend upon the permeability of the last lift of dredged material placed in the CDF. Excellent composite action is achieved with soils having permeabilities as large as 1×10^{-5} cm/s. This permeability level can be achieved by may dredged materials. Thus an effect cover composite barrier may be constructed of a GM overlying a lift of low-permeability dredge material. The dredged material must have sufficient strength to work on and thus may require some form of preloading to preconsolidate it.

Water-balance barrier—A water-balance barrier system relies on Richards' Effect (Richards 1931) to store surface water infiltration in the upper soil layer and evapotranspiration to remove it from that same layer. A significant amount of DOE research has focused on development of water-balance barriers for long-term (1000-year) isolation of nuclear waste (Wing 1993). Richards' Effect is based on the observation that a fine-grained soil layer that overlies a coarse grained soil layer will not give up moisture to the deeper layer until it is completely saturated. This is due to the strong capillary forces acting on the liquid within the fine-grained soil. A water-balance barrier has a fined-grained vegetative layer that has sufficient hydraulic storage capacity to contain the maximum anticipated infiltration and relies on plants to remove the water from this layer.

A water-balance cover over a CDF would require that a portion of the available dredged materials be coarse sands so that the capillary break could be established beneath the fine-grained cover. If sands are available, the water-balance cover has a very low cost.

Erosion Control/Vegetative Layer Design—The erosion of a final cover on a CDF is limited by the gentle slopes associated with these facilities due to the nature of the dredged materials. By limiting the slopes to less than 5%, even minor rill erosion is eliminated. A significant design concern results if a barrier layer is incorporated into the cover system. As water percolates through the vegetative layer it mounds on the barrier layer unless drained. Allowing the water to mound increases leakage through the barrier layer and may kill the surface vegetation by "drowning" the roots. A drainage system must therefore be part of a barrier design unless a water-balance analysis indicates that the mounding is limited.

Summary and Conclusions

This paper has summarized alternative methods for limiting contaminant pathways from CDFs for problematic dredge materials. These alternatives include the addition of engineered barrier and/or water-balance components in the dikes, basin, and cover. Additionally, this paper includes operational alternatives for establishing the pathway barriers. Thus the basin of the CDF could be lined using and engineered CCL or it could be sealed by placing an initial layer of clean fine-grain dredged material in the CDF. Either barrier layer could be effective in limiting the movement of leachate from the dredged material into the ground water beneath the CDF.

This paper also attempted to portray the overlapping federal regulations that may govern the design of a CDF depending upon site hydrogeologic conditions and the concentration and partitioning potential of the contaminants (see Table 1). CDF designs can be conceptualized into several categories that depend on the extent of partitioning of the contaminant of concern to water. The need for pathway controls should reflect the nature and extent of the contamination associated with the sediments.

References

Chen, K. Y., Gupta, S. K., Sycip, A. Z., and Lu, J. C., 1978, "Confined Disposal Area Effluent and Leachate Control (Laboratory and Field Investigations)," University of Southern California, Los Angeles, and Department of Army Waterways Experiment Station, Vicksburg, MS.

COE, 1987, *Confined Disposal of Dredged Material, Engineer Manual EM 1110-2-5027*, Department of the Army, Washington, DC, September.

EPA, 1979, "Design and Construction of Covers for Solid Waste Landfills," EPA/600/2-79/165, Municipal Environmental Research Laboratory, Cincinnati, OH.

EPA, 1985, "Covers for Uncontrolled Hazardous Waste Sites," EPA/540/2-85/002, Hazardous Waste Engineering Laboratory, Cincinnati, OH.

EPA, 1986, "Test Methods for Evaluating Solid Wastes: Physical/Chemical Methods," EPA SW-846, Office of Solid Waste and Emergency Response, Washington, DC.

EPA, 1989, "Technical Guidance Document: Final Covers on Hazardous Waste Landfills and Surface Impoundments," EPA/530-SW-89-047.

EPA, 1990, "Report on Great Lakes Confined Disposal Facilities," EPS/905/9-90/003, Environmental Review Branch, Planning and Management Division, Chicago, IL, August.

EPA, 1991, "Conducting Remedial Investigations/Feasibility Studies for CERCLA Municipal Landfill Sites," EPA/540/P-91/001, Office of Solid Waste and Emergency Response, Washington, DC.

EPA, 1992, "Solid Waste Facility Criteria," *Federal Register*, Vol. 57, No. 124, EPA/OSW-FR-92-4146-6.

Giroud, J. P. and Boneparte, R., 1992, "Rate of Leakage through a Composite Liner Due to Geomembrane Defects," *Geotextiles and Geomembranes*, Vol. 11.

Kelly, M., 1988, *Mining and the Freshwater Environment*, Elsevier Applied Science, New York.

Krizek, R. J. and Fitzpatrick, J. A., 1976, *Investigation of Effluent Filtering Systems for Dredged Material*, Northwestern University, Evanston, IL.

Pranger, S. A. and Schroeder, P. R., 1986, "Dye Tracer Studies at the Kenosha, Manitowoc, Milwaukee, and Kewaunee Harbors Confined Disposal Facilities," Misc. Paper D-86-4, Department of Army Waterways Experiment Station, Vicksburg, MS.

Richards, L. A., 1931, "Capillary Conduction of Liquids in Porous Mediums," *Physics*, Vol. 1, pp. 318–333.

Schroeder, P. R., 1983, "Chemical Clarification Methods for Confined Dredged Material Disposal," Technical Report D-83-2, Department of Army Waterways Experiment Station, Vicksburg, MS.

Schroeder, P. R., 1987, "Verification of the Hydrologic Evaluation of Landfill Performance (HELP) Model Using Field Data," EPA/600/2-87-050, Hazardous Waste Engineering Laboratory, Cincinnati, OH.

Schroeder, P. R., McEnroe, B. M., Peyton, R. L., and Sjostron, J. W., 1988, "Hydrologic Evaluation of Landfill Performance (HELP) Model: Documentation for Version 2," Department of Army Waterways Experiment Station, Vicksburg, MS.

Selisker, D. and Gallagher, J., 1983, *The Ecology of Tidal Marshes of the Pacific Northwest Coast: A Community Profile*, Division of Biological Services, Fish and Wildlife Series, Dept. of the Interior, Washington, DC.

Thackston, E. L. and Palermo, M. R., 1990, "Field Evaluation of the Quality of Effluent from Confined Dredged Material Disposal Areas: Supplemental Study-Houston Ship Channel," Technical Report D-90-9, Department of Army Waterways Experiment Station, Vicksburg, MS.

Wing, N. R., 1993, "Permanent Isolation Surface Barrier: Functional Performance," prepared for U.S. Department of Energy, Westinghouse Hanford Co., WHC-EP-0650, UC-702, October.

Zimmie, T. F. and LaPlante, C., 1990, "The Effect of Freeze-Thaw Cycles on the Permeability of Fine-Grained Soil," *Proceedings, 22nd Mid-Atlantic Industrial Waste Conference*, Philadelphia.

Bruce I. Collingwood,[1] Marco D. Boscardin,[1] and
Richard F. Murdock[1]

Abating Coal Tar Seepage into Surface Water Bodies Using Sheet Piles with Sealed Interlocks

REFERENCE: Collingwood, B. I., Boscardin, M. D., and Murdock, R. F., "**Abating Coal Tar Seepage into Surface Water Bodies Using Sheet Piles with Sealed Interlocks,**" *Dredging, Remediation, and Containment of Contaminated Sediments, ASTM STP 1293*, K. R. Demars, G. N. Richardson, R. N. Yong, and R. C. Chaney, Eds., American Society for Testing and Materials, Philadelphia, 1995, pp. 220–226.

ABSTRACT: A former coal tar processing facility processed crude coal tar supplied from manufactured gas plants in the area. Coal-tar-contaminated ground water from the site was observed seeping through an existing timber bulkhead along a tidal river and producing a multicolored sheen on the surface of the river. As part of a short-term measure to abate the seepage into the river, a 64-m long anchored sheet pile wall with sheet pile wing walls at each end was constructed inland of the timber bulkhead. The sheet piles extended to low-permeability soils at depth and the interlocks of the sheet piles were provided with polyurethane rubber seals. Based on postconstruction observations for leakage and sheens related to leakage, the steel sheet piles with polyurethane rubber interlock seals appeared to provide a successful seal and abate coal-tar-contaminated ground water seepage into the river. The tie rod penetration sealing proved to be a more problematic detail, but through several postconstruction grouting episodes, an effective seal was produced.

KEYWORDS: anchored sheet pile wall, sheet piling, coal tar, site remediation

Prior to 1960, a coal tar processing facility (facility) located near Boston, MA operated on a 3.3 hectare parcel adjacent to a tidal river. The facility processed crude coal tar supplied from manufactured gas plants in the area. The facility was razed circa 1960. Coal-tar-contaminated ground water from the site was observed seeping through an existing timber bulkhead and producing a multicolored sheen on the surface of the river.

As part of a short-term measure to stop the migration of the coal-tar-contaminated water into the river, a 64-m long anchored sheet pile wall with 9-m long sheet pile wing walls at each end was constructed inland of the timber bulkhead. Refer to Fig. 1 for a plan view of the sheet pile wall layout. The sheet piles extended to low-permeability soils at depth and the interlocks of the sheet piles were provided with polyurethane rubber seals. After construction of the sheet pile wall, the timber bulkhead was cut off at the mudline and removed from the site. An observation trench was installed about 7 m inland from the anchored sheet pile wall, and the site was capped with bituminous concrete pavement to reduce infiltration.

---

[1] Geotechnical engineer, project manager, and principal, respectively, GEI Consultants, Inc., Winchester, MA 01890-1970.

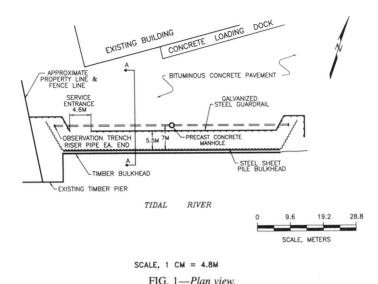

FIG. 1—*Plan view.*

Excavated materials were stockpiled on the site until each material could be characterized and disposed off site.

Background

Crude coal tar is a black, viscous liquid or semisolid substance derived from the distillation of bituminous coal during the production of coke. Coke was used as a fuel source for the manufacture of steel, whereas coal tar typically was a byproduct. Coal tar products have many uses including wood preservatives, road stabilizers, and solvents. Coal tar products are also used in the manufacture of plastics, resins, dyes, paints, synthetic fibers, perfumes, pesticides, explosives, and pharmaceuticals. The facility manufactured finished products as well as intermediate products that were used in the manufacture of other products. Coal tar is only slightly soluble in water and has a specific gravity between 1.18 and 1.23. Coal tar is considered a dense nonaqueous-phase liquid.

The facility was constructed on filled land that was previously a tidal flat. GEI performed a comprehensive environmental assessment of the site and identified a zone of coal-tar-contaminated soil (source area) inland of a timber bulkhead that was built circa 1900. At this site, the coal tar typically consists of polycyclic aromatic hydrocarbons, aromatic compounds, and phenolics. The bulkhead was in poor condition and there was ongoing erosion of the inland backfill soil through openings in the bulkhead. Historical records indicated that the area adjacent to the bulkhead had been the site of several buildings for the facility. Buildings and structures in this part of the site included warehouses, a cooper's shop, storage buildings, oil storage tanks, asphalt stills, and pitch cooling bays. Insufficient facility records are available to describe the facility's operations or to explain how contaminants were released onto the site. Samples from the soil borings located immediately inland of the timber bulkhead indicated that hydraulic fill had been used to backfill behind the bulkhead.

A fill and dredging license dated 1899 indicated that the timber bulkhead was supported by a batter pile-supported deadman constructed 3.7 m inland from the bulkhead. The bulkhead piles were attached to the deadman piles by 0.04-m diameter tie rods at 1.5-m spacings. The piles supporting the deadman are on 1.5-m centers. The permit drawing indicated that the inland support piles for the deadman were connected to a single level of timber batter piles.

Sheet Pile Wall Design

The main design goals for this project included abating the ongoing seepage of coal-tar-contaminated ground water into the river, removing the source material from the area immediately inland from the timber bulkhead, capping the inland area to minimize storm water infiltration, and designing a system that would reduce worker exposure to coal tar contaminants during construction. This project is a short-term measure that has been implemented ahead of the final site remediation. Work continues on the overall site remediation investigation including investigation of other sources and contaminant migration.

A watertight wall was needed to replace the timber bulkhead, and sheet piling was selected for the wall. Due to the very close proximity of the wall location to the river, other wall alternatives such as slurry walls were not considered. GEI's review of available sheet piles lead to HOESCH of West Germany, who was capable of factory installing an interlock seal that effectively seals the interlock, providing a hydraulic barrier. HOESCH uses polyurethane rubber to form the sheet pile interlock seal. Polyurethane rubber commonly is known as a very durable material which, for this site, is compatible with the environmental weathering and chemical conditions. The sheet piles are delivered as double sheets which are preassembled in the factory. The assembled interlock is die-punched on the interlock along the length of the sheet pile to fasten the sheets together then the polyurethane rubber is factory injected into the punched interlock. A preformed interlock seal was also factory-installed in the open interlock of the double sheet. The seal is designed to be in compression once the adjacent sheet pile is installed in the field. This sealing system has been used on many projects in Europe and, according to HOESCH, this project is the first American project using this interlock seal system.

The soil profile at the site consisted of 2 m of miscellaneous fill, 8.8 m of soft organic silt, and 8 m of medium stiff clay, then dense glacial till as shown on Fig. 2. The ground water level was located at the top of the organic silt stratum. The organic silt stratum appeared to be a barrier to the vertical migration of the coal tar. Several soil borings were performed to establish the extent of contamination. The borings were advanced through the source and into the organic silt which consistently showed that only the top few inches were stained black by coal-tar-contamination. In addition, ground water samples collected from the organic silt stratum for chemical testing showed no detection of coal tar constituents. HOESCH 175 19.5-m long sheet piles were used for the entire length of the wall. The sheet piles were made from marine grade steel (see ASTM A 690, Specification for High-Strength Low-Alloy Steel in H-Piles and Sheet Piling for Use in Marine Environments) and coated with coal tar epoxy [8 mils (0.20 mm)] on each side of the sheet pile for the top 10 m in accordance with the Massachusetts Building Code. The code states that corrosion protection is needed to protect the steel surface from the pile cutoff grade to a grade 4.6 m below the zone of objectionable material. The zone of objectionable material was set at 5.5 m below the top of the sheet pile wall. Coal tar epoxy is a standard coating used to protect sheet piles. The bottom 9.5 m of the sheet piles were not coated. The polyurethane seal was installed in the top 9.1 m of the sheet pile interlocks. The design life of the wall is 25 years.

The difference between the final grade inland of the sheet pile wall to the mudline varied from 1 to 5 m. To minimize lateral displacements of the wall during backfilling, an anchored system was designed. The area inland of the timber bulkhead had remnants of old founda-

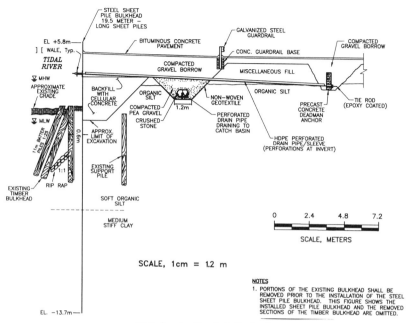

FIG. 2—*Section A-A (typical)*.

tions and pipes, so pre-excavation was necessary to clear obstructions before the sheet piles and anchorage could be installed. Precast concrete deadman anchors were selected for the anchorage which avoided formwork and reduced the time workers were in the excavation. Obstructions were removed during the installation of the anchors. The precast concrete anchors were cast off the site and coated with coal tar epoxy for additional corrosion protection. The dimensions of each anchor were 0.3 m thick by 1.2 m high by 12.5 m long. Each anchor was connected to the sheet pile wall by four tie rods spaced 3.2 m apart. The bottoms of the anchors were installed 2.1 m below the final grade.

Dywidag tie rods were used to connect the sheet pile wall to the deadman anchors. The tie rods were No. 14 DYWIDAG Grade 60 Threadbar (see ASTM A 615, Specification for Deformed and Plain Billet-Steel Bars for Concrete Reinforcements) and factory-coated with an electrostatically applied powdered-epoxy coating. Each tie rod was 18.9 m long and was encased in a 0.15-m diameter High Density Polyethylene (HDPE) perforated drain pipe from the deadman anchor to the sheet pile wall. The purpose of the HDPE pipe was to allow settlement around the tie rod without increasing the load on the tie rod. The tie rod was installed at the invert of the pipe. The tie rods sloped down to the deadman anchors. The tie rods penetrated through the sheet pile wall to the waler on the seaward side of the wall where it was locked off. The waler consisted of two C-channels (C12X20.7) made of ASTM A588 steel [see ASTM A 588, Specification for High-Strength Low-Alloy Structural Steel with 50 ksi (345 MPa) Minimum Yield Point to 4 in. (100 mm) Thick]. The waler was positioned 1 ft (127 cm) above Mean High Water.

Because there could be a tendency for ground water to mound behind the wall and then find a route around the wall to the river, a ground water observation trench was designed as shown on Fig. 2. The observation trench was located 7 m inland from the wall parallel to the sheet pile wall. The trench consisted of perforated HDPE pipe bedded in crushed stone.

The open trench was lined with filter fabric to the top of the organic silt stratum. The HDPE pipe extended to the ground surface via riser pipes at each end of the trench to permit cleaning of the pipe if necessary.

The coal-tar-contaminated soil immediately behind the sheet pile wall was considered a source zone and was removed. The source zone extended to about 5.5 m below the top of the sheet pile wall. Due to potential tidal effects on the sheet pile system, the excavation and backfilling were coordinated with the tidal cycle. Cellular concrete fill expedited the backfilling and added an additional low-permeability barrier between the sheet pile wall and the coal-tar-contaminated ground water. Cellular concrete is a lightweight concrete produced by adding a foaming agent to a sand and cement mix in a concrete truck. Also, the cellular concrete was self-leveling and reduced the time that laborers would have to work in the excavation by avoiding the use of soil backfill.

Excavated material from the work was visually characterized, based on observed staining or presence of mobile contaminants, into one of four categories: Group I (coal-tar-contaminated soil with greater than 50% coal tar by volume); Group II (coal-tar-contaminated soil with less than 50% coal tar by volume); Group III (construction rubble such as bituminous concrete, concrete, timber piles, steel, tires, etc.); or Group IV (reusable on-site fill). Each of these materials was stockpiled on the site. GEI coordinated the disposal of these materials after they were characterized and acceptable disposal facilities were identified.

The sheet pile wall was designed to be continued further north or south along the river, if necessary. At each corner of the wall, male ends of sheet piles were welded along the full length of the corner sheet piles so the wall could be extended while maintaining the continuity of the wall.

After the anchored wall was installed, the site was regraded and 0.10 m of bituminous concrete was placed on the gravel borrow subbase. The proposed site grading routed storm water to collection points where the storm water exited into the river.

Construction

Prior to the start of the construction, applications were submitted for several permits including Notice of Intent to the Conservation Commission for an Order of Conditions, Chapter 91 Waterways License from the Massachusetts Department of Environmental Protection (DEP) Waterways Regulation Program, Environmental Notification Form to the Massachusetts Environmental Policy Act Unit for a Certificate of the Secretary of the Executive Office of Environmental Affairs, Army Corps of Engineers Permit from the Army Corps of Engineers, and Water Quality Certificate from the DEP Division of Water Pollution Control, and Federal Consistency Concurrence from the Massachusetts Coastal Zone Management. All the permits were received within seven months after the applications were submitted.

The contractor mobilized to the site on 10 September 1992. The overall construction sequence was as follows: (a) preparation of the soil containment area for stockpiling excavated materials, (b) excavation of soil from the area immediately inland from the timber bulkhead, extraction of the inland timber support piles and batter piles, (c) installation of the observation trench and precast concrete manhole, (d) installation of the precast concrete deadman anchors, (e) installation of the sheet piles inland of the timber bulkhead, (f) excavation of the source material, (g) placement of cellular concrete in the source zone, (h) installation of the tie rods, (i) backfilling of the inland area with gravel borrow, (j) removal of the timber bulkhead, (k) paving of the site with bituminous concrete, and (l) disposal of the excavated materials. The contractor had originally planned to install the sheet piles as the first step of the installation of the wall, but due to delays in the sheet pile manufacturing and delivery to the site, it was necessary to install the precast concrete deadman anchors and interceptor trench first to avoid temporarily shutting down the job.

The contractor provided a full-time site safety officer to monitor the work at the site. Site personnel wore protective clothing, boots, gloves (if necessary), hard hats, and safety glasses to protect themselves from dermal contact with the excavated soils. The contractor had provisions on the site to upgrade to the use of respirators and air monitoring was regularly performed in work areas.

The HOESCH sheet piles were manufactured in West Germany and transported to the United States by ship. The sheet piles were coated in the United States with coal tar epoxy. It is important to note that the lead time required to actually take delivery of the sheets was underestimated and forced a rescheduling of activities at the site to continue construction.

Obstructions were encountered in almost all excavations. Obstructions included concrete foundations, fuel tanks, timber piles, and steel piping. All pipes were cut and plugged with concrete or bentonite paste, or both. Other unanticipated obstructions included twice as many inland timber support piles and a second level of battered timber piles. In all, four times the estimated number of timber piles had to be extracted for a total of 196 piles. The inland battered piles crossed the alignment of the proposed sheet pile wall and had to be extracted. The pile driving subcontractor used a vibratory hammer to clamp onto the tops of the exposed piles, then vibrated the piles out of the ground. Only four piles broke off and could not be completely extracted. The broken timber piles did not impact the installation of the steel sheet piling.

The pile-driving subcontractor used a double acting diesel impact hammer (ICE 520-30) to drive the double sheets. A vibratory hammer could not be used because it would void HOESCH's warranty for the sheet piles. HOESCH indicated that the interlocks are so tight that the rapid movements caused by a vibratory hammer would increase the temperature at the interlocks, which would adversely affect the polyurethane rubber interlock seal. A 54-metric ton crane with fixed leads was used to align the sheet piles and hammer during installation. The top of each double sheet was driven to 0.6 m above the planned top of wall elevation, and then the previous double sheet was redriven to the planned top of wall elevation. By using this procedure, the driving of a double sheet would not drag the previous sheet down beyond the proposed top of wall elevation. To minimize damage to the polyurethane seal and reduce the friction between the rubber and the advancing steel end of the sheet pile, the contractor brushed a water-insoluble lubricant (Burolan-GM), supplied by HOESCH, on the polyurethane rubber interlock seals prior to installation.

Inland form sheet piles were installed perpendicular to the wall to form a series of cells for the removal of the source zone material and placement of cellular concrete fill. Each cell was excavated to about 3.5 m deep, about 13 m wide along the wall, about 2 m inland from the wall at the bottom of the excavation, and about 5 m at the top of the cell. The inland form sheet piles were covered with polyethylene sheeting, then the cells were backfilled with cellular concrete with a mean unit weight of 1090 kg per m^3 (slightly denser than sea water). Every other cell was excavated and backfilled, then the inland form sheet piles were extracted. The other cells were excavated and backfilled with the adjacent walls of the previously placed cellular concrete acting as the form. Bentonite panels were placed against both sides of exposed sides of the previously cast cellular concrete to act as seals along the potential pathways that may form between the cells due to cold joints and concrete shrinkage.

A total of 713 m^3 of cellular concrete was placed. The cellular concrete was mixed on the site by injecting Mearl Geocell foam into concrete trucks containing Portland Type II cement, sand, and water. The cellular concrete was placed so that the top of concrete was just below the tie rod elevation at the wall. The tie rods and waler were installed according to the design without any unanticipated problems. The tie rod penetrations through the wall were coated with bentonite paste mastic to prevent sea water from passing through the annular space around the tie rods. An acetylene torch was used to burn holes through the sheet piles to install the tie rods through the wall. The inland area above the concrete fill was backfilled with a compacted gravel borrow.

The removal of the timber bulkhead was performed after completion of the sheet pile wall by using an excavator equipped with a LaBounty Shear to cut the timber bulkhead at the mudline. The shear was lowered over the sheet pile wall to the timber bulkhead which was about 0.6 to 1.8 m horizontal out-board from the sheet pile wall as measured at the mudline. The timbers of the bulkhead were cut and removed by the shear. A second excavator was used to remove the remaining coal-tar-contaminated soils, a potential source material, that was between the timber bulkhead and the sheet pile wall. This operation was performed very efficiently with minimal disturbance to the sediments on the seaward side of the wall.

About 36.3 metric tons of Group I material, 5050 metric tons of Group II material, about 971 metric tons of Group III material, and about 816 metric tons of Group IV material were excavated and stockpiled on the site. Groups I, II, and III materials were disposed off the site. Group IV material was used on the site as on-site fill. The Group I material was disposed at an out-of-state lined landfill. The Group II material was characterized as a Massachusetts Special Waste and disposed in a Massachusetts landfill. The Group III materials were generally sent to recycling facilities except for 151 metric tons of rubble and trash which was disposed at a local landfill.

Performance and Conclusions

The sheet pile wall was installed in December 1992 and accomplished the design goals set for the project. The coal-tar-contaminated ground water seepage was abated, source material was removed, and the inland area was capped. During the installation of the cellular concrete in one cell near the center of the wall, a delay in placing the concrete allowed enough time for the tide to reach its highest level on the outside of the wall, which was about 3.4 m above the bottom of the excavation. No visible seepage of sea water was observed coming through the exposed section of wall, indicating that the urethane interlock seals appeared to be watertight. To date, GEI has not observed seepage from the sheet pile interlocks on the seaward side of the wall. Minor seepage, without a sheen, was observed at several tie rod penetrations and was eliminated by injection of polyurethane grout through drilled ports above each tie rod. Apparently the annular space around several of the tie rods was too large for the bentonite mastic to perform as designed. The polyurethane grout expands upon contact with water. In addition, no coal tar has been observed in the manhole. The authors believe that the observation trench will provide a preferred pathway for any DNAPL compounds present.

The steel sheet piles with polyurethane rubber interlock seals appeared to be a successful seal in abating coal-tar-contaminated ground water seepage into the river at the former coal tar processing facility. The tie rod penetration sealing proved to be a more problematic detail, but through several postconstruction grouting episodes an effective seal was produced.

Acknowledgments

T. Ford Company, Inc. of Georgetown, MA was the general contractor for the installation of the anchored sheet pile wall project.

K. T. Valsaraj,[1]* L. J. Thibodeaux,[1] and D. D. Reible[2]

Modeling Air Emissions from Contaminated Sediment Dredged Materials

REFERENCE: Valsaraj, K. T., Thibodeaux, L. J., and Reible, D. D., **"Modeling Air Emissions from Contaminated Sediment Dredged Materials,"** *Dredging, Remediation, and Containment of Contaminated Sediments, ASTM STP 1293*, K. R. Demars, G. N. Richardson, R. N. Yong, and R. C. Chaney, Eds., American Society for Testing and Materials, Philadelphia, 1995, pp. 227–238.

ABSTRACT: Volatilization rates for hydrophobic organic compounds from a confined disposal facility (CDF) containing contaminated dredged material are presently unknown. The primary purpose of this study is to indicate the availability of theoretical models for the evaluation of volatile emissions from a CDF. Four emission locales are identified and modeled: the sediment relocation (dredging) locale, the exposed sediment locale, the ponded sediment locale, and the vegetation-covered sediment locale. Rate expressions are derived to estimate the volatile organic chemical (VOC) emission from each locale. Emission rates (in mass of total VOCs per unit time) are primarily dependent on the chemical concentration at the source, the surface area of the source, and the degree to which the dredged material is in direct contact with air. The relative magnitude of these three parameters provides a basis upon which a tentative ranking of emission rates from the different locales can be given. Exposed sediment results in the greatest estimated emissions of volatiles followed by water with high levels of suspended sediments, such as might occur during dredging or during placement in a CDF. Expected to be lower in volatile emissions are dredged materials covered by a quiescent water column or vegetation.

KEYWORDS: volatile organic compounds (VOCs), emissions, contaminated sediments

Contaminated sediments in lakes and waterways represent a source of aquatic pollution. There are several alternatives for remediation of these contaminated sediments, of which dredging and disposal is one such scheme. Contaminated dredged materials are generally placed in confined disposal facilities (CDFs) either permanently or temporarily before introduction to final disposal. A CDF is a diked area that allows gravity drainage and storage of dredged materials. Traditionally, the primary water quality concerns associated with a CDF were suspended solids losses in the effluent during placement. Volatile organics and metals, however, can also enter the air either during or after disposal. The emission rates of volatile organic compounds (VOCs) and metals from CDFs are presently unknown. Validated mathematical models to predict the emissions are needed to evaluate confined disposal alternatives for dredged materials. This paper deals with the development of models that, in principle, are capable of addressing the release of volatiles, and in particular VOCs, from the water and dredged material compartments of a CDF. The proposed models form the basis for

---

[1] Associate professor and professor, respectively, Department of Chemical Engineering, Louisiana State University, Baton Rouge, LA 70803-7303.
* To whom all correspondence should be addressed.
[2] Professor of chemical engineering, Louisiana State University, and Shell Professor of Environmental Engineering, University of Sydney, New South Wales, Australia.

subsequent experimental testing and validation leading to appropriate methods for volatile emission estimation. The primary objectives of this paper are the following:

1. To identify the primary VOC generation processes for CDFs,
2. To develop appropriate models for each pathway, and
3. To illustrate the application of the algorithms to a hypothetical CDF containing dredged material contaminated with PCBs from New Bedford Harbor.

One would further combine these algorithms with a proper air dispersion model to determine downwind concentrations and air plume sizes. Neither these aspects of air modeling nor experimental validation of the theoretical models are addressed in this paper.

Emission Locales and Models

The volatile emissions from a CDF can occur via a variety of pathways and mechanisms. Hence it is useful to classify them into four general VOC emission locales. Figure 1 illustrates the general locations of each major locale within the CDF. Figure 1 represents (a) those during filling of the CDF and (b) those after the CDF filling operation has ceased. The first locale involves CDF operations concerned with sediment relocation (e.g., dredging, transporting, discharging), and includes the dredge site which is not in the CDF. The second locale is the exposed and drying sediment devoid of any vegetative cover. This locale covers the region from the edge of the water to the dike or to the vegetation line that commences the upland or marsh region adjacent to a CDF. The third locale is the water which also includes the area of sedimentation during disposal. The fourth locale includes the region that

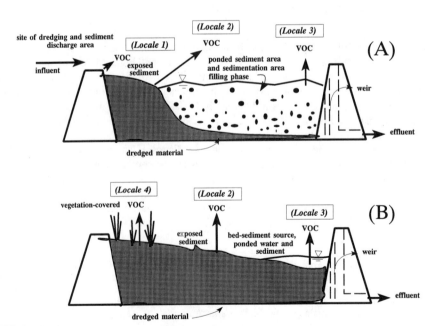

FIG. 1—*VOC emission locales (a) during dredging of sediments and filling of a CDF and (b) after filling of the CDF and vegetation-cover.*

is covered by vegetation (both grasses and trees). Each of these locales contributes to a VOC flux (expressed in mass per unit time per unit area).

Locale 1: Sediment Relocation Devices

During a dredging and filling operation, sediment plumes remain suspended in the water column. The degree of resuspension will obviously depend upon the type of dredge used. VOC emissions may be enhanced as a result of the increased load of contaminants in the water column and the turbulence imparted to the air-water interface. Contaminants will be released from the suspended sediments and via dispersal of the interstitial porewater. The contaminants so placed in solution will subsequently volatilize to the air. Consider an area A m^2 encompassed by a suspended sediment plume of solids concentration ρ_s g/m^3. Assuming that the rate of volatilization is small in comparison to the total contaminant mass associated with the solids, an equilibrium model can be used to obtain the concentration of A in water

$$C_w = \frac{w\ \rho_s}{1 + K_d \rho_s} \tag{1}$$

where w is the mass of A per kg of solid (sediment) and K_d is the equilibrium partition coefficient of A between water and solid (m^3/kg). Notice that the above equation applies to all concentrations of A less than its equilibrium aqueous solubility (C_A^*). The assumption of equilibrium is predicated upon long residence times for the suspended solids which will be obviously realized only for small particles. The VOC emission rate at the air-water interface is

$$n_A = K_w (C_w - C_w^*) \tag{2}$$

where K_w is the overall liquid phase mass transfer coefficient for A (m/s) and C_w^* is the hypothetical concentration of A in water in equilibrium with the concentration C_a in air (g/m^3). $C_w^* = C_a / H_c$ where H_c is Henry's constant (molar concentration ratio, dimensionless). The use of Eq 2 depends on knowledge of the local suspended sediment concentration from which C_w can be estimated from Eq 1 and a mass transfer coefficient, K_w. The coefficient K_w can be computed from known correlations using the two-resistance theory (Thibodeaux 1979)

$$\frac{1}{K_w} = \frac{1}{k_w} + \frac{1}{H_c k_g} \tag{3}$$

This coefficient depends on the degree of turbulence at the air-water interface. Hence the presence of dredge machinery will likely influence K_w. The contribution from mechanical dredges, for example, are three-fold and include influences on the water surface under the bucket that is being hoisted or lowered through the water surface, the surface of the bucket with the sediment heap, and the surface of the receiving vessel. The following equation represents the flux from a bucket carrying sediment being transmitted through the water surface:

$$n_A = k_g (C_a^* - C_a) \tag{4}$$

where $C_a^* = w H_c/K_d$ is the effective concentration of the pollutant in air originating from the sediment in the dredge bucket. k_g in the above equation can be obtained by assuming that correlations for turbulent mass transfer from an object are applicable, such as (Welty et al. 1984)

$$\frac{k_g D}{D_{Aa}} = 2 + 0.6 \left(\frac{Dv_\infty}{v}\right)^{1/2} \left(\frac{v}{D_{Aa}}\right)^{1/3} \quad (5)$$

where D is the characteristic length of the bucket (here its diameter, m), D_{Aa} is the diffusivity of the solute in air (m²/s), v_∞ is the wind speed across the bucket (m/s), and v is the kinematic viscosity of air (kg/m.s).

For the case where the receiving vessel is an open top container that acts as a continuous VOC source, the same expression for flux as given in Eq 4 is applicable. However, it has to be modified to consider the evaporating water surface in the vessel that is z m below the top of the vessel of diameter D. This modification is (Thibodeaux 1989)

$$\frac{k_g D}{D_{Aa}} = 0.036 \left(1 - \frac{z}{D}\right)\left(\frac{Dv_\infty}{v}\right)^{1/1.25} \left(\frac{v}{D_{Aa}}\right)^{1/3} \quad (6)$$

Locale 1 also includes emissions resulting during the sediment slurry flowing into the water or onto the sediment delta of the receiving CDF. The slurry discharge can be of two types: *nonsubmerged* or *submerged*. For the nonsubmerged discharge, the following equation gives the quantity of VOC approaching the discharge point that is volatilized:

$$n_A = Q F C_w \quad (7)$$

where Q is the volumetric flow rate of water (solids free, m³/s) and F is given by

$$F = \frac{0.033\, ab\, (1 + 0.046(T - 273))\, H_d \left(\frac{D_{Aw}}{D_{O2,w}}\right)^{1/2}}{1 + 0.033ab\, (1 + 0.046(T - 273))\, H_d \left(\frac{D_{Aw}}{D_{O2,w}}\right)^{1/2}} \quad (8)$$

where a is the water quality factor (=1 for polluted water), b is 0.6 for a round broad-crested curved face spillway, T is the water temperature (K), H_d is the height through which the water falls (m), D_{Aw} is the molecular diffusivity of VOC (m²/s), and $D_{O2,w}$ is the molecular diffusivity of oxygen (m²/s). The equation for n_A is predicated upon the assumption of equilibrium near the exit which is easily realized owing to long residence time and high turbulence levels in the pipe.

For the submerged discharge, the above equation is not applicable. In this case, the sediment is covered by overlying water that provides an additional resistance to mass transfer. The methodology to be used in this case is similar to the one for locale 3 discussed later in this paper.

VOC emissions occurring during the overland flow of water and sediment from the discharge point to the edge of the CDF can be modeled as a rivulet. Along the entire stretch of the rivulet, sediment deposition occurs and the redeposited sediment (with concentration w of pollutant) is a source of VOC. The suspended solids concentration along the rivulet is given by

$$\rho_s = \rho_s^\circ \exp\left(-\frac{v_d\, xy}{Q}\right) \qquad (9)$$

where ρ_s° is the concentration of particles at the head of the rivulet (g/m³), v_d is the net particle deposition velocity (m/s), x is the distance from the rivulet head (m), and y is the width of the rivulet (m).

A plug flow model devised by Thibodeaux (1989) for the pollutant concentration in the rivulet can be combined with the above equation to give the following for dissolved pollutant concentration at any point x

$$C_w = \alpha + (C_w^\circ + \alpha) \exp\left(-(K_{sw} + K_w)\frac{xw}{Q}\right.$$

$$\left.+ \left(\frac{K_{sw} + K_w}{v_d}\right) \ln\left[\frac{1 + K_d \rho_s^\circ}{1 + K_d \rho_s^\circ \exp\left(-\frac{v_d\, xw}{Q}\right)}\right]\right) \qquad (10)$$

where

$$\alpha = \frac{K_{sw}\dfrac{w^*}{K_d} + K_w C_2^\infty}{K_{sw} + K_w} \qquad (11)$$

K_{sw} is the water-side mass transport coefficient for the pollutant above the bed (m/s) and C_w° is the initial VOC concentration at $x = 0$ (g/m³). C_w obtained thus can now be integrated with respect to x and substituted in the equation for flux.

Locale 2: Exposed Sediment

This particular sediment locale is the one that is exposed directly to air, and is the most complex of the four VOC emission locales. This locale can go through cycles of wet and dry regimes. Since VOC sorption on sediments and effective diffusivity is heavily influenced by moisture content, the evaporation of water can have a significant impact on the VOC emission from this locale.

In the upper layers of the exposed sediment, the water transport is driven by advection processes. It percolates downward and moves up by capillary forces. The movement of water also contributes to the movement of VOC dissolved in it. This coupled transport of VOC can be described by the following equation:

$$n_A = \left[\frac{\left[y_{H_2O}^i \left(\dfrac{M}{M_{H_2O}}\right)^{1/2} - 1\right] + 1}{y_{H_2O}^i} - 1\right] k_{gs}\, y_w^i \qquad (12)$$

where $y_{H_2O}^i$ is the mol fraction of water at the interface, y_w^i is the mol fraction of VOC at the interface, M and M_{H2O} are the molecular weights of the VOC and water, respectively, and k_{gs} is the sediment to air mass transfer coefficient for the VOC (m/s).

Wet exposed sediments are the largest source of VOC emissions in that the water is essentially saturated with respect to the sediment contaminants. Surface depletion of the contaminant, however, is likely to occur quickly. Ultimately, the VOC evaporation must occur via a series of steps: (a) diffusion from the particle surface into porewater, (b) diffusion through the water film, (c) desorption from the water film into the air boundary layer, and finally (d) diffusion in the air. Since the water film is fairly thin, it provides minimal resistance. If, further, it is assumed that the water film and the particle are in equilibrium, then Eq 4 is adequate in estimating the VOC emission rate with the sediment/air-side mass transfer coefficient, k_{gs}, replacing k_g. Thibodeaux and Scott (1985) have developed equations to predict the air-side mass transfer coefficient, k_{gs}, under these conditions of evaporation from surface soils and sediments. This coefficient needs to be corrected for the water vapor transport enhancement factor described in Eq 12. As evaporation from the sediment surface continues, the top layers quickly become depleted of the VOC. Further losses will come via diffusion from within the sediment pore spaces. At some point during this process, the resistance to mass transfer will change from being air-side controlled to sediment-side diffusion controlled. The critical depth h to which the VOC evaporation plane recedes before sediment-side controls the mass transfer is given by

$$h = \frac{19 \, D_{Aa} \, \epsilon_a^{10/3}}{k_{gs} \, \epsilon_T^2} \tag{13}$$

where ϵ_a is the air-filled porosity of the sediment (m³/m³) and ϵ_T is the total soil porosity (m³/m³). The coefficient 19 assumes that 95% of the mass transfer resistance is in the sediment-side when this criterion is met.

For the VOC emission from particles below the sediment-air interface, the solute has to diffuse through the air-filled sediment pore spaces prior to emerging into the air boundary layer. This is the rate limiting step, for which the flux through the air-sediment interface is given by

$$n_A = \left[\frac{D_{eff} \left(\frac{\epsilon_a H_c + K_d \rho_b}{H_c} \right)}{\pi t} \right]^{1/2} \left[\frac{w H_c}{K_d} - C_{As}^i \right] \tag{14}$$

where $D_{eff} = D_{Aa} \, \epsilon_a^{10/3} / \epsilon_T^2$ is the effective diffusivity within the sediment pore spaces (m²/s), ρ_b is the bulk density of the sediment (g/m³), and C_{As}^i is the VOC concentration at the soil surface (g/m³). t is the time since the sediment system has been in place (s).

The overall equation for flux from the exposed sediment can be obtained by combining the above equation for sediment-side control with that for the air-side control. The final equation is

$$n_A = \cfrac{\left(\cfrac{wH_c}{K_d} - C_a\right)}{\left[\left[\cfrac{\pi t}{D_{eff}\left(\cfrac{\epsilon_a H_c + K_d \rho_b}{H_c}\right)}\right]^{1/2} + \cfrac{1}{k_{gs}}\right]} \qquad (15)$$

where all terms have been defined previously.

The basic assumption in the above VOC transport model is that the process is rate-limited by molecular diffusion of chemical species through the sediment pore spaces and that the particles, covered by moisture, are the chemical source, always maintaining an equilibrium concentration with the pore gas. The mechanisms that dominate the VOC emission process over the long term are likely to be as demonstrated above. The wet and dry cycles will have a secondary effect in enhancing the transport rate. These cyclic conditions will transport the VOC upward in clay and low-permeability sediments. Water evaporation at the surface and its capillary rise transport the soluble VOC upward where it readsorbs onto the cleaner surface and awaits dry-out so it can revaporize. In sandy, high-permeability sediments downward percolation is likely to occur. Surface cracking of sediments will enhance the overall sediment porosity and will increase the VOC emission rate. As a first approximation, the solid geometric shapes that make up the cracked sediment surface can be modeled as vertical spines with cross-sectional area, A_c m^2 and perimeter, P_c m. Heat transfer analogies for such shapes are available in the literature which by way of analogy theories can be converted to mass-transfer expressions and used for this purpose (Welty et al. 1984). Evaluation of this and other secondary effects on the VOC emission rate in the upper soil layers, such as capillary water movement, awaits further development of similar models.

In short, this sediment locale is likely to be a significant source of VOCs from a CDF. In general, the pathway is short and the surface area is extensive, so one would expect a relatively large emission rate.

Locale 3: Ponded Sediment

This VOC emission locale is one in which the sediment is either suspended in or at the bottom of a water body. The VOCs have to diffuse through the water surface to enter the atmospheric boundary layer. Figure 1 illustrates this locale during filling of the CDF and after the disposal has been completed. The emission during these two phases have to be considered separately.

Filling Phase—For this case, the model for emissions from a rivulet discussed earlier is applicable. Equations 2, 9, and 10 provide the necessary algorithm for this. The significant difference between those for the rivulet and for the ponded sediment (lake) is that the mass transfer coefficients K_w and K_{sw} will be smaller for this case. The overall settling velocity v_d will also be different for the quiescent conditions of the lake than the turbulent conditions of the rivulet. The settling of particles in a lake is by flocculation or zone settling and is a function of many different variables (Montgomery and Poindexter 1978). For an unstratified lake K_{sw} can be calculated from the expression derived by Thibodeaux and Becker (1982), whereas for the stratified lake the expression given by Thibodeaux (1979) should be used.

Bed Sediment Source—Once the discharge of dredged materials into the CDF has ceased, the bed sediment is the source of VOCs to the atmosphere. The VOC emission then occurs in steps: (a) desorption from the bed surface, (b) molecular diffusion through the benthic boundary layer, (c) diffusion through the water column, and (d) volatilization through the atmospheric boundary layer. If steady-state conditions are assumed then the following equation holds for the VOC concentration in water

$$C_w = \frac{\left(w \dfrac{K_{sw}}{K_d} + K_w C_w^\infty\right)}{K_{sw} + K_w} \quad (16)$$

In combination with Eq 2, the flux can be obtained.

The ponded sediment locale is likely to be a significant emission source near the point where the sediment/slurry enters the CDF. At this point, since high VOC and suspended sediment concentrations are likely, large emissions are possible. With solids settling out as the water flows through the CDF, the vaporization potential diminishes. As the CDF fills, the ponded sediment locale decreases in importance and hence do the emissions. There are both laboratory and field demonstrations of the bed sediment as a source of VOCs to the atmosphere (Karickhoff and Morris 1985; Larsson 1985). However, more work needs to be done in this area to validate the mathematical models such as presented above.

Locale 4: Vegetation-covered Sediment

The fourth emission locale in a CDF is the area that is covered by vegetation. It encompasses the region that commences with the line of grass at the edge of the exposed sediment and extends into that region of the CDF containing older sediments. The existence of the vegetation-cover exerts a significant influence on the soil environment, as compared to the exposed locale. Increased vegetation makes the soil more porous which in turn increases both the air-filled soil porosity and total soil porosity. This will likely increase the VOC diffusivity through the soil. On the other hand, plant-derived natural organic matter in soil can retain a large portion of the VOC through adsorption, and retard the VOC emission rate. The presence of plants in the air boundary layer provides a further resistance to VOC transport and hence decreases the value of k_g. This reduces the water vapor and VOC transport rates directly. Indirectly it also cools the air above the soil and reduces the rate of VOC emissions. A reduction in moisture evaporation also helps maintain a high soil water content which in turn reduces the air-filled soil porosity and leads to decreased effective diffusivity of VOC through the soil. The net effect of the vegetation-cover is to reduce the VOC emission compared to an exposed sediment locale. The models developed for the exposed sediment will generally apply here. Thus Eq 4 with k_g reduced due to vegetation cover will give us the flux of VOC from this locale. Since this locale appears to be the lowest source of VOC emissions to the air, existing vignette models may be appropriate at this stage. With more field data, one can justify developing refined models for this locale.

Table 1 is a ranking of VOC emission locales based on the aerial extent, concentration and the mass transfer coefficient. The most significant one is locale 2 followed by locale 1. Both locales 3 and 4 are of low priority and are ranked 3. Based on this ranking the following example will consider only locales 1 and 2 for estimation of emissions for a representative CDF.

Application to a New Bedford Harbor CDF—In order to demonstrate the applications of the theoretical models developed above, calculations were made based on the conditions of the pilot scale CDF proposed for the New Bedford harbor. Detailed operational, physical, and chemical aspects of the site are available in Otis (1987).

TABLE 1—*Tentative ranking of CDF chemical emission locales.* $R = (Area, A) \bullet (Concentration, C) \bullet (Mass\ Transfer\ Coefficient, K)$.

Locale No.	Description	A	C	K	Rank
1	Sediment relocation devices	L	H	H	2
2	Exposed sediment	H	H	H	1
3	Ponded sediment	H	L	L	3
4	Vegetation-covered sediment	H	L	L	3

NOTE: H ≡ high, L ≡ low.

The primary cell has a capacity to hold approximately 19 000 m³ of slurry. The surface area is approximately 5776 m². The two chemicals of concern are Aroclor 1242 (A-1242) and Aroclor 1254 (A-1254). Their properties are summarized in Table 2. The partition constants were estimated from Otis (1987), the solubility and vapor pressures are from the U.S. Environmental Protection Agency (EPA) priority pollutant data list. The Henry's constants were obtained from the solubility and vapor pressure data and are dimensionless ratios (molar concentration in air/molar concentration in water). Table 3 details the site specific information obtained from Otis (1987).

Emissions during Filling—Filling is assumed to be through discharge of the slurry into the CDF using a submerged diffuser. The suspended solids concentration was assumed to be that obtained using the standard elutriate test as reported in Otis (1987). Since this is a ponded sediment locale Eqs 1 and 2 should apply. Combining these equations yields

$$n_A = K_w \left[\left(\frac{w}{K_d + \frac{1}{\rho_s}} \right) - C_w^* \right] \quad (17)$$

Since PCB emission is water-side controlled, $K_w \approx k_w$. The appropriate equation for k_w is the following (Springer et al. 1986)

$$k_w = 19.6\ v_x^{2.23}\ D_{Aw}^{2/3} \quad (18)$$

TABLE 2—*Physicochemical properties of Aroclor at 298 K.*

Property	Aroclor 1242	Aroclor 1254
Sediment-water Partition Constant, K_d (m³/kg)	188	304
Henry's Constant (dimensionless)	0.0249	0.0337
Aqueous Solubility (kg/m³)	2.4×10^{-4}	3.0×10^{-5}
Vapor Pressure, (mm Hg)	4.06×10^{-4}	7.71×10^{-5}
Molecular Weight (kg/mol)	0.267	0.238
Aqueous Phase Diffusion Constant, D_{Aw} (m²/s)	4.6×10^{-10}	4.8×10^{-10}
Air Phase Diffusion Constant, D_{Aa} (m²/s)	3.6×10^{-6}	3.8×10^{-6}

TABLE 3—*New Bedford harbor pilot-scale CDF site-specific information and data.*

Total PCB Concentration in the Bed Sediment, w	4.32×10^{-4} kg/kg
Aroclor Ratios, wt%	48% A-1242 and 52% A-1254
CDF Suspended Solids Concentration, ρ_s	0.49 kg/m³
Temperature, T	298 K
Wind Speed (average), v_x	40 km/h
Dredged Material Air Porosity, ϵ_a	0.3
Dredged Material Total Porosity, ϵ_T	0.7
Dredged Material Bulk Density, ρ_b	1.2×10^{-3} kg/m³
Water Depth	1.2 m
Area of CDF	76 m × 76 m

where v_x is the wind velocity over the pond (assumed to be 40 km/h). For A-1242, $k_w = 0.07$ m/h. Assuming that no PCBs exist in the air over the CDF gives the maximum flux from the pond. Hence with $C_w^* = 0$, we have, for A-1242, $C_w = w/(K_d + 1/\rho_s) = 1.09 \times 10^{-3}$ kg/m³. Note that this is less than C_A^∞ for A-1242. Substituting this in Eq 17 above we obtain the flux of A-1242 as 7.63×10^{-8} kg/m² · h. For the entire area the emission rate is $N_A = n_A A = 4.43 \times 10^{-4}$ kg/h. For A-1254 the emission rate is obtained as 3.11×10^{-4} kg/h.

Emissions from Exposed Sediment—Once the CDF is filled and the water removed, the solid dredged material will be exposed directly to the atmosphere. Equation 15 is the appropriate one for ascertaining the flux of chemical under these conditions. The effective concentration of A-1242 in soil pore space is given by the term $w H_c/K_d = 2.75 \times 10^{-8}$ kg/m³. Note that this is less than the vapor density of A-1242 which is 5.8×10^{-5} kg/m³. As in the previous calculation, assuming $C_a = 0$ gives the maximum flux. The term $\epsilon_a + K_d\rho_b/H_c = 9.6 \times 10^6$ and $D_{eff} = 1.33 \times 10^{-7}$ m²/s. The appropriate equation for the soil-air side mass transfer coefficient is (Thibodeaux and Scott 1985)

$$k_{gs} = 0.036 \, Re^{4/5} \, Sc^{1/3} \, \frac{D_{Aa}}{L} \qquad (19)$$

where Re is the Reynold's number $= v_x L/v_a$, Sc is the Schmidt number $= v_a/D_{Aa}$, L is the fetch of the pond ($= 76.2$ m), and v_a is the kinematic viscosity of air $= 1.5 \times 10^{-5}$ m²/s. Assuming that $v_x = 40$ km/h, we obtain $k_{gs} = 15.8$ m/h. Substituting these values in Eq 15 we get for the flux of A-1242

$$n_A = \frac{4.4 \times 10^{-7}}{1 + 0.58\sqrt{t}} \left(\frac{kg}{m^2 \cdot h}\right) \qquad (20)$$

For a total area of 5.8×10^3 m², the emission rate for A-1242 is

$$N_A(t) = \frac{2.552 \times 10^{-3}}{1 + 0.58\sqrt{t}} \left(\frac{kg}{h}\right)$$

with t in hours.

For A-1254, the emission rate is

$$N_A(t) = \frac{2.32 \times 10^{-3}}{1 + 0.4\sqrt{t}} \left(\frac{kg}{h}\right)$$

with t in hours.

Figure 2 represents the emission rate of A-1242, A-1254, and total aroclor emissions as a function of time. It can be seen that the flux rapidly decreases from its maximum value and reaches a plateau in about 3000 h. The initial flux from this locale is much larger than during the filling stages.

Conclusions

Much general information is available in the technical literature on the subject of volatile chemicals in water and on solids relative to contact with the air phase. The basic theory of volatilization from such sources is in place. We have developed an assemblage of vignette models and associated equations plus guidance for their use in estimating VOC emission from CDFs. The models are general so that they can be applied to any CDF; however, features of proposed dredging and disposal operations at New Bedford harbor were used as a specific example. Four principal VOC emission locales were identified to exist in any CDF: sediment relocation devices (dredging and associated activities), exposed sediment, ponded sediment, and vegetation-covered sediment.

Emission rates (in mass of total VOCs per unit time) are primarily dependent on the chemical concentration at the source, the surface area of the source and the degree to which the dredged material is in direct contact with air. The relative magnitude of these three parameters provide a basis upon which a tentative ranking of emission rates from the different locales can be given. On this basis the exposed sediment locale ranks first. The sediment locale with a high suspended solids concentration (as occurs with sediment relocation devices) ranks second. Low in rankings are the ponded sediment with the sediment covered by a quiescent water column and the vegetation-covered sediment locale.

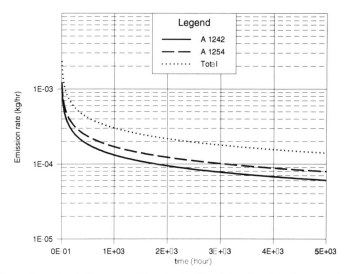

FIG. 2—*Estimated emission rates of Aroclor 1242 and Aroclor 1254 from a typical CDF.*

Although the models are physically based and components of the models have been validated in a variety of environmental applications, there exists no data to test their ability to predict vaporization rates from a confined disposal facility (CDF). Some aspects of the models presented here are based on very crude approximations, and further development is needed. The algorithms given in this paper are to be tested against actual experimental data either in the field or pilot scale CDFs.

The recommendations on laboratory and field testing reflect general research activities that must be performed to build a higher degree of confidence in the predictive capability of these models.

Acknowledgment

Although this work was supported in part by funds from the U.S. Environmental Protection Agency (Grant R819165-01) to the LSU Hazardous Substance Research Center (South and Southwest), it has not been subjected to the Agency's required peer review and hence does not necessarily reflect the views of the Agency.

References

Karickhoff, S. W. and Morris, K. R., 1985, "Impact of Tubificid Oligachaetes in Pollutant Transport in Bottom Sediments," *Environmental Science and Technology*, Vol. 19, No. 1, pp. 51–56.

Larsson, P., 1985, "Contaminated Sediments of Lakes and Oceans Act as Sources of Chlorinated Hydrocarbons for Release to Water and Atmosphere," *Nature*, Vol. 317, No. 6035, pp. 347–349.

Montgomery, R. L., 1978, "Methodology for Design of Fine-grained Dredged Material Containment Areas for Solids Retention," *Technical Report D-78-56*, U.S. Army Engineers Waterways Experiment Station, Vicksburg, MS.

Otis, M. J., 1987, "Pilot Study of Dredging and Dredged Material Disposal Alternatives-Superfund Site, New Bedford Harbor, MA," U.S. Army Engineer Division, New England, Waltham, MA.

Thibodeaux, L. J., 1979, *Chemodynamics*, John Wiley and Sons, New York.

Thibodeaux, L. J., 1989, "Theoretical Models for Evaluation of Volatile Emissions to Air during Dredged Material Disposal with Applications to New Bedford Harbor, MA," Miscellaneous paper EO-89-3, U.S. Army Engineers Waterways Experiment Station, Vicksburg, MS, pp. 1–44.

Thibodeaux, L. J. and Becker, B., 1982, "Chemical Transport Rates Near the Sediment in Wastewater Impoundments," *Environmental Progress*, Vol. 1, No. 4, pp. 296–304.

Thibodeaux, L. J. and Scott, H. D., 1985, "Air/Soil Exchange Coefficients," *Environmental Exposure from Chemicals*, Vol. 1, W. B. Neely and G. E. Blau, Eds., CRC Press, Boca Raton, FL, pp. 65–89.

Welty, J. R., Wicks, L. E., and Wilson, R. E., 1984, *Fundamentals of Momentum, Heat and Mass Transfer*, John Wiley and Sons, New York, p. 266.

Nader S. Rad,[1] *Robert C. Bachus,*[2] *and Brian D. Jacobson*[1]

Soil-Bentonite Design Mix for Slurry Cutoff Walls Used as Containment Barriers

REFERENCE: Rad, N. S., Bachus, R. C., and Jacobson, B. D., **"Soil-Bentonite Design Mix for Slurry Cutoff Walls used as Containment Barriers,"** *Dredging, Remediation, and Containment of Contaminated Sediments, ASTM STP 1293,* K. R. Demars, G. N. Richardson, R. N. Yong, and R. C. Chaney, Eds., American Society for Testing and Materials, Philadelphia, 1995, pp. 239–251.

ABSTRACT: Soil-bentonite slurry cutoff walls (SBSCWs) have been used for decades by geotechnical engineers in conjunction with soil excavation to control ground water movement. In recent years, soil-bentonite slurry cutoff walls have been increasingly used as containment barriers around contaminated soils to impede or, in some cases, nearly eliminate the off-site migration of contaminated ground water or other potentially hazardous liquids. Because of this recent application, the conventional design methodologies for the soil-bentonite slurry cutoff walls, specifically the requirements for the soil-bentonite slurry mix acceptance, have been adjusted. Low hydraulic conductivity, workability, and physical stability commonly used as the main acceptance criteria for conventional soil-bentonite slurry mixes are no longer the only governing parameters. Compatibility of the soil-bentonite slurry mix with the ground water/ liquid is as important and should be considered a major factor in the soil-bentonite slurry mix design.

The paper presents the procedures used and the results obtained during an extensive laboratory testing program performed to select varying soil-bentonite slurry mix components for a soil-bentonite slurry cutoff wall constructed around an old landfill at a former oil refinery. The landfill is underlain to varying depths by a coarse granular soil that has been exposed to oil-products. Compatibility of three commercially available bentonite products with the free oil-products and the oil-contaminated ground water found at some locations in the landfill was initially investigated. Based on the test results, one of the bentonite products was selected for use in the soil-bentonite slurry testing program. A clayey soil from a borrow source, potable water from the site, and subsurface soils from the proposed soil-bentonite slurry wall alignment were used to form different soil-bentonite slurry mixes. Slump tests were performed to evaluate the workability of the mixes. Based on the test results, a single mix was selected for further study, including permeability/compatibility testing. The results of the compatibility testing program are presented and discussed in the paper. A specific design mix methodology for evaluating the chemical compatibility of soil-bentonite slurry mixes with permeants is proposed.

KEYWORDS: soil-bentonite, containment, contaminant, compatibility, slurry cutoff wall, laboratory testing, oil-products, design mix methodology

Soil-bentonite slurry cutoff walls (SBSCWs) are increasingly used as hydraulic containment barriers around existing landfills and contaminated subsurface soils to isolate the contaminated soils and ground water from the surrounding subsurface environment. Using various construction techniques (Case International Company, 1982; Spooner et al. 1985),

---

[1] Laboratory director and assistant program manager–environmental testing, respectively, GeoSyntec Consultants, Atlanta, GA 30342.

[2] Principal, GeoSyntec Consultants, Atlanta, GA 30342.

SBSCWs are commonly constructed using the subsurface soils from the alignment of the slurry wall, a commercially available bentonite, water available at the site, and, if necessary, a clayey soil (herein referred to as borrow clay) from a potential borrow source. A laboratory testing program is commonly performed to select a soil-bentonite design mix (SBDM) that satisfies the following criteria:

- Workability. The SBDM should be of a consistency that could be easily mixed and placed in trenches using conventional construction equipment and techniques (Case International Company 1982; Spooner et al. 1985).
- Low hydraulic conductivity. The SBDM should have a hydraulic conductivity less than a specified value (typically, 1×10^{-10} to 1×10^{-8} m/s).
- Compatibility. The SBDM should be resistant to chemical alteration by the liquid contaminants (herein referred to as contaminants) and contaminated ground water; test results should indicate that the hydraulic conductivity of the SBDM would not change with time when in contact with the contaminants and/or the contaminated ground water.

The goal of this paper is to outline and briefly discuss the laboratory testing program the authors use to rationally select a SBDM. A case history is also presented.

Laboratory Testing Program

The laboratory testing program performed to select a SBDM typically consists of the following five steps: (a) soil index testing, (b) soil-contaminant compatibility testing, (c) bentonite-contaminant compatibility testing, (d) workability testing, and (e) permeability/compatibility testing. To simulate the worst-case scenario, free-contaminant (if applicable) is used in the compatibility evaluations (i.e., in steps b, c, and e); in other cases the contaminated ground water from the site is used. Each of these five steps is briefly discussed in the following sections.

Soil Index Testing

Soil index testing is performed to evaluate the physical characteristics of the subsurface soils, and, if applicable, the borrow clay, and the resulting composite soil mix (SM) formed by blending the subsurface soils and the borrow clay. Soil index tests are often used as the initial screening of the soil components anticipated for use in the SBDM. These tests are used to provide information regarding: (a) potential variations of the subsurface soil along the alignment of the wall, and (b) a rough estimate of the expected hydraulic conductivity of the subsurface soil or, when applicable, the soil mix (e.g, hydraulic conductivity of clayey soils may vary from 1×10^{-10} to 1×10^{-7} m/s, while hydraulic conductivity of silty soil may vary from 1×10^{-8} to 1×10^{-5} m/s) before the addition of any bentonite. The information regarding stratigraphic variations provides the design engineer with the knowledge needed to decide: (a) if more than one SBDM should be selected for a given SBSCW, and (b) if the subsurface soil should be amended with a borrow clay.

It is not uncommon to use two to three different SBDMs along a given wall. This is usually the case when the wall is relatively long and the subsurface soil conditions vary considerably along the alignment of the wall. Similarly, as mentioned previously, it is not uncommon to form a soil mix by blending a borrow clay with the subsurface soils. The hydraulic conductivity of the subsurface soils or, when applicable, the soil mix provides an estimate of the amount of bentonite needed to achieve the specified hydraulic conductivity.

In general, the higher the expected hydraulic conductivity of the soil mix, the higher the percentage of bentonite additive necessary to achieve the specified hydraulic conductivity.

The index testing program commonly includes:

- Sieve and hydrometer analysis (ASTM D 422, Method for Particle-Size Analysis of Soils)
- Atterberg limits (ASTM D 4318, Test Method for Liquid Limit, Plastic Limit, and Plasticity Index of Soils)
- Soil classification (ASTM D 2487, Classification of Soils for Engineering Purposes (Unified Soil Classification System)).

The soil used in the SBDM should ideally be a well-graded soil with a fines content (i.e., percent passing a standard No. 200 sieve) of not more than 30 to 40% (D'Appolonia 1980; Spooner et al. 1985). A fines content of less than 30 to 40% is desirable since the clay portion of soils is commonly more susceptible to chemical alteration than the coarse materials. Therefore, a SBDM with a high fines content is typically more susceptible to chemical alteration than one with a low fines content. A high fines content, however, may be specified by the design engineer, particularly when the subsurface soil naturally contains a high percentage of fines.

If the soil is suspected of containing organic matters, a loss-on-ignition (LOI) test, ASTM D 2974 (Test Methods for Moisture, Ash, and Organic Matter of Peat and Other Organic Soils) should be performed to determined the organic contents of the soil. Based on very limited available information and personal experience, the authors recommend that the soils used in the SBDM contain less than 5 to 10% organic contents.

Soil-Contaminant Compatibility Testing

It is important to determine if the soil particles used in the SBDM are compatible with the contaminant(s) and/or the contaminated ground water. Thus, it is important to perform compatibility testing on the soil used in the SBDM using the contaminant and/or the contaminated ground water. Testing is performed separately on both the coarse and the fine particles (i.e., particles larger or smaller than Standard No. 200 sieve opening, respectively) of the soil mix.

Coarse Particles—The testing program on the coarse particles commonly consists of particle-size distribution analyses (i.e., ASTM D 422). Generally, a representative sample of the soil used in the SBDM is washed over a standard No. 200 sieve and subsequently oven-dried at 230°F (110°C) for at least 24 h. The specimen is then examined under a microscope to record the initial physical appearance (e.g., angularity, surface roughness, etc.) of the particles. Following this initial examination, the specimen is subjected to a particle-size distribution analysis. The specimen is then submerged in the contaminant and/or the contaminated ground water for a period of approximately 30 days. Afterwards, the specimen is removed, washed over a standard No. 200 sieve, and oven-dried. Once again, the specimen is examined under the microscope to establish possible variations in the physical appearance of the particles. The specimen is then subjected to a particle-size distribution analysis. Deterioration of the surface of the particles, any weight loss, or a change in the particle-size distribution are all indicative of incompatibility of the coarse soil particles and the contaminant and/or the contaminated ground water.

It should be noted that the aforementioned 30-day time period is arbitrary and may vary considerably depending upon the mineralogy of the soil particles and the chemistry of the contaminant and/or the contaminated ground water. Based on the authors' past experiences, the 30-day residence period appears to be sufficient for potential chemical reactions to occur between the soil and the contaminant and/or the contaminated ground water. The authors suggest to perform triplicate testing to establish testing deviation and uncertainty. Triplicate testing would help clarify deteriorations caused by incompatibility between the coarse particles, and the contaminant and/or the contaminated ground water from possible testing deviation.

Fine Particles—Soils exhibiting low plasticity indices are generally more resistant to chemical alteration (Spooner et al. 1985). Thus, whenever possible, the authors try to use low plasticity soils in the design of soil bentonite mixes. Additionally, to provide a more direct indication of the compatibility of the fine particles with the contaminant and/or the contaminated ground water, the authors commonly perform Atterberg Limit tests (i.e., ASTM D 4318) and/or hydrometer analyses (i.e., ASTM D 422) before and after submergence of the fines in the contaminant and/or the contaminated ground water for a period of approximately 30 days. Once again, this period is arbitrary and based on the authors' past experience.

Variations in the plasticity (i.e., Atterberg results) or the settlement time (i.e., hydrometer results) of the fines before and after soaking in the contaminant and/or the contaminated ground water are indications of potential incompatibility between the fines and the contaminant.

As a time-saving and somewhat cost-saving alternative, a double hydrometer test may be performed on the fines using ASTM D 4221, Test Method for Dispersive Characteristics of Clay Soil by Double Hydrometer, to evaluate the dispersivity of the clay portion of the soil. Note that in this case the fines are not soaked in the contaminant, and thus, the residence soaking period (e.g., 30 days) is not required. A high dispersion potential is an indication of dispersive soils which are commonly more prone to chemical alteration.

Bentonite-Contaminant Compatibility Testing

Addition of small quantities of bentonite (e.g., 4 to 6%) commonly has a significant effect on the hydraulic conductivity of the SBDM. Therefore, it is important to confirm that the bentonite used in the SBDM is compatible with a specific contaminant. If the bentonite is incompatible with the contaminant, the hydraulic conductivity of the SBDM may increase significantly when SBSCW is subjected to exposure to the contaminant.

To the best knowledge of the authors, there is little or no information available in the literature about the compatibility of varying available bentonite products and different contaminants. Considering the lack of available information and the wide range of contaminants that may be encountered, the authors strongly recommend that a bentonite-contaminant (or contaminated ground water) compatibility testing program be performed in the initial stages of the laboratory testing program. Such a compatibility testing program commonly includes:

- Marsh funnel viscosity (American Petroleum Institute (API) 13B, Standard Procedure for Field Testing Drilling Fluids; Marsh Funnel Viscosity Test)
- Filtrate loss (API 13B, Standard Procedure for Field Testing Drilling Fluids; Filtration Low Temperature/Low Pressure Test)
- Slurry pH (API 13B, Standard Procedure for Field Testing Drilling Fluids; pH Test).

Since the bentonite is usually used in slurry form to stabilize the walls of the excavated trench (Case International Company 1982; Spooner et al. 1985) prior to construction of the SBSCW, the bentonite-water slurry used in the laboratory compatibility testing should, preferably, satisfy the following index properties (Case International Company 1982; Xanthakos

1979): (a) a minimum Marsh funnel viscosity of 35 to 40 s; (b) a filtrate loss of less than 30 mL in 30 mins; (c) a pH of 8 to 10; (d) a bentonite content of 4 to 8%; and finally, (e) a total unit weight of approximately 65 pcf (10 kN/m^3).

To study the compatibility of the bentonite used in the SBDM, the contaminant and/or the contaminated ground water is added to the bentonite-water slurry, and the mixture is tested using the aforementioned testing methods. Alteration of the properties of the bentonite-water slurry is indicative of potential incompatibility of the bentonite, and the contaminant and/or the contaminated ground water. In some cases the authors have noted separation of the bentonite from water, indicating an extreme degree of incompatibility.

Workability Testing

The SBDM used in the construction of a SBSCW must be workable (i.e., it must be of a consistency that will be easy to place without any entrapped air pockets or segregation of the particulate materials). The technique used in the construction of the wall dictates the required consistency of the mix. Commonly, a consistency similar to that of wet-concrete is desirable. The workability of the SBDM is evaluated by conducting slump tests. Commonly, ASTM C 173, Test Method for Air Content of Freshly Mixed Concrete by the Volumetric Method, is used.

A slump value of 3 to 6 in. (75 to 150 mm) and a moisture content of 15 to 35% are usually indicative of a workable SBDM. Soil-bentonite slurry mixes with lower slump values are difficult to place and may contain air voids. Higher slump values, depending upon the gradation of the soil mix, may result in segregation of the soil-bentonite slurry mix, causing larger soil particles to fall to the bottom of the trench.

Permeability/Compatibility Testing

The ultimate test of the SBDM is its ability to maintain low hydraulic conductivity while in contact with the contaminant and/or the contaminated ground water. The purpose of the previous tests has been to provide indications of the potential success of this test. The hydraulic conductivity of the SBDM and its compatibility with the contaminant and/or the contaminated ground water are evaluated using the general guidelines provided in the following standards:

- ASTM D 5084 (Test Method for Hydraulic Conductivity of Saturated Porous Materials Using a Flexible Wall Permeameter) or
- United States Environmental Protection Agency (USEPA) Method 9100 SW-846, Revision 1, 1987, Standard Test Method for Measurement of Hydraulic Conductivity of Saturated Hydraulic Conductivity, Saturated Leachate Conductivity and Intrinsic Permeability.

The aforementioned standards provide the necessary guidelines on the size of the specimens, and the method used to saturate, consolidate, and permeate the specimen. The consolidation pressure is specified by the project engineer and is commonly a function of the height of the wall. It is recommended to use at least two consolidation pressures corresponding to the expected range of the effective stress along the height of the wall. However, usually only one effective stress is specified by the project engineer. In such cases, the specified consolidation pressure is usually equal to the calculated effective stress for the mid-height of the wall. It should be noted, however, that due to arching effects (especially in narrow walls) the actual effective stress may be less than the calculated value.

Commonly, both the SBDM and the soil used in the SBDM are tested. The soil used in the SBDM is used as a reference test to study the worst-case scenario where the bentonite may be attacked/chemically altered by the contaminant. The test specimens (i.e., the soil and the SBDM) are first permeated with water to establish the baseline hydraulic conductivity value. Commonly, the water from the site that will be used in preparing the bentonite-slurry mix is used at this stage of the testing. After consistent hydraulic conductivity values are obtained, the specimens are permeated with the contaminant and/or the contaminated ground water until: (a) consistent hydraulic conductivity values are obtained, (b) the inflow and outflow rates are approximately equal, and (c) two to three pore volumes of the permeant have passed through the specimens. To expedite the testing program, hydraulic gradients larger than the values recommended by the applicable standards (e.g., ASTM 5084) may be used provided that the hydraulic gradient is: (a) increased incrementally and only during the baseline hydraulic conductivity study (e.g., during the first stage of the testing where site water is used as the permeant), and (b) immediately reduced if a pronounced variation in the baseline hydraulic conductivity is observed.

In general, an increase by an order of magnitude in the hydraulic conductivity value when the specimens are permeated with the contaminant and/or the contaminated ground water indicates that the contaminant and/or the contaminated ground water may adversely affect the hydraulic conductivity of the SBDM. It should be noted, however, that in some cases, even if there is an increase in the hydraulic conductivity relative to the baseline value, the final and stable hydraulic conductivity value may still be significantly lower than the required/specified value. As a conservative approach, in such cases, the authors assume that the SBDM is acceptable only if the final hydraulic conductivity of the reference test (i.e., the hydraulic conductivity of the soil used in the SBDM, without any addition of bentonite, after permeated with contaminant and/or contaminated ground water) is lower than the specified hydraulic conductivity value. Commonly, a hydraulic conductivity value of less than 1×10^{-10} to 1×10^{-8} m/s is specified by the regulatory agencies.

Case History

The remainder of this paper describes a case history of one project. This includes a brief project background, a discussion of the test materials, and the results of the testing program performed to select SBDMs.

Background

A SBSCW was constructed around the perimeter of a landfill at a former oil refinery located in the northeastern United States. The landfill is underlain by a coarse granular soil to a depth varying from approximately 30 to 40 ft (9 to 12 m) below the ground surface. A thick deposit of clay with a hydraulic conductivity of approximately 2×10^{-10} to 2×10^{-9} m/s underlies the coarse granular soil. The granular soil has been exposed to refinery by-products (herein referred to as free-product) since the 1950s. Volatile and semivolatile organic compounds (e.g., xylene and trimethylbenzene) as well as heavy metals (e.g., arsenic and lead) exist at the site.

A SBSCW, which is keyed into the underlying clay, was to be constructed around the landfill to isolate the free-product and the potentially contaminated ground water from the surrounding subsurface environment. The subsurface granular soil from the alignment of the slurry wall, clay from a borrow source, a commercially available bentonite, and the water from a municipal water supply available at the site were used to select a SBDM for the construction of the SBSCW.

Test Materials

Subsurface Soil—Subsurface soil samples were obtained at various depths within the coarse granular soil. The samples were relatively consistent and composed of well-graded gravel with silt and sand. Thus, all the samples were blended together to obtain a composite subsurface soil sample. General information about the composite subsurface soil is presented in Table 1 and Fig. 1.

Borrow Clays—Three different borrow clay samples (herein referred to as BC1, BC2, and BC3) were tested. Table 1 and Fig. 1 present the results of the index tests and particle-size analysis tests, respectively, performed on the samples. BC3 was selected for further testing since: (a) it was less plastic than BC1; and (b) it was less expensive than BC2.

Test Liquids—The city water available at the site (herein referred to as the site water), the free-product, and the contaminated ground water were used in the testing program.

Bentonite—Three commercially available bentonite products (herein referred to as B1, B2, and B3) were used in the testing program to select the bentonite most resistant to chemical and physical alteration when in contact with the free-product and/or the contaminated ground water.

SBDM Selection Approach

Soil Mix—Based on the project specifications, a soil mix (i.e., only soil without any bentonite) with a well-graded particle-size distribution and 30 to 40% fines was to be designed using the composite subsurface soil sample and the selected borrow clay. The composite subsurface soil and the borrow clay were to be blended at their as-received moisture contents.

Considering the particle size distributions of the composite subsurface soil and the selected borrow clay, a soil mix was formed which comprised of (by dry weight) 67% of composite subsurface soil and 33% of BC3. As presented in Table 1 and Fig. 1, the soil mix had a fines content of approximately 32%.

Soil-Contaminant Compatibility—Figure 2 presents the particle-size distributions for the coarse fraction of the soil mix before and after soaking in the free-product for a period of 30 days. There was no weight loss and, as presented in Fig. 2, there was little or no change

TABLE 1—*Soil index test results.*

Material Source	Percent Passing #200 Sieve ASTM D 1140	Atterberg Limits ASTM D 4318			Classification
		Liquid Limit (%)	Plastic Limit (%)	Plasticity Index (-)	
Subsurface Soil	7.7	NP	NP	NP	GM-GW; well-graded gravel with silt and sand
Borrow Clay 1 (BC1)	99.8	45	25	20	CL; lean clay
Borrow Clay 2 (BC2)	48.0	23	16	7	SC; clayey sand with gravel
Borrow Clay 3 (BC3)	56.0	23	15	8	CL; sandy lean clay
Soil Mix	32.0	NA	NA	NA	Silty, clayey sand with gravel[a]

[a] Based on a visual observation.
NA Not available.

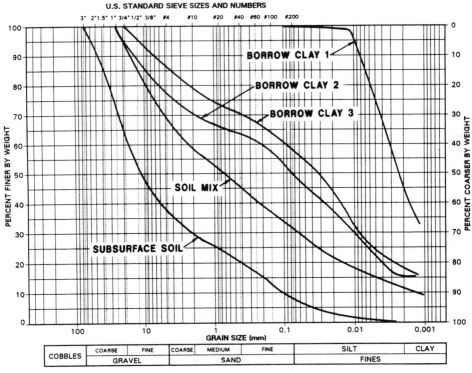

FIG. 1—*Particle-size distribution of the soils used in the testing program.*

in the particle-size distribution of the soil mix. Additionally, visual examinations of the coarse particles before and after soaking showed no visual surficial deterioration due to soaking in the free-product. Based on these results, it was concluded that the free-product had no adverse chemical effects on the coarse soil particles.

A double hydrometer test was performed on the fine fraction of the soil mix. The results indicated that the soil mix had a dispersion value of 59%, and thus was dispersive in nature. This indicates that the fines content of the soil mix may be susceptible to chemical alteration.

Bentonite-Contaminant Compatibility—To select the bentonite most compatible with the free-product and the contaminated ground water, a sample of each bentonite was mixed with the site water to form a hydrated slurry with a bentonite content of approximately 10% by total weight. The slurry samples were allowed to hydrate for a period of 24 h. Afterwards, 1.1 lb (500 g) specimens of the slurry samples were diluted with each of the test liquids (i.e., the free-product, the contaminated ground water, and the site water) to obtain bentonite slurry specimens with a bentonite content of 5% by total weight. The newly formed bentonite slurry specimens were allowed to age for a period of seven days, as specified by the project engineer, prior to testing. The test results are present in Table 2. Referring to this table, it may be observed that: (a) the three bentonite-water slurry specimens were relatively similar, (b) the contaminated ground water did not have a strong effect on the behavior of the bentonite-water slurry specimens, and (c) addition of the free-product to the bentonite-water slurry specimens resulted in gelling and formation of bentonite clots. Based on availability and cost considerations, B3 bentonite was selected for the testing program.

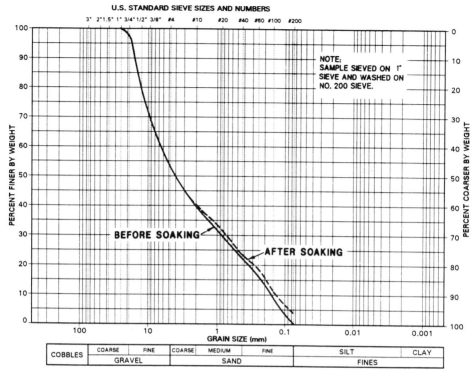

FIG. 2—*Particle-size distribution of the coarse fraction of the soil mix before and after soaking in the free-product for 30 days.*

Workability Testing—A bentonite-water slurry, composed of 95% site water and 5% B3 bentonite, was formed and hydrated for a period of approximately 24 h. A SBDM was formed by incrementally adding the hydrated bentonite-water slurry to the soil mix. After each bentonite-water slurry addition, a slump test was performed on the resulting soil ben-

TABLE 2—*Bentonite compatibility test results.*

Slurry Sample No.	Test Liquid	Unit Weight[a] (pcf)	Marsh Viscosity (s)	Filtrate Loss (mL/30 min)	Slurry pH (-)	Temperature (°C)
B1	Site Water	65	43	17.8	9.4	21.1
	Ground Water	62	47	25.8	7.6	21.7
	Free-Product	58	[b]	5.6	7.3	20.8
B2	Site Water	65	NA	17.8	9.1	20.9
	Ground Water	64	58	24.8	7.7	21.5
	Free-Product	63	[b]	11.8	7.3	21.3
B3	Site Water	62	41	15.6	9.4	22.1
	Ground Water	64	40	14.8	7.7	21.7
	Free-Product	60	[b]	5.8	7.6	22.3

[a] 62.4 pcf 9.81 kN/m³.
[b] Sample gelled; unable to test.

tonite mix until a slump of approximately 4 in. (100 mm) was achieved. As presented in Table 3, the selected SBDM had a moisture content of approximately 17.4%, a dry unit weight of approximately 115 pcf (18 kN/m^3), and a bentonite content of only 0.7% (i.e., 7 g for 1000 g of dry soil mix).

Permeability/Compatibility Testing—Permeability tests were performed on the soil mix (i.e., the reference test) as well as the SBDM. An effective stress of approximately 5 psi (35 kPa), corresponding to the average effective stress at the mid-height of the wall, was used to consolidate the test specimens. The specimens were initially permeated with the site water to establish the baseline hydraulic conductivity values. Permeation continued until consistent hydraulic conductivity values were obtained and the inflow and outflow rates were equivalent. The test results are presented in Table 3 and Figs. 3 through 6. Referring to the figures, the initial hydraulic conductivity of the soil mix was approximately 5×10^{-10} m/s. The SBDM had an initial hydraulic conductivity of approximately 3×10^{-10} m/s. To expedite the remainder of the testing program, the hydraulic gradient was incrementally increased for both of the specimens to approximately 400. As shown in the figures, no adverse effects on the recorded hydraulic conductivity values were observed even after approximately 4 to 6 pore volumes of the site water had been passed through the specimen.

The specimens were then permeated with the free-product. Free-product, and not the contaminated ground water, was used as a means to simulate the worst-case scenario. At least 2 to 3 pore volumes of the free-product were permeated through the test specimens before the tests were terminated. As shown in Table 3 and Figs. 3 through 6, introducing the free-product had little or no effect on the hydraulic conductivity of either the soil mix or the SBDM specimen. It was thus concluded that the SBDM is compatible with the free-product and/or the contaminated ground water and can be used in the construction of the SBSCW at the site.

Summary

Soil bentonite slurry cutoff walls are constructed around contaminated soil to isolate the contaminant and prevent further contamination of the surrounding subsurface environment. A soil bentonite mix must be designed for the wall which has a hydraulic conductivity less than a specified value, is resistant to the contaminant, and is workable. The subsurface soil from the alignment of the wall, a clay from a borrow source (when needed), and bentonite are used in the SBDM. The SBDM should preferably comprise of a well-graded soil mix

TABLE 3—*Design mix composition, workability, and permeability/compatibility test results.*

| | Soil Mix Composition | | | Physical Properties | | | Hydraulic Conductivity USEPA 9100 | | | |
| | | | | | | | Test Specimen Initial Conditions | | Hydraulic Conductivity (m/s) | |
Sample ID	Subsurface Soils (%)	BC3 Borrow Clay (%)	Bentonite (%)	Fines (%)	Moisture Content (%)	Slump (mm)	Moisture Content (%)	Dry Unit[a] Weight (pcf)	Water	Free Product
Soil Mix	33	67	0	32	...	...	17.9	119	4.8E-10	4.2E-10
SBDM	33	67	0.7	33	17.4	102	17.4	115	2.7E-10	2.4E-10

[a] 62.4 pcf 9.81 kN/m^3.

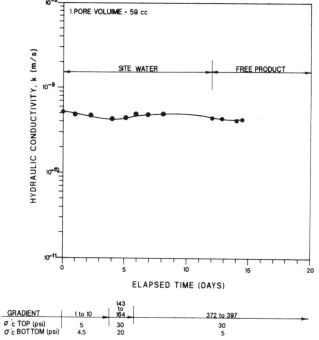

FIG. 3—*Hydraulic conductivity of the soil mix versus time.*

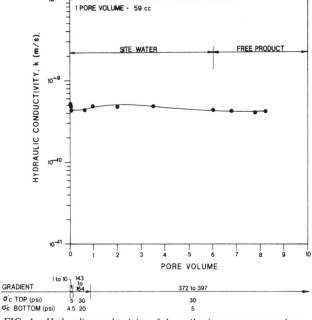

FIG. 4—*Hydraulic conductivity of the soil mix versus pore volume.*

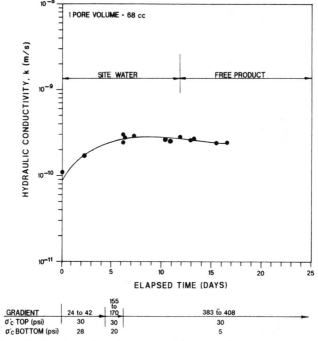

FIG. 5—*Hydraulic conductivity of the soil-bentonite design mix versus time.*

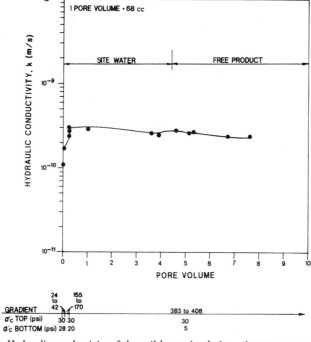

FIG. 6—*Hydraulic conductivity of the soil-bentonite design mix versus pore volume.*

with 30 to 40% fines. Both the soil and the bentonite used in the SBDM must be compatible with the contaminant. The final and the most important test of the SBDM is its ability to sustain a low hydraulic conductivity when permeated with the contaminant for an extended period of time.

This paper presented a recommended laboratory testing program developed by the authors to design a soil bentonite mix compatible with a contaminant and/or contaminated ground water. Results from a testing program where a soil bentonite mix was designed for a soil bentonite slurry wall that was to be constructed around the perimeter of a landfill at a former oil refinery were presented.

Acknowledgment

The authors are grateful to Mr. David Christensen of Atlantic Richfield Company for permission to use the data presented in the case history section.

References

Case International Company, 1982, *Case Slurry Wall Notebook*, Manufacturers' Data, Case International, Houston, TX.
D'Appolonia, D. J., 1980, "Soil-Bentonite Slurry Trench Cutoffs," *Journal of the Geotechnical Engineering Division of ASCE*, Vol. 106, No. 4, pp. 399–417.
Spooner, P., Wetzel, R., Spooner, C., et al., 1985, *Slurry Trench Construction for Pollution Migration Control*, Noyes Publication, Park Ridge, NJ.
Xanthakos, P. P., 1979, *Slurry Walls*, McGraw-Hill Book Company, New York.

Jean-Hugues Deschênes,[1] *Michel Massiéra,*[2] *and Jean-Pierre Tournier*[3]

Testing of a Cement-Bentonite Mix for a Low-Permeability Plastic Barrier

REFERENCE: Deschênes, J.-H., Massiéra, M., and Tournier, J.-P., **"Testing of a Cement-Bentonite Mix for a Low-Permeability Plastic Barrier,"** *"Dredging, Remediation, and Containment of Contaminated Sediments, ASTM STP 1293,* K. R. Demars, G. N. Richardson, R. N. Yong, and R. C. Chaney, Eds., American Society for Testing and Materials, Philadelphia, 1995, pp. 252–270.

ABSTRACT: A laboratory investigation was carried out to determine the physical properties of a suitable mix of cement-bentonite for a low-permeability cutoff wall implemented in 1991 under the north dyke at the LG-1 hydroelectric site, which is part of the James Bay Development Project in Northern Quebec. The project specifications required the mix to have a hydraulic conductivity of less than 10^{-8} m/s and be able to sustain a plastic deformation of 6% without fissuring, as measured in a triaxial compression test with a lateral pressure of 100 kPa on a specimen cured for 90 days. A two-phase laboratory study was carried out. Phase 1 involved the determination of the mix design, while the second phase consisted of the laboratory control of the manufactured mixture. During phase 1, laboratory samples were prepared with different proportions of cement, bentonite, sand, and water. The permeability, triaxial compression, and unconfined compression tests were performed after curing periods ranging from 8 to 120 days. Once the appropriate mix was selected, it was tested to check if it met the specifications. During phase 2, or during actual production or construction, density and viscosity measurements of the cement-bentonite slurry were made routinely at the plant and at the discharge of the slurry pipe. Cylindrical specimens were molded using slurry from the discharge of the pipe and from the trench. Permeability, triaxial compression, and unconfined compression tests were performed on the specimens after curing periods ranging from 3 to 103 days.

KEYWORDS: slurry trench, confinement, plastic barrier, impermeable cutoff wall, mix design, laboratory testing and control, triaxial compression, unconfined compression, hydraulic conductivity

Environmental regulations concerning contaminated soils and wastes are becoming more stringent. As a result, technical solutions to deal with the contamination are becoming more costly. It is generally accepted that there exists two types of practical solutions to counteract the negative effects of contaminated soils and wastes. The contaminants can either be excavated and disposed or they can be contained and isolated. Among the latter type, the cement-bentonite slurry wall is an attractive solution that offers impermeability, safety, and economy, and if properly designed is easy to implement. These slurry cement-bentonite walls are currently used as barriers for water-retaining structures. Because the cement-bentonite

---

[1] Senior engineer, Géoconseil Inc., Montréal, Québec, Canada H1X 1W6.
[2] Professor, Department of Civil Engineering, Université de Moncton, Moncton, New Brunswick, Canada E1A 3E9.
[3] Senior engineer, Société d'Energie de la Baie James, Montréal, Québec, Canada H2Z 1W7.

slurry is self-hardening, it first serves as support for the trench during its excavation, then, once it has hardened, acts as a low-permeability cutoff wall. This technique is relatively recent in North America, and was first applied to the building of cutoff walls for embankment dykes approximately 20 years ago.

Few results of tests conducted on various self-hardening cement-bentonite slurry mixes for the building of cutoff walls have been published, so far. Limited information on the behavior (mechanical strength and deformation) of cement-bentonite slurry mixes with respect to their use in the construction of cutoff walls has been published within the last ten years. This has mainly been reported by Millet and Perez (1981), Chapuis et al. (1984), and Gill and Christopher (1985).

Such a cutoff wall was constructed in 1991 at the hydroelectric site of LG-1, which is part of the James Bay Development Project in Northern Quebec. The owner of this project is the James Bay Energy Corporation (SEBJ). This cutoff was implemented over a length of 1 km along the north dyke, which was seated on a recent geological deposit consisting of a mixture of sand and silt containing coarse sand lenses. The depth of this permeable layer ranged from 5 to 25 m, before reaching a thick clay deposit. The slurry wall barrier that now controls the permeability under the north dyke is formed by a 0.6 m wide wall of a hardened cement-bentonite slurry mix, 1030 m long, with a depth ranging from 5.7 to 27.7 m.

It can be seen that in this particular project, the slurry wall barrier was not used to contain wastes, but mainly as a cutoff to stop underground water movement. However, such a barrier could be used to isolate wastes. For environmental use, the mix should not only be subjected to the verification of its mechanical behavior, but also to its chemical compatibility between the wastes, and its leachates and the barrier materials. This verification is needed to determine if the wastes or the leachates could chemically alter the slurry wall materials which could lead to an increase in hydraulic conductivity (Ryan 1987).

The aim of this paper is to present results of tests carried out to determine the physical, mechanical, and hydraulic properties of the cutoff wall materials. These tests were performed on the slurry before and during the construction of the cutoff wall, in order to verify a cement-bentonite mix meeting the specifications for a plastic cutoff wall. A testing program was extended to verify proposed slurry mix formulae using the same ingredients (cement, bentonite, water) as in the field. Another testing program was established to control the quality of the slurry during the actual construction of the cutoff wall. It will be shown that the properties of these mixes depend essentially upon the quality and the type of cement and bentonite being used.

Requirements for Impermeable Cutoffs

Cutoff walls must be composed of plastically deformable materials which, at various stages during and after construction, can retain their permeability properties, while being sufficiently flexible to conform to the deformations of the surrounding soils without fissuring. During construction of the barrier, the material has to be sufficiently fluid in order to excavate the trench easily and sufficiently dense to withstand the lateral pressure of the *in-situ* soils and stabilize the excavation sides. Furthermore, for the case of an environmental cutoff, the barrier materials would have to be sufficiently inert and should not be chemically transformed or altered by the percolation of the leachates yielded by the wastes.

To achieve adequate behavior, it is necessary that the barrier meet the following criteria:

- mechanical criterion: strength, perennity, deformability and permanence;
- hydraulic criterion: low hydraulic conductivity;

- chemical criterion: compatibility

Numerous factors influence the hydraulic conductivity of a barrier (Millet and Perez 1981). These include continuity and integrity of the barrier during the excavation of the trench, the properties of the material composing the barrier, and the integration of the barrier with surface structures such as dykes or other water-retaining structures. Aside from the properties of the slurry materials, all other factors are directly related to the design and mainly to the construction techniques. As leachates are the result of particular wastes, it is impossible in this paper to discuss all eventualities. We refer the reader to papers by Ryan (1985, 1987) which discuss the possible chemical attack of leachates on the barrier materials. The behavior of the barrier with respect to hydraulic conductivity and durability is termed "compatibility." The leachates may in certain cases have detrimental effects, while in other cases, they may be totally insignificant, as found by Manassero and Viola (1992) and Jefferies (1992). Thus, it is imperative that future design of cement-bentonite cutoff walls be developed with compatibility being assessed for the particular leachates produced by the wastes. In this paper, only the mechanical and hydraulic properties affecting the slurry materials have been studied, not the chemical properties.

In this project, the selection of a cement-bentonite slurry trench was based on the variable depth (5.7 m to 27.7 m) of the permeable layer and the presence of an underlying thick deposit of clayey silt and silty clay which is normally consolidated and which easily reaches 30 m in thickness. The construction of a 6 m-high dyke directly on the permeable layer, which would cause approximately 200 mm of settlement, had to be taken into account. The requirements of the tender document were the following:

- Six percent plastic deformation without fissuring as measured after a curing of 90 days, in a triaxial compression test under a lateral pressure of 100 kPa at a deformation rate of 0.1% per minute. This test was devised by SEBJ according to Chapuis et al. (1984) where it was considered that the most critical site condition is when the slurry has been released in the trench when the slurry is not completely hardened, and where full drainage has not been achieved.
- A coefficient of permeability less than or equal to 10^{-8} m/s. This value was judged by SEBJ as the maximum acceptable after numerous flow net studies.

The triaxial test required in the specifications is nonconventional. It is a "hybrid" test where the sample is not completely drained during the compression. This test was devised in order to represent the most unfavorable conditions of the *in-situ* behavior where full drainage is not completely achieved. The test is called, in this paper, a consolidated isotropically partially drained test (CIPD). The hydraulic conductivity was also measured in a triaxial cell after full saturation of the samples.

The specifications also required a minimal compressive strength of the slurry mixture after 28 and 91 days of curing. This allowed the determination of the undrained shear strength of the self-hardening slurry and its undrained modulus of elasticity. In this project, the unconfined compression tests were conducted on a mix design done prior to construction and were also run on samples from the cement-bentonite slurry manufactured in the field and taken at the discharge of the pipe in the trench. The specifications called for the consistency of the hardened slurry to be stiff to very stiff. In other words, the unconfined compression strength should range between 100 to 400 kPa.

Laboratory Testing Prior to Construction

Cement-Bentonite Slurry Mix Design

Prior to construction, the laboratory hired by the contractor conducted a series of tests in order to develop a proper mix design, satisfying the above-mentioned requirements.

Materials Used and Composition of the Trial Mixes—In this testing program, four mixes were achieved with bentonite and cement used in the actual construction. The bentonite was Bara-Kade 90 from the Belle Fourche plant of Baroïd Minerals and Chemicals in South Dakota. This bentonite meets API-13M specification of the American Petroleum Institute (1988). It is a powder bentonite, having an average pH of 9.3 and 76% particles finer than the No. 200 sieve (0.074 mm). The cement used was a portland type 20 which originated from the Saint-Constant (Quebec) area and was produced by the Lafarge Corporation. This cement contained 0.7% of free calcium oxide (CaO). This cement meets the requirements of ASTM C 150, Specification for Portland Cement. A retarding agent was used so that the initial set of slurry would not hinder the construction procedure. This agent is a calcium-sodium lignosulfate named Marasperse C-21 manufactured by Lignotech Inc. (U.S.).

Four initial mixes, BJ-1 to BJ-4, were subjected to testing in the laboratory of the contractor's consultant. The mixes BJ-1, BJ-2, and BJ-3 were manufactured using water from the laboratory tap. Mix BJ-4 which contained the same solid materials as mix BJ-2, was done with water taken from the site. The various compositions of the mixes are shown on the left side of Table 1. This mix was done to check if the results were the same using water from different sources. The BJ-3 mix had the same composition as BJ-2 except sand was added. This was carried out in order to measure the influence of sand which could eventually contaminate the cement-bentonite mixture and modify its mechanical and hydraulic characteristics. The composition of each mix was based on proportions that were already tested by SEBJ and reported by Chapuis et al. (1984).

Mix Preparation—For each mix, the bentonite-water solution was mixed using a three-blade industrial mixer at a rotary speed of 700 r/min for 4 min to obtain a uniform slurry. The initial density and viscosity (in seconds on the Marsh cone) of the slurry was measured. After a hydration period of 5 h (mixes BJ-1 and BJ-3) and 24 h (mix BJ-4), the bentonite-water slurry was remixed (700 r/min during 90 s) and its physical characteristics (density and viscosity) were measured again.

Cement-bentonite mixes were prepared by adding cement to the hydrated bentonite. This slurry was mixed by using the same mixer at a rotation speed of 750 to 900 r/min during 90 s. For each mix, a small quantity of retarding agent was added to the cement-bentonite-water slurry. The density and the viscosity were again measured. To obtain mix BJ-3, sand

TABLE 1—*Proportion of Trial Mixes of Cement-Bentonite.*

	Tests by Contractor's Laboratory				Tests by SEBJ's Laboratory		
Mix No.	BJ-1	BJ-2	BJ-3	BJ-4	ME-5	ME-6	ME-7
B/W, %	3.72	3.68	3.68	3.68	3.69	3.69	3.69
C/W, %	35.0	30.0	30.0	30.0	31.9	31.9	31.9
S/W, %	0	0	26.7	0	0	36.0	60.0
R/W, %	0.17	0.11	0.11	0.11	0.10	0.10	0.10

NOTES: B = bentonite, C = cement, W = water, R = retarding agent; all percentage by weight.

from the site was added to the BJ-2 mix; the quantity of sand was 20% of the cement-bentonite's total weight, which corresponds to 26.7% of the weight of water. For all mixes, the slurry was molded into 51 mm in diameter by 127 mm high cylindrical samples. All samples were closed and sealed after molding and stored in a curing room.

These different samples were subjected to triaxial compression and permeability tests to determine a suitable mix. The final mix recommended by the consultant of the contractor adopted for the field work consisted of the following quantities of material: a bentonite-water (B/W) ratio of 0.037, a cement-water (C/W) ratio of 0.32 with the retarding agent being 1 to 4 kg/m^3. The results are presented hereafter.

Verification Tests Prior to Construction—To verify the mix proportions recommended by the contractor, SEBJ carried out verification trial mixes in its field laboratory. In this testing program, three mixes, ME-5 to ME-7, were prepared using the mixed proportions selected by the contractor's consultant. Mix ME-5 was identical to the one recommended by the contractor. Mixes ME-6 and ME-7 contained, respectively, aside from the same ingredients as the adopted mix, 36% and 60% of sand with respect to weight of water. These mixes were prepared to check the influence of the sand content on the mechanical and hydraulic characteristics of the cement-bentonite-water mix. These different sand contents were chosen based on the experience reported for the Yacyreta Hydro-electric Project (Larsen et al. 1991). In that project, the sand content was, on average with respect to weight of water, 35.5% or 11.2% of the total volume of the slurry. This addition of sand was made to represent the possible contamination of the slurry by the collapse of sand from the side walls. The composition of the mixes is shown on the right half of Table 1. The sample preparation procedure was identical to the one previously described. The samples were mixed into 75 mm in diameter by 150 mm cylindrical molds. These were stored for testing and future use.

Testing Procedures

The determination of the coefficient of permeability and the triaxial compression was carried out in the triaxial cell. For the permeability test, the procedure required the use of a confining pressure and a back-pressure for sample saturation. The effective confining pressure used in all compression tests was 100 kPa and the back-pressures were 690 kPa (contractor) and 207 kPa (SEBJ), respectively. After saturation at 98% and consolidation of the samples, a hydraulic gradient was applied between the bottom and the top ends of the samples. This gradient varied from 12 to 24 for the contractor, while it was maintained at 30 for the SEBJ test program. The triaxial compression tests (CIPD) were carried out using the Bishop and Henkel (1962) procedure. The axial rate of deformation was maintained at 0.1% per min. All saturation and permeability tests were carried out using site water. Finally, all unconfined compression tests were carried out according to the ASTM D 2166, Test Methods for Unconfined Compressive Strength of Cohesive Soil.

Results

Characteristics of the Mixes in their Fluid State—During the mixing of the different batches, the density of the bentonite slurry was measured using the mud scale (contractor) and in a 0.2 L graduated test tube (SEBJ). The viscosity of the bentonite slurry was measured using the Marsh cone according to the API-RP-13B (1988) test procedure. The test values obtained are relatively constant at 1.022 for the density and 32 s for the viscosity. The density of the cement-bentonite slurry without sand varied between 1.20 and 1.23 with a C/W ratio of 0.30 to 0.35, respectively. These values are in good agreement with the theoretical calculated density given by the expression proposed by Chapuis et al. (1984).

Characteristics of the Mixes in their Solid State—A summary of the results obtained from the permeability, unconfined compression, and triaxial compression tests carried by both the contractor and SEBJ are shown in Tables 2 and 3, respectively.

It can be observed from the contractor's results that the measured hydraulic conductivity slightly decreases with time (mixes BJ-1 and BJ-4) or stays approximately constant (mix BJ-2). In the case of the SEBJ tests, the hydraulic conductivity is approximately constant (ME-6 and ME-7) or increases (ME-5) with time. In all cases, the coefficient of permeability of the mixes after 90 days is on the order of 10^{-8} m/s. The highest and lowest values of the coefficient of permeability after 90 days are obtained respectively for mix BJ-1 which contains highest cement content, and mix ME-7 which contains the highest sand content.

The strength of the mixes is dependent mainly on C/W ratio and varies greatly with different C/B ratio (Tornaghi 1972; Caron 1973; Chapuis et al. 1984). The unconfined compression tests were carried out on samples cured for 3, 7, 28, and 90 days. The summary of these results are given in Tables 2 and 3. Figure 1 shows the influence of the C/W ratio. Mixes BJ-1 (C/W = 0.35) and ME-5 (C/W = 0.32), which have the highest cement content, have obviously the highest strength.

The influence of the sand content on the unconfined compression strength and on the E modulus, for a C/B ratio of 0.32 is shown in Figs. 2 and 3. The addition of sand tends to increase the strength and the modulus of the mix. The E modulus, here, is taken as the tangent-modulus on the stress-strain curve.

As shown in Fig. 4, the type of test has an important influence on the stress-strain curve and on the deformation at peak strength. It is generally accepted that the unconfined compression tests are not appropriate to determine the deformations of a hardened slurry without fissuring. According to the unconfined compression tests, the axial deformation to obtain the maximum strength varied from 0.8 to 1.3% depending on curing time (Tables 2 and 3). However, it is generally accepted (Gill and Christopher 1985) that these triaxial compression tests are more adapted to determine the deformation under which a slurry wall can be subjected without fissuring. The samples in the triaxial cell subject to lateral stress in drained or partially drained conditions can be subject to higher lateral and axial deformation before reaching maximum strength. The CIPD triaxial tests were carried out on the different mixes. In all tests, the samples showed axial deformation above 8% at maximum strength. In all cases, no fissures were observed prior to 10% axial deformation. It should be noted that the tests carried out on mixes ME-5 to ME-7 at 90-day curing were stopped after 6% of triaxial deformation. The samples were dismantled and inspected for fissures and microfissures according to the specifications. All other triaxial tests were carried out to failure or to an axial deformation of 20%. A summary of these results is shown in Tables 2 and 3. Figure 5 gives examples of the increase of the triaxial compression strength with time.

Mixes BJ-1 (C/W = 0.35) and ME-5 (C/W = 0.32), which have the highest proportion of cement, have the highest strength (Fig. 5). It should be noted that the C/W ratio was only slightly between 0.30 and 0.35. The influence of the sand content on the triaxial strength for a C/B ratio is shown in Fig. 6. The addition of sand to the mix alters the strength only slightly.

The initial water content can influence the mechanical and hydraulic characteristics of the different mixes. The water contents vary little with curing time. For mixes without sand, the water contents are a function of the C/W ratio. These range from 192 to 200% for mix BJ-1 (C/W = 0.35); from 192 to 221% for mix ME-5 (C/W = 0.32); and from 210 to 239% for mixes BJ-2 and BJ-4 (C/W = 0.30). For mixes with sand, the water contents are less.

TABLE 2—Characteristics of the mixes in their solid state (contractor's laboratory).

Mix No. (C/W, B/W, S/W)	Time, days	Permeability Tests			Unconfined Compression Tests				Triaxial CIPD Tests				
		Initial Density	w_i, %	k, m/s	Initial Density	w_i, %	q_u, kPa	ε_f, %	Initial Density	w_i, %	$(\sigma_1 - \sigma_3)_f$, kPa	ε_f, %	Cracking Development, %
BJ-1 (0.350, 0.037, 0.000)	7				1.248[a]	200[a]	190[a]	1.0[a]					
	8								1.252	193	495	11.1	10.8
	16	1.252	200	1.9×10^{-8}									
	18				1.248	197	350	1.0					
	28	1.242	194	8.1×10^{-9}					1.252	195	560	9.1	10.8
	29	1.247	197	6.5×10^{-9}					1.263	192	657	9.1	10.8
BJ-2 (0.300, 0.037, 0.000)	7	1.258	195	4.4×10^{-9}									
	15				1.230[a]	233[a]	122[a]	1.2[a]					
	21	1.216	227	1.5×10^{-8}									
	28	1.216	230	1.2×10^{-8}	1.226	227	281	0.8					
	29	1.216	229	1.9×10^{-8}									
	90								1.221	232	382	19.7	not observed
	92												
BJ-3 (0.3, 0.037, 0.267)	28	1.337	143	1.5×10^{-8}					1.356	141	429	14.6	17.2
	89								1.325	149	458	14.5	17.5
BJ-4 (0.300, 0.037, 0.000)	7				1.214[a]	239[a]	123[a]	1.0[a]					
	20	1.220	210	1.2×10^{-8}									
	22												
	28	1.216	233	9.1×10^{-9}	1.220	234	272	1.1	1.216	228	422	18.1	not observed
	35	1.223	232	6.2×10^{-9}									
	98								1.231	231	459	16.9	not observed
	120								1.223	229	503	15.5	not observed

[a] Mean of two results.

TABLE 3—*Characteristics of the mixes in their solid state (SEBJ's laboratory).*

Mix No. (C/W B/W S/W)	Time, days	Permeability		Initial Density	Unconfined Compression Tests[a]					Triaxial CIPD Tests			Cracking Development, %
		w_i, %	k, m/s		w_i, %	q_u, kPa	ε_f, %	E, MPa		w_i, %	$(\sigma_1 - \sigma_3)_f$, kPa	ε_f, %	
ME-5 (0.320 0.037 0.000)	0			1.210									
	3	193	4.3×10^{-9}		221	92	1.1	11.9		203	441	13.2	≈10.0
	7	203	2.9×10^{-8}	1.252	207	175	1.2	23.4		207	526	8.0	≈10.0
	28	192	1.5×10^{-8}	1.356	205	367	1.2	50.0		199	635	6.3	not observed[b]
	90				214	503	0.9	83.8					
ME-6 (0.320 0.037 0.360)	0												
	3	106	1.3×10^{-8}		110	97	1.3	14.8		121	459	9.3	≈10.0
	7	114	2.4×10^{-8}	1.416	107	181	1.0	25.5		121	508	7.1	≈10.0
	28	126	1.3×10^{-8}	1.437	107	367	1.0	57.8		129	575	6.5	not observed[b]
	90				117	521	0.9	108.5					
ME-7 (0.320 0.037 0.600)	0												
	3	88	3.0×10^{-8}		88	126	1.3	17.3		85	466	9.5	≈10.0
	7	93	3.1×10^{-8}		85	206	1.0	37.6		85	525	8.0	≈10.0
	28	84	3.7×10^{-8}	1.512	83	439	0.9	76.3		88	609	5.5	not observed[b]
	90				81	577	0.9	118.8					

[a] Tests stopped at around 6% of strain.
[b] Mean of three results.

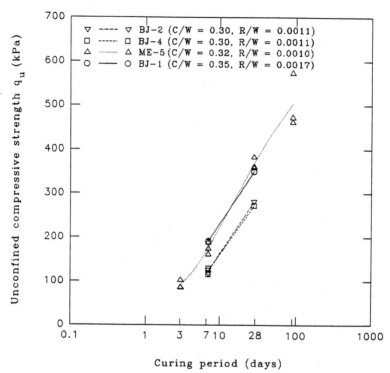

FIG. 1—*Increase of the unconfined compression with time.*

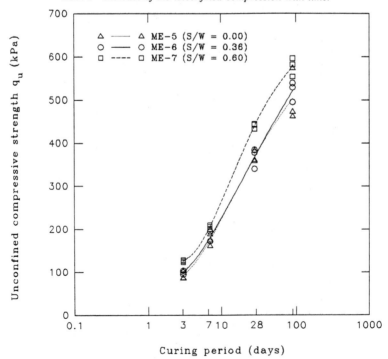

FIG. 2—*Increase of the unconfined compression with time (C/W = 0.32).*

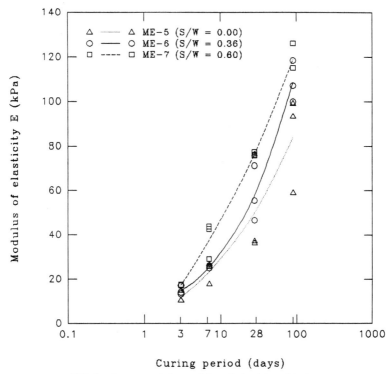

FIG. 3—*Increase of the modulus of elasticity with time.*

Testing During Construction

Control of the Cement-Bentonite Slurry in its Fluid State

Bentonite Slurry or Mud—During the preparation of this slurry, the physical properties were controlled each day by the contractor by measuring the slurry's density and its viscosity with the Marsh cone according to API RP 13B (1988). The average values obtained on 38 tests were, respectively, 1.026 for the density and 32 s for the viscosity. SEBJ carried out the same tests on 15 samples. The results were identical to the ones obtained by the contractor.

Cement-Bentonite Slurry—The physical characteristics of the slurry (density and viscosity) were controlled at the plant at least twice a day by the contractor. The mean values obtained on 39 tests were, respectively, 1.225 for the density and 50 s for the viscosity.

SEBJ also carried out independent tests on a regular basis at the discharge of the slurry pipe. The mean values are lower than those obtained by the contractor. On a total of 39 tests, the density is 1.199, while the viscosity is 46 s.

The density and the viscosity of the slurry in the trench was measured once a day by the contractor. However, in 70% of the measurements, the viscosity was higher than 65 s, because the contamination with the sand from the excavation walls thickens the slurry. The average density on 31 tests was 1.30, but these are much more scattered than the ones measured at the plant. The SEBJ did six routine tests and obtained values of 1.33.

Bleeding occurs when water rises to the surface in the slurry after sedimentation of the particles prior to the setting of the mix. In practice, a bleeding of 3% of the volume is

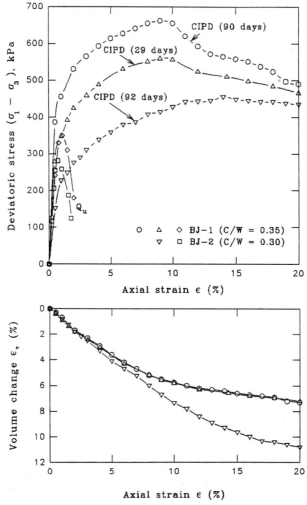

FIG. 4—*Typical stress-strain curves.*

considered acceptable, according to Chapuis et al. (1984). A total of 52 tests was carried out by SEBJ on the slurry prior to being discharged into the trench. The measured values range from 0.5% to 3.1% with an average value of 1.6%. The sand content of the slurry sample at mid-height of the trench was measured at least once a day by SEBJ. These sand contents are expressed with respect to the total volume of the slurry according to the API RP 13B (1988) test procedure. The average sand content was measured to be 10% by volume of slurry. This sand content corresponds to sand/water weight ratio of 33%. The histogram in Fig. 7 indicates the quantity of sand in the slurry at mid-level of the trench. It shows the sand content by volume to vary from 2 to 22%, which was less than expected.

Control of the Cement-Bentonite Slurry in its Solid State—Summary of the permeability, unconfined compression, and triaxial CIPD compression test results, as performed by the contractor's laboratory and by the SEBJ laboratory, respectively, are presented in Tables 4 and 5. Testing procedures were identical to those previously described.

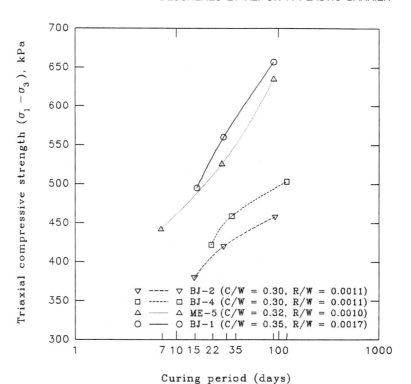

FIG. 5—*Increase of the triaxial compression strength with time.*

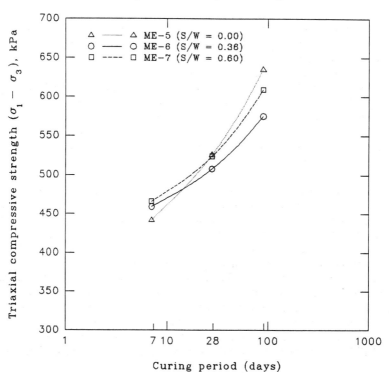

FIG. 6—*Increase of the triaxial compression strength with time (C/W = 0.32).*

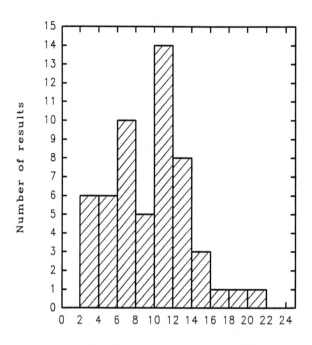

FIG. 7—*Histogram of sand content.*

Permeability tests were done by the contractor on samples molded from slurry collected at the exit of the mixer (slurry CBM), or directly in the trench (slurry CBT). These tests were carried out on the molded samples after a 90-day curing period. The hydraulic conductivity was measured at levels ranging between 0.3×10^{-8} m/s and 1.8×10^{-8} m/s. The CBM-3 and CBT-1 samples with the lowest water contents have higher values of hydraulic conductivity. The tests done by SEBJ on samples molded from slurry collected at the discharge of the pipe indicate a measured hydraulic conductivity ranging between 0.4×10^{-8} m/s and 2.0×10^{-8} m/s. It is important to note that tests CB-08 and CB-19 were done on molded samples having the highest and the lowest density, respectively. The coefficient of permeability of the hydraulic conductivity is at the most 10^{-8} m/s after a 90-day curing period.

The unconfined compression tests performed by SEBJ laboratory were done after 3, 7, 28, and 91 day curing periods on samples molded from slurry collected at the discharge of the pipe. Figures 8 and 9 show all the results obtained. For identical curing periods, the unconfined compression strength is approximately a function of the inverse of the water content (Fig. 10). This relation also exists with the initial density of the slurry. The unconfined compression strength is approximately a direct function of the initial density of the slurry. In all practical cases, these two factors (density and water content) are for a particular slurry function of the C/W ratio. This explains the variations observed on the slurry sampled at the discharge of the pipe. It should be noted that the cement-bentonite in the trench is more homogeneous because it is continuously mixed by the equipment.

The triaxial CIPD tests have been carried out by the contractor on slurry sampled from the plant (CBM-4) or in the trench (CBT-3) and by the SEBJ at discharge of the slurry pipe

TABLE 4—Control of the cement-bentonite slurry in its solid state (contractor).

Sample No.	Permeability Tests				Triaxial CIPD Tests					
	Time, days	Initial Density	w_i, %	k, m/s	Time, days	Initial Density	w_i, %	$(\sigma_1 - \sigma_3)_f$, kPa	ε_f, %	Cracking Development, %
CBM-1	103	1.255	199	3.6×10^{-9}						
CBM-2	100	1.255	198	3.8×10^{-9}						
CBM-3	94	1.277	150	1.8×10^{-8}						
CBM-4	90	1.237	209	3.0×10^{-9}	91	1.237	208	675	6.2	6.6
CBT-1	100	1.288	157	5.8×10^{-9}						
CBT-2	94	1.277	179	4.2×10^{-9}						
CBT-3	90	1.300	162	3.3×10^{-9}	91	1.300	162	630	6.6	8.4

TABLE 5—*Control of the cement-bentonite slurry (SEBJ's laboratory).*

Sample No.	Time, days	Density	Viscosity Marsh Cone (s)	Permeability		Unconfined Compression[a]				Triaxial CIPD		
				w_i, %	k, m/s	w_i, %	q_u, kPa	ε_f, %	E, MPa	w_i, %	$(\sigma_1 - \sigma_3)_f$, kPa	ε_f, %	Cracking Development
CB-08	0	1.261	50										
	3					218	182	0.9	26.4				
	7					232	235	1.6	28.8				
	28	1.247		209	1.4×10^{-8}		541	1.2	80.0				
	91	1.200	41	209	5.8×10^{-9}	227	665	1.0	110.4				
CB-11	0												
	3					223	164	1.2	21.1				
	7					212	275	1.0	50.1				
	28	1.250		195	1.0×10^{-8}	201	488	0.8	76.7				
	91	1.145	46	198	1.0×10^{-7}	214	546	0.9	96.9				
CB-19	0												
	3					290	75	1.4	12.0				
	7					296	133	1.3	18.9				
	28	1.198		266	1.2×10^{-8}	289	269	1.0	39.0				
	91	1.254	48	274	2.0×10^{-8}	295	342	0.9	56.4				
CB-28	0												
	3					228	179	1.4	30.5				
	7					226	280	1.2	52.7				
	28	1.241				223	580	0.9	92.9				
	91	1.198	48	208	1.1×10^{-8}	227	673	0.7	142.6	212	650	5.5	not observed[b]
CB-31	0												
	3					242	120	1.2	17.1				
	7					235	223	1.2	22.9				
	28	1.238				233	572	0.8	103.9				
	91	1.208	48	216	4.1×10^{-9}	231	648	1.1	81.9	217	694	5.7	not observed[b]
CB-38	0												
	3					232	171	1.2	39.1				
	7					220	269	1.4	26.2				
	28					225	518	1.3	63.8				
	91	1.245		199	5.4×10^{-9}	226	691	1.0	111.4	207	689	5.7	not observed[b]

[a] Tests stopped at around 6% of strain.
[b] Mean of three results.

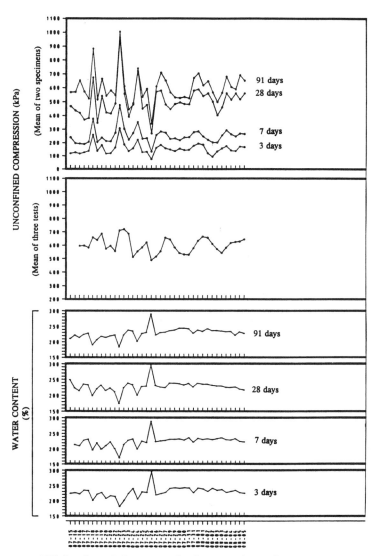

FIG. 8—*Control chart: unconfined compression and water content.*

(CB). All these were carried out after a curing period of 90 or 91 days. In two tests (CBM-4 and CBT-3) carried out by the contractor, the axial deformations corresponding to the maximum strengths were higher than 6%. The three tests (CB-28, CB-31, and CB-38) carried out by SEBJ have been stopped at 6% axial deformation. The samples were dismantled from the triaxial cell to inspect the specimens for fissures or microfissures. A summary of these results is given in Tables 4 and 5. Figure 11 shows typical stress-strain curves.

Conclusion

The cement-bentonite mixes behave like deformable and plastic materials. In this project, the requirement of no fissures after 6% deformation in a sample subject to a triaxial stress

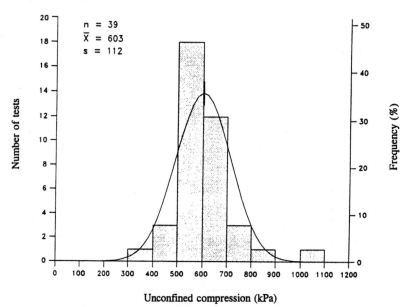

FIG. 9—*Histogram of the unconfined compression tests at 91 days.*

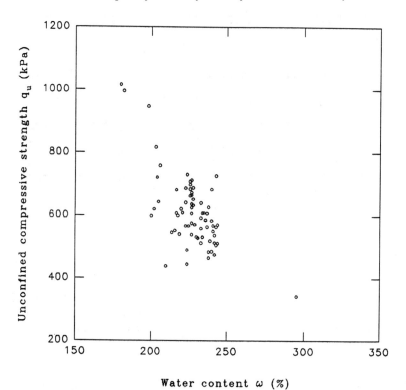

FIG. 10—*Variation of the unconfined compression with water content (91 days).*

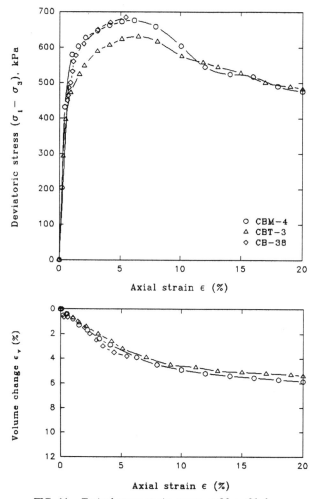

FIG. 11—*Typical stress-strain curves at 90 or 91 days.*

under a lateral stress of 100 kPa at a curing time of 90 days was met in all cases. The measured coefficient of permeability obtained in all tests is at the most 10^{-8} m/s. To decrease hydraulic conductivity to 10^9 m/s, it would be required to increase the bentonite content by 5 or 6% (Gill and Christopher 1985).

From the test results, the following remarks can be made:

- The mix proportions described in the literature usually give an adequate performance.
- Verification tests can be made based on existing results without having to initiate a complete mix design.
- Verification tests should be performed to evaluate the mechanical behavior of mix material with addition of sand, which could contaminate the slurry.
- The unconfined compression tests are relevant control tests that reflect not only the strength of the mix, but the quality of its production.

- To further decrease the hydraulic conductivity, tests tend to show that the bentonite content would have to be increased.
- The determination of the density appears to be the most useful control test during production.

Acknowledgments

The authors wish to thank the SEBJ for authorizing the publication of these data. The design of the slurry trench was achieved through the collaboration of the SEBJ engineering specialists, the board of engineering experts of SEBJ, and the consulting engineers of TECSULT Inc. The construction of the barrier was done by VIBEC who subcontracted the works to INQUIP-SOLCAN. This subcontractor used the laboratory facilities of Woodward Clyde Consultants. Laboratoire de Béton was retained by SEBJ to carry out the control testing.

References

American Petroleum Institute, 1988, "API Specification for Oil-well Drilling Fluid Materials," *API Spec. 13A,* 12th ed., API, Washington, DC.

American Petroleum Institute, 1988, "API Recommended Practice Standard Procedure for Field Testing Drilling Fluids," *API Spec. 13B,* 12th ed., API, Washington, DC, 47 pp.

Bishop, A. W. and Henkel, D. J., 1962, *The Measurement of Soil Properties in the Triaxial Test,* 2nd ed., Edward Arnold and Co., London.

Caron, C., 1973, "Un nouveau style de perforation, la boue auto-durcissable," (A New Type of Deep Cut-off-Plastic Cement), *Annales de l'Institut de bâtiment et des travaux publics,* Paris, No. 311, pp. 2–40.

Chapuis, R. P., Paré, J. J., and Loiselle, A. A., 1984, "Laboratory Test Results on Self-Hardening Grouts for Flexible Cutoffs," *Canadian Geotechnical Journal,* Vol. 21, pp. 185–191.

Gill, G. A. and Christopher, B. R., 1985, "Laboratory Testing of Cement-Bentonite Mix for a Proposed Plastic Diaphragm Wall for Complexe La Grande Reservoir Caniapiscau, James Bay, Canada," *Hydraulic Barriers in Soil and Rock, ASTM STP 874,* American Society for Testing and Materials, Philadelphia, pp. 75–92.

Jefferies, S. A., 1992, "Contaminant-Grout Interaction," *Grouting Soil Improvement and Geosynthetics,* Vol. 2. *ASCE Geotechnical Special Publication No. 30.* Vol. 2, New Orleans, pp. 1393–1402.

Larsen, W. R., Hopton, N. R., Ramirez, F. R., and Varde, O., 1991, "Yacyreta Cement-Bentonite Slurry Cutoff," *International Commission on Large Dams,* 17th Congress, Q-66, R. 80, Vienna, pp. 1491–1518.

Manassero, M. and Viola, C., 1992, "Innovative Aspects of Leachate Containment with Composite Slurry Walls: A Case History," *Slurry Walls: Design, Construction and Quality Control, ASTM STP 1129,* D. B. Paul, R. R. Davidson, and N. J. Caralli, Eds., American Society for Testing and Materials, Philadelphia, pp. 181–193.

Millet, R. A. and Perez, J. Y., 1981, "Current USA Practice: Slurry Wall Specifications," *Journal of the Geotechnical Division,* Vol. 107, No. GT8, ASCE.

Ryan, C. R., 1985, "Slurry Cutoff Walls: Applications in the Control of Hazardous Waste," *Hydraulic Barriers in Soil and Rock, ASTM STP 874,* A. I. Johnson, R. K. Friskel, N. J. Caralli, and C. B. Patterson, Eds., American Society for Testing and Materials, Philadelphia, pp. 9–23.

Ryan, C. R., 1987, "Vertical Barriers in Soil for Pollution Containment," *Conference on Geotechnical Practice for Waste Disposal 1987, ASCE Special Publication No. 13,* R. D. Woods, Ed., June, pp. 15–17, 182–204.

Tornaghi, E., 1972, "Parois souples moulées dans le sol avec un coulis de perforation bentonite-ciment," (Plastic Slurry Trenches in Soils Using Cement Bentonite), *5th European Conference on Soil Mechanics and Foundation Engineering.* Madrid, pp. 577–583.

Reto Hollenweger[1] and Lisa Martinenghi[1]

Design and Performance Verification of a Soil-Bentonite Slurry Wall for the Hydraulic Isolation of Contaminated Sites

REFERENCE: Hollenweger, R. and Martinenghi, L., **"Design and Performance Verification of a Soil-Bentonite Slurry Wall for the Hydraulic Isolation of Contaminated Sites,"** *Dredging, Remediation, and Containment of Contaminated Sediments, ASTM STP 1293,* K. R. Demars, G. N. Richardson, R. N. Yong, and R. C. Chaney, Eds., American Society for Testing and Materials, Philadelphia, 1995, pp. 271–286.

ABSTRACT: The polluted ground of contaminated sites is often encapsulated by diaphragm walls, with concrete or concrete-stabilized earth being common construction materials in Europe. At the Institute of Geotechnical Engineering at the Swiss Federal Institute of Technology Zurich, new material mixtures that contain a high percentage of clay minerals have been developed over the past few years. Two experimental diaphragm walls, one in Switzerland and one in Germany, were constructed using such materials. A diaphragm wall 55 m deep using this mixture is currently being placed to encapsulate a contaminated landfill site in Germany. In this paper, an evaluation of the new mixture and results from both the experimental and full-scale walls, as well as complementary laboratory investigations, are presented and examined.

KEYWORDS: slurry walls, diaphragm walls, site encapsulation, diffusion, permeability, *in situ* tests, wall construction technique

The Malsch landfill is located about 20 km south of Heidelberg, Germany. Between 1971 and 1984, approximately 700 000 tons (7×10^8 kg) of hazardous waste were placed in a disused clay pit. The exact composition of the waste placed before 1976 is completely unknown. The general landfill layout is presented in Fig. 1. Due to its unfavorable inclined position, the landfill was subject to East-West ground water flow. Pertinent landfill information is summarized in Table 1.

Ground water monitoring was carried out simultaneously with waste depositing, but it was in 1978 that increased chloride was first measured in the ground water. The contaminant concentration sharply increased with time and in 1983 contaminant values comparable to those in the landfill were measured at a distance of more than 200 m downstream from the fill.

Extensive site investigations, necessitated by the extremely complex geological and hydrological situations, were initiated at that time and continued until 1990. These studies yielded the following information:

- The more or less dense landfill floor exhibited very few disturbed zones through which leachate could flow out.

---

[1] Institute of Geotechnical Engineering (IGT), Swiss Federal Institute of Technology Zurich (ETHZ), 8093 Zurich, Switzerland.

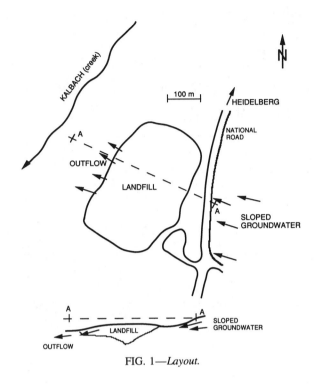

FIG. 1—*Layout.*

- The largest water inflow occurred over the surface.
- Contamination from leachate reached more than 200 m in the western direction.

A surface cap and deep drains were installed between 1987 and 1990 as a first remediation measure (Fig. 2). These construction measures led to a reduction from 25 m^3/day to 6 m^3/day of collected effluent.

The planning and construction of a vertical diaphragm wall was begun in 1990 on the basis of the results of the extensive site investigations. The wall was to enclose the entire

TABLE 1—*Landfill data.*

Landfill Management Company	Sondermüllbetriebsgesellschaft mbH, Fellbach-Schmiden
Site	Former Reimschloch clay pit
Landfill Type	Pit construction with surelevation
Total Surface Area	Approx. 70 000 m^2
Filled Area	Approx. 50 000 m^2
Service Period	1971-1984
Types of Waste Deposited	Salt slag, heavy metal hydroxides, contaminated soils, industrial waste including chlorinated compounds
Deposited Quantities	Approx. 700 000 tons ≈ 570 000 m^3

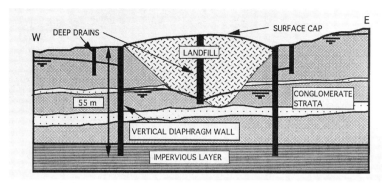

FIG. 2—*Site section with geological data and remedial constructions.*

landfill, seal the existing conglomerate strata, and key itself into the clay layer found at a depth of up to 55 m (Fig. 2).

Diaphragm Wall Material

For the remediation project of the Malsch landfill, one of the primary requirements for the chosen sealing material was longevity in the presence of high leachate concentrations and under complicated and unfavorable geological conditions.

Additional requirements for the diaphragm wall were:

- The wall material must have small pore volume.
- A large portion of active material should be used (for example, clay) in order to ensure high contaminant adsorption.
- The material should have the highest possible density in order to replace the excavation slurry completely (two-phase construction method).
- Large quantities of material are necessary for the construction of such a wall. This implies material workability over long periods of time.
- Two geological faults are present at the landfill site. This necessitates that the wall should not be too brittle.

Mineral sealing with a high portion of solid particles was chosen, which is only possible for two-phase diaphragm walls. A new wall material was developed and the construction of an experimental wall was supervised in order to check construction using this material for a Swiss landfill site. Here, the material requirements, but not the geological conditions, were very similar to those at the Malsch site (Hermanns 1993). A slightly modified version of this experimental wall material was then used at the Malsch site, where a more flexible wall was required due to the difficult geology. This was obtained through a reduction in cement and an increase in clay content. The composition of the new material, which was developed in the laboratory and is able to be used directly in this form, is presented in Table 2.

Clay

Clay minerals have shown themselves to be very stable materials. The danger that clay minerals be dissolved under "normal" leachate conditions has been demonstrated by Madsen

TABLE 2—*Composition of the wall material.*

Clay (Opalinus Clay or Opalite), kg/m^3	650
Cement (Portland), kg/m^3	150
Electrofilter Ash, kg/m^3	100
Water, kg/m^3	670
Fluidizing Agent (Sikament), kg/m^3	4.5
Water Content, %	74
Wet Density, t/m^3	1.56
Dry Density, t/m^3	0.90
Grain Density, t/m^3	2.72
Porosity, %	67

(1990) to be very improbable. New investigations on different clays have shown that both the adsorption and diffusion behavior in the presence of various organic and inorganic substances may be positively influenced through a judicious choice of clay minerals (Stockmeyer and Kruse 1991; Hermanns 1993).

Cement

The wall material must attain a certain stiffness as rapidly as possible in order to counter any effects of consolidation settlement and "hanging up" of the wall on the surrounding ground. This necessitates the addition of an hydraulic binding agent. High temperature furnace cement (HOZ) with special low hydration temperature (NW) and high sulfate resistance (HS) was used.

EFA-Filler

An attack on the cement structure was expected from the inorganic contents of the Malsch leachate (especially chloride and sulphate). In order to limit the cement content of the wall material, a filler with latent hydraulic properties and pozzolanic activity was used to replace some of the cement. A filter ash resulting from a coal-burning electropower plant was chosen, referred to as EFA-Filler (*e*lectro*f*ilter *a*sh).

Fluidizing Agent

The construction of a two-phase vertical diaphragm wall necessitates a wall material with a relatively high density in order to replace the excavation slurry completely. A fluidizing agent must be mixed in with the other components so that such a viscous wall material remains workable. The product Sikament 10 (Sika 1991) permitted better cement dispersion, and resulted in a fluid material in the first phase, without decrease in final density and resistance.

Water

Normal tap water, both in the laboratory and on site, was used in this mixture.

Laboratory Investigations

The new wall material was prepared and thoroughly tested in the laboratory. The investigations included the following tests:

- permeability tests with tap water as well as with synthesized Malsch leachate, called M 100% (tap water + 120 g/L chloride + 4 g/L sulfate + 8 g/L ammonium)
- compressive strength tests
- determination of representative pore radius using a mercury porosimeter
- diffusion tests.

Wet and dry densities, void ratio, and water content were also determined.

Sample containers (cylindrical $d/h \approx 10$ cm/12 cm) were filled with the material suspension after the completion of mixing and were kept in water at 18°C until the tests were carried out.

Permeability Values

The permeability to tap water and landfill leachate was tested in a triaxial cell with a gradient of $i \approx 30$. Under these conditions the confining pressure was approximately 0.3 bars higher than the flow pressure. The quantity of water that flowed through the sample was measured. As it is known that leachate in a landfill is constant neither with respect to position nor to time, the synthetic leachate (Leachate M 100%) was used. The target permeability value was $k \leq 5 \times 10^{-10}$ m/s. The permeability test apparatus is shown in Fig. 3.

Compressive Strength

The determination of the unconfined compressive strength of cylindrical samples was carried out according to the German standard DIN 18 136. The sample dimensions were $d/h \approx 10$ cm/10 cm, and the deformation velocity (for unrestrained side deformation) was 1.27 mm/min. After the test the sample was oven dried at 105°C and wet and dry density and water content values were measured.

Determination of Representative Pore Radius

The determination of the representative pore radius was carried out with a mercury porosimeter. The test procedure is described in the literature (Bucher et al. 1982). The mass

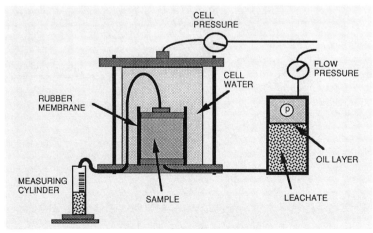

FIG. 3—*Test apparatus for the determination of* k *values.*

276 CONTAMINATED SEDIMENTS

of the mercury that penetrates the probe as a function of the pressure was measured until a pressure value of approximately 4000 bars arrived, which corresponds to a pore radius of ≈2 nm.

Diffusion Tests

Diffusion is a compensation process, in which atoms or molecules move from an area of high concentration to one of low concentration. Diffusion can be a significant transport mechanism for pollutants in barrier materials with low hydraulic permeability (Madsen and Kahr 1993). At the present time, no standard test method exists for the determination of diffusion constants for two-phase diaphragm walls, and it is not possible to determine a diffusion profile (for pure diffusion) over a specific sample height. In this case, the sample preparation was long and arduous because of the low strength of the wall material (it was not possible to divide the samples into lengths of less than a centimetre) and the lack of organic substances in the material. This was due on the one hand to the volatility of the substances used, and on the other hand to the low concentrations in the test solution and the test procedure, which is mainly suitable for softer materials. A special test method was developed in which the samples do not need to be cut into slices (Hermanns 1993).

The complete test apparatus, shown in Fig. 4, consists of impermeable glass components that permit essentially no diffusion. The fresh wall material suspension is poured into the inner glass tube to a height of approximately 2 cm. The sample is covered with water and left to harden for 28 days, the inner tube being temporarily closed at the bottom with tape during the setting period so that the suspension remains inside. This 2 cm thick layer then divides the liquid of the inner glass tube from that of the outer glass tube. At the beginning of the test, the water in the outer tube is replaced with fresh distilled water and the inner one is filled with chlorophenol. It is necessary to make sure that both liquid levels have the same height in order to avoid hydraulic gradient, which would affect pure diffusion. Both solutions are then changed at regular intervals and the chlorophenol concentration in the distilled water is measured. The breakthrough time for a certain concentration and barrier material may be extrapolated with the aid of a nonstationary diffusion model. A breakthrough

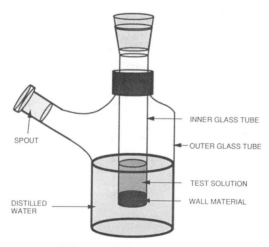

FIG. 4—*Diffusion test apparatus.*

time of approximately 450 years was determined for the prepared laboratory samples (Hermanns 1993).

Mineralogical Composition

The mineralogical composition of the opalinus clay was investigated by X-ray diffraction (Madsen and Nüesch 1990). Opalinus clay consists mainly of the clay minerals illite, mixed-layer illite, smectite, kaolinite, and chlorite (total amount 50 to 60%), quartz, and carbonates. The specific surface area of opalinus clay is about 70 m^2/g, and the cation exchange capacity is about 10 meq/100 g.

Experimental Wall at the Malsch Landfill Site

Even though a certain experience with the slightly different material used for the experimental wall at the Swiss site had been acquired, it was decided to construct a new experimental wall at the Malsch site in order to ascertain the effects of the local geology and the modified material parameters (lower density and compressive strength). Such a test was designed to clarify the following points:

- mixability and pumpability of the material
- completeness of the replacement of the excavation slurry with the wall material
- construction of the wall with a view to setting behavior (crack development)
- formation of joints
- excavation verticality
- excavation method (hydraulic grab or drilling head)
- test of a new contractor's installation method using a "harp" filling device (fully described later in text)
- use of laboratory results during wall construction
- *in situ* permeability.

In September/October 1992 the experimental wall was constructed to a depth of ≈30 m. Problems were encountered during the excavation of soil panels using an hydraulic grab. Neither the necessary construction output nor verticality requirements (deviation from plumb ≤0.5% of the wall depth) could be attained. A drilling head was then used for the excavation, which provided much better results. In addition to laboratory investigations carried out on samples from the experimental wall, the following *in situ* tests were made:

- axial strain in the wall (sliding micrometer measurements)
- setting behavior (pore water pressure and temperature measurements)
- verticality measurements of the excavated trench (sonar measurements)
- permeability tests in boreholes (double packer).

Axial Strain Measurements

The complete distribution of axial strain (shortening and elongation) along a particular measuring line may be determined with sliding micrometres. Sliding micrometer measurements are carried out using a portable probe that measures the change in length between two markers along a measuring line. The precision of a single measurement is $\pm 3 \times 10^{-6}$ m. Material deformation behavior during setting is of great importance for the suitability of a diaphragm wall. The deformation behavior of the test wall was evaluated in two panels, in which sliding micrometer tubes were placed with measuring markers every metre over the entire depth. The construction depth was approximately 30 m with 28 measuring positions. Initial measurements were carried out about 1 h after placement of the wall material.

Figure 5 presents measurements carried out at both 9 and 20 days after the initial measurements in one panel. An overall settlement of approximately 43 mm may be seen after 9 days, which can be explained by water transmission into the ground during wall consolidation. After 20 days, additional uniformly distributed settlement values of 1.2 mm were measured; the first measurements had not been uniform with depth.

Temperature Measurements

The temperature development in the diaphragm wall gives information concerning the setting behavior of the wall material. Heat is given off during the setting process, which diffuses in the surroundings. At the end of the setting process, the temperatures asymptotically approach those of the soil. The temperature development of the wall material was measured continuously in two panels of the test wall with electric temperature sensors of the PT100 type built into the wall (Solexperts 1983). The measurements were recorded using a SOLEXPERTS GEO monitor (Solexperts 1983). Temperature measurements as a function of time are given in Fig. 6.

The maximum setting temperatures (25 to 30°C) were attained during the first 50 h, followed by a rapid, continuous temperature decrease.

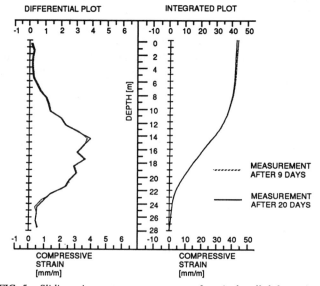

FIG. 5—*Sliding micrometer measurements of vertical wall deformation.*

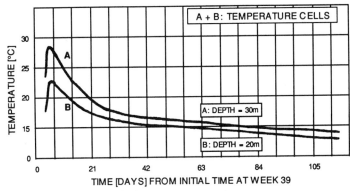

FIG. 6—*Temperature evolution.*

Pore Water Pressure Measurements

Pore water pressures were monitored using electric pore water pressure cells. Measurements carried out before the replacement of the slurry by the wall material permitted an *in situ* determination of the slurry density, and similar measurements carried out during the wall construction permitted an appreciation of the setting speed of the wall material (buildup of grain structure). Pore pressure measurements are presented in Fig. 7.

The pore water pressure measurements showed an increase at the beginning of the installation which mirrored the construction advancement. The measured pressures began to decrease approximately 4 h after the beginning of installation when the cells were immersed in the slurry. This shows that the formation of the grain structure, and therefore the development of strength, took place very quickly, and that the overburden pressure of the fluid was taken up by this grain structure. The pore water pressure decreased after approximately 12 to 18 h to the pressure level of the artesian ground water.

Sonar Measurements

It was possible to measure the vertical deviation of the excavation from the desired position using sonar measurements in the slurry. An oriented ultrasonic sensor was lowered with

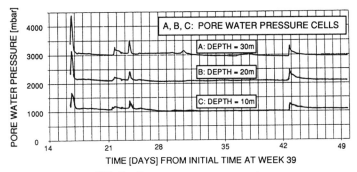

FIG. 7—*Pore pressure measurements.*

constant velocity into the section, thus enabling the distance from the sensor to both opposite wall limits to be known. The measurements were continuously recorded on a paper band. The results of this procedure at the panel ends permitted the determination of the overlap between two panels.

It may be seen in Fig. 8 that the necessary verticality was not achieved using the hydraulic grab, but rather with the drilling head.

In Situ Permeability

Full-core boreholes with the possibility for sample extraction were carried out in order to permit permeability tests of the wall *in situ* as well as of the resulting samples. However, some problems were encountered during this procedure.

For example, it was not possible to measure permeability values with the double packer, as leaks around the packers could not be avoided despite an increase in packer pressure. Pumping and recharge tests were then carried out, with poor results, followed by fluid logging investigations. The fluid logging results revealed water flow between the boreholes at several levels, as well as hydraulic flow paths under different boreholes. A direction-oriented camera inspection in several of the boreholes was ordered when this behavior could not be explained. Marked horizontal cracking was discovered which, however, did not correspond with the sliding micrometer measurements. The experimental wall at Malsch differed from the Swiss experimental wall in that it was more flexible and had less strength. It is assumed that the wall was damaged during the boring or the packer tests. The leaks during the packer tests and the poor results from the pumping and recharge tests also help to explain these cracks. The wall was later excavated over a depth of 6 m and a width of 2.5 m at two different locations in order to confirm this hypothesis. One location was around two boreholes, and the other in an undisturbed section of the wall. These excavations and subsequent

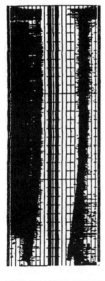
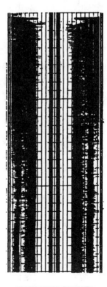

EXCAVATED PANEL   DRILLED PANEL

FIG. 8—*Sonar measurements (SOLEXPERTS)*.

visual examinations revealed that the wall was damaged and cracked near the boreholes and completely crack-free at the undisturbed location. One crack ran along a borehole, and others could be traced to it. With respect to the Swiss experimental wall, the strength reduction of the Malsch experimental wall was such that a limit had been reached after which full-core boreholes and packer tests could no longer be carried out without damaging the wall.

Conclusions from the Construction of the Experimental Wall at the Malsch Landfill Site

Tests on the full-core samples, as well as those from filled samples from the mixing plant, showed basically positive results. Complete replacement of the excavation slurry with the wall material, homogeneity of the wall material, as well as sufficient density at panel interfaces, were also able to be confirmed.

The construction of the experimental wall showed that the new wall material could be mixed, pumped and worked with on site. The low density of the mixture due to the addition of clay material necessitated precise examination of the installation conditions regarding replacement of the bentonite slurry. Control boreholes in the experimental wall were no longer feasible, as explained above. If a wall section must be rebored, as is the case during complete filling, then overdrilling and refilling of the section should be carried out.

Construction of a Vertical Diaphragm Wall at the Malsch Landfill Site

Large primary panels were constructed in order to limit the number of overlaps and permit the largest possible construction output. These panels were 9.3 m wide and up to 55 m deep. The secondary panels were approximately 2 m wide. The thickness of the wall was kept constant at 0.8 m, which corresponded to a filling volume of almost 410 m^3 for the largest panel. In order to guarantee the complete filling of such a wide panel, it was necessary to develop a new system. This system was composed of a filling "harp" installed in common filling tubes. The harp itself expands over the entire excavation section and has up to five openings distributed over the excavation width (additional tubes may be added according to the contractor's experience). The openings may be pneumatically opened one at a time and guarantee uniform filling. During the filling, it must be made sure that the harp, whose upper edge is located 1.5 m above the filling openings, is submerged at least 1.5 m in the wall material. This system proved itself to be highly efficient during the construction project.

Since boreholes in the wall could no longer be used for quality control based on the reasons stated above, only filled samples of fresh wall material could be tested. These only provide information concerning the quality of the filled mass, not concerning if the filling was correctly carried out. Therefore, a precise quality control program was followed during the wall construction in order to minimize the risk of a poorly constructed wall. Uptake of the excavation slurry permitted a degree of filling of less than unity only to be detected (degree of filling = the ratio of installed to theoretical volume). As the density difference between the excavation slurry and the wall material increased, the risk of poor material replacement decreased. Therefore, the excavation slurry (IBECO-HT-X-Bentonit) must be desanded before filling until its density is inferior to the limit value of 1.2 t/m^3. A lower limit for the slurry density was fixed at a value of 1.15 t/m^3, with a permitted variation of ± 0.05 t/m^3, in order to ensure adequate excavation stability. The excavation slurry was part of a large closed cycle (storage basin-panel-desanding machine-storage basin) and was recirculated and reused.

The mixing process took place in two permanently installed, identically built, computer-controlled mixing units. The raw materials were delivered by tank trucks in powdered form and pumped directly into the upright storage silos. The powder was then taken from the silos to the mixer by spiral conveyor. From there the powder was poured into the mixer and was weighed by pressure measuring cells. Mixing was carried out by two circulating pumps and one forced flow mixer. After the addition of all of the components, the wall material was pumped through a pipe line into a storage tank where it was kept in continual movement. The storage basin was an open metal reservoir fitted with an agitating mechanism. The wall material was then pumped from the storage basin to the filling apparatus. This apparatus had a drum, around which the connecting tubes of the harp were coiled. The harp was continuously pulled upwards during the filling operation.

The separate components of the wall material used were, in order: water, opalinus clay, filler, cement, and Sikament. The weighed quantities were computer recorded, and these data were used for the calculation of the degree of filling. The minimum mixing time amounted to 1 min, as Sikament is only fully mixed after 20 to 30 s. The material was assumed to be homogeneous and lump-free after the mixing procedure.

The various activities of the following quality control program (QCP) which concerned the material, the geometry, and the filling of the diaphragm wall were scrupulously carried out. The in-house monitoring was carried out by the firm Hochtief AG, and the external monitoring by the Swiss Federal Institute of Technology Zurich (supervisor of the external monitoring) in collaboration with the engineering consultants GBA in Darmstadt, Mühltal. The geotechnical boundary conditions and the stability of the diaphragm wall were the responsibility of a third party, and were not included in the QCP.

Quality Control Program (QCP)

Input Checks:
Cement

- taking of reserve samples
- grain size tests on random samples in a cement laboratory

EFA-Ash

- taking of reserve samples
- water content and uptake tests using the Enslin-Neff method

Opalinus Clay

- taking of reserve samples
- water content and uptake tests
- check of mineralogical composition by X-ray diffraction

Excavation Slurry:

- check of mineralogical composition
- slurry density before replacement by wall material (≤ 1.2 t/m^3)

Diaphragm Wall Material:

- verification of density 1.56 ± 0.02 g/cm^3
- checking of composition and processing of the components
- determination of unconfined axial strength of cylindrical samples

- determination of water permeability value k
- determination of representative pore radius
- determination of diffusion behavior

Diaphragm Wall Geometry:

- checking of the excavation depth
- verification of excavation width ≥ 0.8 m
- verification of overlap of neighboring panel ≥ 0.08 m
- verification of minimum width of the wall at the foot in the joint region ≥ 0.8 m
- verification of deviation of two vertical panel axes in the foot region ≤ 0.4 m
- verification of total deviation from vertical $\leq 0.5\%$ of the wall depth measured by inclinometer
- the depth of the wall was determined according to the information received during the wall construction to confirm that the waterbearing conglomerate layers were fully sealed
- the verticality of the panels was measured by a computer-controlled inclinometer attached to the drilling head and all information was computer recorded. The inclinometer precision was checked weekly by angle and distance measurements using theodolite.

Degree of Filling of the Panels:

- the degree of filling was calculated for each panel
- measurement of the filled volumes using two methods: (1) mass balancing during mixing (scales), (2) Inductive flow meter (IDM).

Problems Encountered during the Handling of the Wall Material

Problems in handling the fresh wall material were encountered at the site during construction in the summer months. The material exhibited altered flow behavior, even though the mixture and the production procedure had not been modified in any way. This led, in extreme cases, to a clogging of the mixers. The first hypothesis was that this was brought about by the summer temperatures; however, detailed investigations showed no correlation between the material behavior and the outside temperatures. It was shown that different samples of freshly delivered opalinus clay differed both in their water uptake velocity and water uptake capacity after 30 min (Enslin-Neff test) (Neff 1988). It was surprising that samples with lower clay content could have a higher water uptake capacity. This was discovered to be caused by a fabric effect. Clay fabric may be damaged after ultrasonic treatment of the suspension for 6 min at a frequency of 20 kHz. The samples that were so treated, dried, and then ground for 10 min with a microgrinder showed that, as expected, the water uptake capacity depends on clay content. However, the rate of the water uptake is strongly influenced by the clay fabric (Nüesch 1994). This showed that such complex material questions cannot be adequately answered without the aid of complementary laboratory investigations.

In winter, problems were encountered with the material temperature during filling. Temperatures lower than 10°C led to a waiting period longer than the minimum 5 days prescribed by the QCP for the overdrilling of one panel during the excavation of the next. It was decided to heat the mixing water, and thus, it was possible to maintain minimal material temperatures of 14°C throughout the winter.

Comparisons of Results from Laboratory and Malsch Site Samples

Samples (cylindrical d/h 10 cm/12 cm) for the testing of the panels were taken from the experimental wall, as well as during the construction of the full-scale wall at the tube drum of the harp, and were kept in water at 18°C until the tests were carried out. The prepared laboratory samples had been stored under identical conditions.

Permeability Tests

Results from different test samples are represented in Fig. 9 for prepared laboratory samples, fresh and full-core samples from the experimental wall and fresh samples from the full-scale wall taken during construction. It should be stated that these were arbitrarily taken samples; thus, a certain deviation in the test results could occur. However, it may be stated overall that the results from all of the samples are comparable. The permeability values of the core samples from the experimental wall were approximately one order of magnitude larger than those of the fresh material samples from the full-scale wall, as was expected. All of the tests showed a decrease in permeability with time. For tests of longer duration, these results would be expected to approach a limiting value asymptotically.

Unconfined Compression Tests

Large differences in unconfined compressive strength for the different samples may be seen in Table 3.

The strength values of the prepared laboratory samples were not reached by the samples from the experimental wall or from the full-scale wall. In accordance with the QCP the limit value at 28 days was 1000 kN/m^2, which was, however, reached by all of the samples.

Pore Radius Distribution

A difference in the results from the various test series was also seen in pore radius distribution (Table 3), but the results confirmed the overall tendency of the other laboratory investigations. The favorable values measured for the prepared laboratory samples were not attained by the other samples. However, the results from the full-scale wall were more favorable than those from the experimental wall.

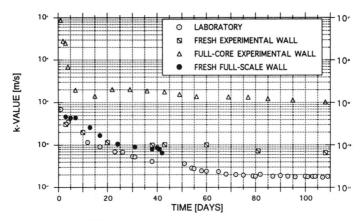

FIG. 9—*Evolution of permeability k-values.*

TABLE 3—*Average compressive strength and representative pore radius measurements.*

Sample Description	Avg. Compressive Strength at 28 days, kN/m^2	Avg. Representative Pore Radius, nm
Laboratory	2000	280
Fresh Experimental Wall	1100	440
Fresh Full-scale Wall	1300	350

Conclusions

The results of the investigations on the wall material originally developed in the laboratory may be favorably compared to those of the experimental wall at the Malsch site. The handling of the material is a difficult and highly delicate operation, in which it is necessary to tread a fine path between conflicting requirements:

- high clay content for improved wall properties (e.g., increased sealing capacity)
- high wall material density to ensure complete replacement of the excavation slurry with the wall material
- high cement content for more rapid wall strength development and lower consolidation sensitivity
- low wall material density to ensure better workability with the fresh material
- high wall material strength to permit testing
- low wall material strength due to low ground strength.

It was seen that a lower strength limit had been reached in the experimental wall, and thus tests in the wall itself could not be carried out. Tests were limited to checking the components and the behavior of the wall material during installation. Because of this, construction and monitoring had to be carried out with extreme precision.

Continuous filling of the panels must be able to be guaranteed, since the fresh wall material remains workable for only about 2 h after it is mixed.

The 32 000 m^2 full-scale diaphragm wall at the Malsch site appears to have been carefully constructed according to the specifications; however, it will be possible to judge the quality of the landfill encapsulation operation fully only with the passage of time.

Acknowledgments

The authors would like to express sincere thanks to Dr. F. T. Madsen and the IGT-Claylab, as well as to Hochtief Contractors for their invaluable collaboration. The review of this paper by Professors H.-J. Lang and P. Amann was also greatly appreciated.

References

Bucher, F., Jeger, P., Kahr, G., and Lehner, J., 1982, Technischer Bericht 82-05, NAGRA, IGT-ETHZ.
Documentation Sika, 1991, "Sikament 10," Sika Chemie.
Documentation Solexperts AG, 1983, "Temperature Sensor PT 100," "GEO Monitor," Solexperts.
Hermanns, R., 1993, "Sicherung von Altlasten mit vertikalen mineralischen Barrierensystemen im Zweiphasen-Schlitzwandverfahren," Doctoral dissertation, Institut für Geotechnik, ETHZ.
Madsen, F. T., 1990, "Chemical Effects on the Hydraulic Conductivity of Waste Containment Barriers," Contaminated Soil '90, Third International KfK/TNO Conference on Contaminated Soil, Karlsruhe.

Madsen, F. T. and Kahr, G., 1993, "Diffusion of Ions in Compacted Bentonite," *Proceedings of the 1993 International Congress on Nuclear Waste Management and Environmental Remediation.*

Madsen, F. T. and Nüesch, R., 1990, "Das Langzeitquellverhalten von Ton- und Anhydritgesteinen" (The Long-Term Swelling Behavior of Clay and Anhydrite Rock), *Beiträge zur Geologie der Schweiz, Geotechnische Serie.*

Neff, H. K., 1988, Der Wasseraufnahme-Versuch in der boden-physikalischen Prüfung und geotechnische Erfahrungswerte, Bautechnik 65, Heft 5.

Nüesch, R., 1994, DTTG Regensburg (D), Beitrag zur Jahrestagung, Vol. 10.

Stockmeyer, M. and Kruse, K., 1991, "Adsorption of Zn and Ni Ions and Phenol and Diethylketones by Bentonites of Different Organophilicites," *Clay Minerals,* Vol. 26.

Management Strategies

Michael R. Palermo[1] *and Jan Miller*[2]

Strategies for Management of Contaminated Sediments

REFERENCE: Palermo, M. R. and Miller, J., **"Strategies for Management of Contaminated Sediments,"** *Dredging, Remediation, and Containment of Contaminated Sediments, ASTM STP 1293,* K. R. Demars, G. N. Richardson, R. N. Yong, and R. C. Chaney, Eds., American Society for Testing and Materials, Philadelphia, 1995, pp. 289–296.

ABSTRACT: In 1985, the U.S. Army Corps of Engineers (USACE) developed a Management Strategy for evaluation of dredged material alternatives which focused on contaminant testing and controls. The Environmental Protection Agency (EPA) later initiated development of a similar management strategy focusing on environmental considerations of disposal alternatives. A USACE/EPA work group was subsequently formed for the purpose of developing a joint Technical Framework endorsed by both agencies. In 1992, the USACE and EPA published the Technical Framework as a guidance document for their agencies' personnel to follow in identifying environmentally acceptable alternatives for the management of dredged material. The USACE and EPA are also cooperating in developing strategies for managing sediment remediation projects. This paper provides a summary of the strategies that can be used in management of contaminated sediments, a description of the more detailed guidance documents available, and plans for future developments.

KEYWORDS: contaminated sediment, dredging, dredged material, remediation, management strategies

Background

In many harbors and waterways, sediments have become contaminated with a variety of chemicals because of municipal and industrial discharges and nonpoint sources. Sediments act as a sink for many contaminants, and sediment contamination of some waterways contributes to environmental degradation and regional environmental problems (National Research Council 1989).

Some of these contaminated sediments fall within waterways and harbors which require dredging to maintain and improve navigable depths. Sediment remediation is also being considered in some of the most seriously contaminated areas. For example, cleanup of contaminated sediments is proposed at several Superfund sites, and, in most cases, the cleanup alternatives under consideration involve dredging for sediment removal prior to treatment or disposal. Dredging for cleanup purposes is increasingly looked upon as a tool in managing contaminated sediments (Engler et al. 1992).

The U.S. Environmental Protection Agency (EPA) has broad missions related to contaminated sediment management under the Superfund and other laws, and EPA is now undertaking a comprehensive effort to coordinate and focus resources on contaminated sediment

---

[1] Research civil engineer, U.S. Army Engineer Waterways Experiment Station, Vicksburg, MS 39180-6199.

[2] Environmental engineer, U.S. Army Engineer North Central Division, Chicago, IL 60606-7205.

problems. The overall strategy is composed of elements related to assessment, prevention, remediation, dredged material management, research, and outreach. As a part of this effort, the EPA will promote an integrated federal agency contaminated sediment strategy.

The U.S. Army Corps of Engineers (USACE) has developed a significant technical expertise in dredging and disposal of contaminated sediments because of its navigation mission. The USACE has also provided support to the EPA and other agencies for sediment management activities and has recently been given limited authority related to "environmental dredging" and broader missions with respect to environmental protection (Palermo et al. 1993).

The extent of the contaminated sediment problem nationwide and the recent initiatives by EPA, USACE, and other agencies require that appropriate technical strategies be developed for managing contaminated sediments. These strategies must consider the range of possible alternatives and the applicable legal and regulatory requirements.

Alternatives for Sediment Disposal/Remediation

Research efforts and field experience have provided a substantial knowledge base regarding the technical aspects of dredging and management of contaminated sediments (Engler et al. 1990; Averett et al. 1990). However, a clear distinction should be made between management of contaminated dredged material in the context of navigation projects versus the context of sediment remediation or cleanup projects.

In the navigation project context, alternatives will necessarily involve removal of the sediments to restore or improve navigable depths within the channels or harbors. The principal thrust is therefore evaluation and selection of dredging and disposal alternatives that are environmentally acceptable. Alternatives could include capping contaminated dredged material at open water disposal sites, placement of contaminated dredged material in confined (diked) disposal areas, and consideration of various contaminant control measures or treatment options.

In the sediment remediation or cleanup context, options that should be considered for managing contaminated sediments include no-action, nonremoval, and removal. No-action involves simply allowing natural processes to gradually improve conditions. Nonremoval options are those that involve restricted use of a contaminated area or treatment or isolation of the contaminated sediments in-place. Removal options are those that involve dredging followed by treatment or disposal of the sediments at another location.

There is a large number of options for dredging, placement of sediments, treatment processes, and control measures that can be used in combination. The resulting range of possible alternatives or scenarios is therefore potentially large. For this reason, a strategy for screening and detailed evaluation of alternatives from a technical standpoint is required. Any strategy must be technically sound, consistent, and compatible with applicable laws and regulations.

Management Strategies

Decision-making strategies are pathways for approaching a complex issue or problem in a logical order or sequence. A strategy can usually be represented as a flowchart or framework of activities and decisions to be followed by very specific applications. Several strategies have been developed related to management of contaminated sediments for both navigation and remediation. Two widely accepted and applied strategies are the USACE/EPA Technical Framework for dredged material management and the Superfund Remedial Investigation/Feasibility Study (RI/FS) Framework for remediation projects.

USACE/USEPA Technical Framework

The USACE and EPA share the responsibility of regulating dredged material management activities, and there is often disagreement and confusion within USACE Divisions and Districts, EPA Regions, and other agencies over how to manage dredge material and how to evaluate proposed disposal alternatives from the environmental standpoint. In the mid-1980s, the USACE developed a Management Strategy (Francingues et al. 1985) for evaluation of dredged material alternatives that focused on contaminant testing and controls, and this strategy was adopted as USACE policy in its dredging regulations (33 CFR 236.1). However, the strategy did not provide guidance on how to evaluate the full range of dredged material management alternatives that can be considered. To meet this need, the USACE and EPA have jointly developed a consistent technical framework for evaluation of dredged material management alternatives. The document is entitled "Evaluating Environmental Effects of Dredged Material Management Alternatives—A Technical Framework" (USACE/EPA 1992).

The technical framework is designed to meet the procedural and substantive requirements of the major laws and regulations pertaining to dredged material management (NEPA, the Clean Water Act, and the Ocean Dumping Act) in a technically consistent manner and is intended to be applicable to all proposed actions involving the management of dredged material. This includes both new work construction and navigation project maintenance, both clean and contaminated materials, and the broad array of management alternatives. The technical framework gives equal consideration to the three major disposal alternatives: confined (diked intertidal and upland) disposal, open-water (aquatic) disposal, and beneficial use applications. A flowchart illustrating the Technical Framework is shown in Fig. 1.

Superfund RI/FS Framework

The Comprehensive Environmental Response, Compensation, and Liability Act of 1980 (CERCLA) and Superfund Amendments and Reauthorization Act of 1986 (SARA) established and reauthorized the Superfund program. The decision-making framework for Superfund projects is shown on Fig. 2, and described in detail in the EPA document entitled *Guidance for Conducting Remedial Investigations and Feasibility Studies under CERCLA* (1988).

The Superfund decision-making framework has two major components, the remedial investigation (RI) and feasibility study (FS). For a Superfund site with contaminated sediments, the remedial investigation would determine the character of the sediments and the extent of contamination, among other information. The feasibility study would include an evaluation of all reasonable remediation alternatives, including treatment and nontreatment options.

Parallel versus Sequential Strategies

Either of the decision-making strategies discussed above might be applied to a sediment remediation project with equal success. In the Superfund strategy, remedial alternatives are evaluated in a parallel fashion (Fig. 3a); that is, a wide range of possible alternatives are evaluated simultaneously, and then a selection is made among the leading candidates. The USACE/USEPA Technical Framework also uses the parallel approach, but with the provision of initial screening to eliminate unreasonable alternatives. However, in actual application of both of these strategies, the detailed evaluations would only be conducted on a reasonable number of promising alternatives which survive a screening process.

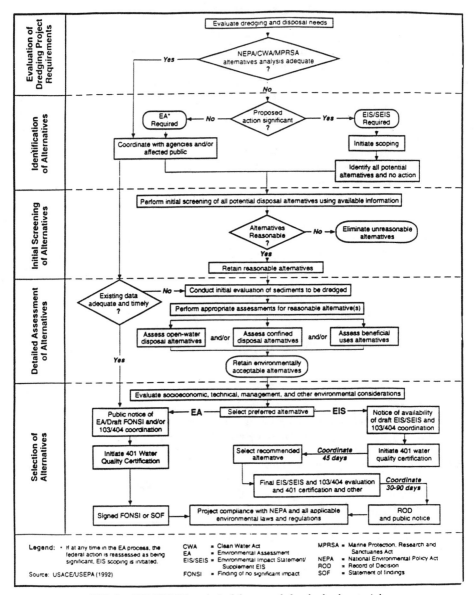

FIG. 1—*USACE/EPA technical framework for dredged material.*

Another possible strategy is a linear or sequential approach to evaluate disposal alternatives (Fig. 3b). With the sequential approach, options are examined in order of increasing complexity. The evaluator leaves the decision path when a suitable alternative is found.

Each of these approaches has its strong and weak points. The advantages of the parallel approach over the sequential approach may be summarized, as follows:

- it has been widely used for RI/FS efforts at Superfund sites and some sites not on the National Priorities List (NPL),

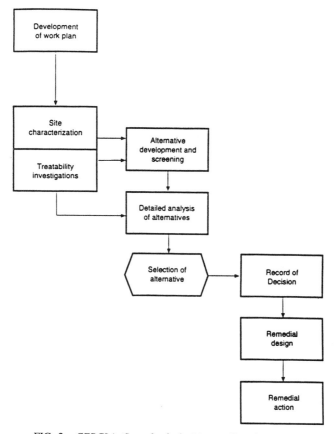

FIG. 2—*CERCLA (Superfund) decision-making framework.*

- it is more familiar to most environmental consultants and regulatory agencies
- it is consistent with the requirements of the National Environmental Policy Act (NEPA), and
- it generally provides principals and stakeholders in remediation projects more than one option for consideration.

The disadvantage of the parallel approach is that the evaluation of numerous alternatives may require significant resources and time.

Projects that are on the National Priorities List (NPL) are required to follow Superfund RI/FS procedures (the parallel approach). However, many (if not most) contaminated sediment sites are not NPL sites. For projects where resources, funding, or time may not allow a detailed evaluation of numerous alternatives, a combined approach may be considered.

A decision-making framework for sediment remediation with a combined approach has been proposed under the EPA Assessment and Remediation of Contaminated Sediments (ARCS) program (EPA 1994) and contains elements of both of the decision-making strategies discussed above (Fig. 4). This framework contains three major activities (boxes) and two decision points (diamonds). The first activity is to define the objectives and scope of the project. The next two activities involve the screening and preliminary design of remediation

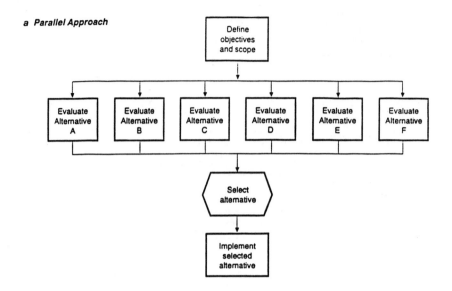

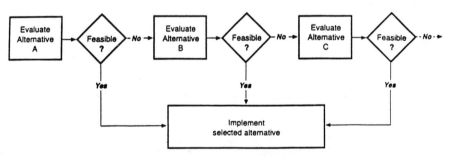

FIG. 3—*(a) Parallel approach; (b) Sequential approach.*

alternatives. The products of these activities are preliminary designs, cost estimates, and estimates of contaminant loss, which are used to determine if there is a feasible alternative that meets the project objectives. If there is more than one alternative that meets these objectives, the selection of the preferred alternative is a second decision point. If there are no feasible alternatives that meet the project objectives, the logic returns to the beginning, where the project objectives, and/or scope must be re-evaluated.

Plans for Future Developments

Both the USACE and EPA are developing additional guidance on management of contaminated sediments within both the navigation dredging and sediment remediation areas. Under the ARCS program, several additional technical guidance documents are planned, to include specific guidance on remediation technology selection and *in-situ* capping. In the area of

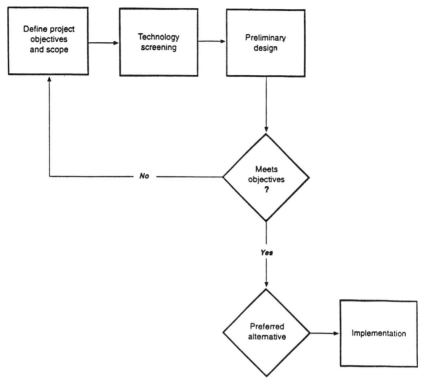

FIG. 4—*ARCS combined decision-making approach.*

dredged material disposal, joint USACE/EPA workgroups are initiating development of technical guidance documents on capping and on confined (diked) disposal. These efforts will hopefully add to the available knowledge regarding contaminated sediment management and will complement the management strategies now being used by both agencies.

Summary and Conclusions

There is no single decision-making strategy for the management of contaminated sediments that fits all purposes. However, established strategies have been applied within the context of both navigation dredging projects and sediment remediation projects. The technical framework developed jointly by the USACE and EPA for environmental evaluation of dredged material disposal alternatives is appropriate for navigation dredging projects. It is consistent with the requirements of the major laws and regulations pertaining to dredged material and provides for evaluation of the full range of dredged material management alternatives that can be considered.

Strategies for evaluation of contaminated sediment remediation projects are not as well established. Although the Superfund RI/FS framework must be followed for remediation projects on the NPL list, this framework does allow for flexibility. Parallel, sequential, and combined approaches for alternatives evaluation have been proposed for remediation projects, and EPA will be providing guidance for these strategies under its ARCS program.

Acknowledgment

This paper summarizes investigations conducted under the various research programs of the U.S. Army Corps of Engineers and the U.S. Environmental Protection Agency. Permission to publish this material was granted by the Chief of Engineers.

References

Averett, D. E., Perry, B. D., Torrey, E. J., and Miller, J. A., 1990, "Review of Removal, Containment, and Treatment Technologies for Remediation of Contaminated Sediments in the Great Lakes," Miscellaneous Paper EL-90-25, U.S. Army Engineer Waterways Experiment Station, Vicksburg, MS.

Engler, R. M., Patin, T. R., and Theriot, R. F., 1990, "Update of the Corps' Environmental Effects of Dredging Programs (FY89)," Miscellaneous Paper D-90-2, U.S. Army Engineer Waterways Experiment Station, Vicksburg, MS.

Engler, R. M., Francingues, N. R., and Palermo, M. R., August 1991, "Managing Contaminated Sediments: Corps of Engineers Posturing to Meet the Challenge," *World Dredging and Marine Construction.*

Environmental Protection Agency, 1988, "Guidance for Conducting Remedial Investigations and Feasibility Studies under CERCLA," EPA/540/G-89/004, U.S. Environmental Protection Agency, Office of Emergency and Remedial Response, Washington, DC.

Environmental Protection Agency, 1994, "Assessment and Remediation of Contaminated Sediments (ARCS) Program—Remediation Guidance Document," EPA 905-R94-003, U.S. Environmental Protection Agency, Great Lakes National Program Office, Chicago, IL.

Francingues, N. R., Palermo, M. R., Peddicord, R. K., and Lee, C. R., 1985, "Management Strategy for the Disposal of Dredged Material: Contaminant Testing and Controls," Miscellaneous Paper EL-85-1, U.S. Army Engineer Waterways Experiment Station, Vicksburg, MS.

National Research Council, 1989, *Contaminated Marine Sediments—Assessment and Remediation,* Marine Board, National Research Council, National Academy Press, Washington, DC.

Palermo, M. R., Engler, R. M., and Francingues, N. R., 1993, "The U.S. Army Corps of Engineers Perspective on Environmental Dredging," *Buffalo Environmental Law Journal,* Vol. 1, No. 2, Buffalo, NY.

USACE/EPA, 1992, "Evaluating Environmental Effects of Dredged Material Management Alternatives—A Technical Framework," EPA842-B-92-008, U.S. Army Corps of Engineers and U.S. Environmental Protection Agency, Washington, DC.

Sandra M. C. Weston[1]

Partnering for Environmental Restoration: The Port Hope Harbour Remedial Action Plan (RAP)

REFERENCE: Weston, S. M. C., "Partnering for Environmental Restoration: The Port Hope Harbour Remedial Action Plan (RAP)," *Dredging, Remediation, and Containment of Contaminated Sediments, ASTM STP 1293*, K. R. Demars, G. N. Richardson, R. N. Yong, and R. C. Chaney, Eds., American Society for Testing and Materials, Philadelphia, 1995, pp. 297–305.

ABSTRACT: A Remedial Action Plan (RAP) is being developed for Port Hope Harbour, one of 43 Areas of Concern (AOCs) identified by the International Joint Commission (IJC). The RAP, when implemented, will lead to the restoration and protection of desirable water conditions in Port Hope Harbour. The environmental concern associated with the harbour can be best viewed as a historical contaminated sediment problem. Approximately 90 000 m^3 of sediment located in Port Hope Harbour's turning basin and west slip are contaminated by uranium and thorium series radionuclides, heavy metals, and PCBs. There are several groups contributing to the development of the RAP. All of these groups have the common goal of developing an environmentally sound plan that reflects the views of the community. Strategic partnerships have been established that recognize the need to integrate and coordinate the efforts of all agencies, stakeholders, and the community. The objective is to develop an environmentally sound remediation plan through an efficient and effective management framework.

KEYWORDS: remedial action plan (RAP), contaminated sediments, low-level radioactive waste, Port Hope, environmental restoration

As a result of past abuses to the Great Lakes, substantial efforts are required to restore conditions in degraded areas (Government of Canada 1993). Intensifying resource constraints have highlighted the need for improved management of remediation projects. A major focus for restoration activities in the Great Lakes basin is the Remedial Action Plan (RAP) program. Forty-two of the 43 Areas of Concern (AOC) in the Great Lakes, for which RAPs are being developed, have contaminated sediments. Of those 42 AOCs, 35 have identified restrictions on dredging activities as a use impairment (International Joint Commission 1991).

Port Hope Harbour is the only AOC for which contaminated sediments is the main environmental concern and the resulting restriction on dredging activities is the only beneficial use impairment. There are several groups contributing to the development of RAPs. All of these groups have the common goal of developing an environmentally sound plan. A challenge for the Port Hope RAP was to establish a management framework that took into consideration the unique context of the environmental restoration problem and provided a distinct and complementary role for all of these groups. To meet this challenge, strategic partnerships were established that defined key roles and responsibilities in the remediation of this AOC.

---

[1] Environment Canada, Toronto, Ontario, Canada.

Background

The International Joint Commission (IJC) was established in 1909 by the Boundary Waters Treaty signed by the governments of Canada and the United States. The commission was formed to help resolve water-related issues along the border between these two countries. In 1912, the two governments referred the matter of pollution of the Great Lakes to the IJC (Great Lakes Science Advisory Board 1993).

Due to increased concern regarding water quality degradation in the Great Lakes, the Great Lakes Water Quality Agreement (GLWQA) was signed by the two governments in 1972. The initial Great Lakes Water Quality Agreement was revised and expanded in 1978 (and again in 1987) to include additional information on more recent concerns and issues. These included toxic substances, nonpoint source pollution, and contaminated sediments. This binational agreement makes far-reaching commitments concerning the restoration and protection of the Great Lakes (International Joint Commission 1994).

The governments of Canada and Ontario work together to implement the responsibilities of the GLWQA through a partnership called the Canada-Ontario Agreement respecting Great Lakes Water Quality (Government of Canada and Government of the Province of Ontario 1990). The most recent Canada-Ontario Agreement (COA) was signed in July 1994.

Areas of Concerns around the Great Lakes

Pursuant to the GLWQA, the IJC has been assessing water quality in the Great Lakes. In 1975, the Great Lakes Water Quality Board (GLWQB) of the IJC began to identify "problem areas." These "problem areas" were sites throughout the Great Lakes and their connecting channels in which water quality objectives of the Great Lakes Water Quality Agreement, or general objectives as expressed by jurisdictional standards or criteria, were exceeded (Port Hope Harbour RAP 1989). In 1981, the GLWQB redefined problem areas as AOCs on the basis of available sediment, water, and biota data. An Area of Concern (AOC) has been defined as:

> "a geographic area that fails to meet the General or Specific Objectives of the Agreement where such failure has caused or is likely to cause impairment of beneficial use or the area's ability to support aquatic life" (International Joint Commission 1988).

Annex 2 of the Canada-U.S. Great Lakes Water Quality Agreement (Canada and the United States 1987) defines "impairment of beneficial use(s)" as a change in the chemical, physical, or biological integrity of the Great Lakes sufficient to cause one or more of the following:

(a) restrictions on fish and wildlife consumption,
(b) tainting of fish and wildlife flavor,
(c) degradation of fish wildlife populations,
(d) fish tumors or other deformities,
(e) bird or animal deformities or reproduction problems,
(f) degradation of benthos,
(g) restriction on dredging activities,
(10) eutrophication or undesirable algae,
(11) restrictions on drinking water consumption, or taste and odor problems,
(12) beach closings,

(13) degradation of aesthetics,
(14) added costs to agriculture or industry,
(15) degradation of phytoplankton and zooplankton populations, and
(16) loss of fish and wildlife habitat.

Forty-three AOCs in the Great Lakes basin have been established. Seventeen of these 43 AOCs are located in Canada. The governments of Canada and Ontario have agreed, under the framework of their agreement, to jointly develop and implement Remedial Action Plans for these Canadian AOCs (Government of Canada and Government of the Province of Ontario 1990).

Remedial Action Plans

The Remedial Action Planning program is a cooperative initiative between Canada and the United States aimed at the cleanup of the 43 AOCs within the Great Lakes basin. A Remedial Action Plan (RAP) is a cleanup management plan that is developed and implemented to help restore and protect the water quality in an AOC.

Each RAP includes (International Joint Commission 1988):

(a) a definition and detailed description of the environmental problems in the AOC, including a definition of the beneficial uses that are impaired, the degree of impairment, and the geographic extent of such impairment;
(b) a definition of the causes of the use impairment, including a description of all known sources of pollutants involved and an evaluation of other possible sources;
(c) an evaluation of remedial measures in place;
(d) an evaluation of alternative additional measures to restore beneficial uses;
(e) a selection of additional remedial measures to restore beneficial uses and a schedule for their implementation;
(f) an identification of the persons or agencies responsible for implementation of remedial measures;
(g) a process for evaluating remedial measure implementation and effectiveness; and
(h) a description of surveillance and monitoring processes to track the effectiveness of remedial measures and the eventual confirmation of the restoration of uses.

The GLWQA requires that each RAP be submitted to the IJC at three stages:

- Stage 1: When the definition of environmental conditions and problems has been completed; as of October 1993, 38 of 43 RAPs have submitted Stage 1 documents to the IJC (International Joint Commission 1994).
- Stage 2: When remedial and regulatory measures have been selected.
- Stage 3: When monitoring indicates that beneficial uses have been restored.

Throughout all three stages, two basic principles underlie the RAP program: (1) the incorporation of an ecosystem approach and (2) public participation (Canada and the United States 1987). The first principle, an "ecosystem approach," focuses upon the interrelationship between all living organisms in the AOC and in the interacting elements of water, air, and land. The second principle, "public participation," is encouraged throughout the RAP process (International Joint Commission 1990).

Port Hope Harbour: Area of Concern

Port Hope Harbour is located in the town of Port Hope (approximately 100 km east of Toronto) on the north shore of Lake Ontario. The turning basin and west slip of the harbor have been designated as one of the AOCs. Port Hope Harbour was listed as an AOC for the following reason:

> "The sediment in the turning basin of Port Hope harbour is contaminated with radium. The dredging and proper disposal of this sediment has been identified as a concern if dredging is undertaken in connection with possible harbour redevelopment" (Port Hope Harbour 1989).

Approximately 90 000 m^3 of sediment located in Port Hope Harbour's turning basin and west slip is contaminated by uranium and thorium series radionuclides, heavy metals (such as iron, lead, zinc, copper, and chromium), and PCBs. Due to the radioactive content, the sediment has been classified as low-level radioactive waste (LLRW) by the Canadian federal government.

By comparison to other AOCs, Port Hope Harbour is a small AOC with regards to:

1. *Local Population.* The current population of the Town of Port Hope is approximately 12 000. The population of the town has grown at an annual average rate of just under 1% from 1961 to 1986 (Ecologistics Ltd. 1992).
2. *Spatial Extent.* The west slip is 40 m in width and 305 m in length. The turning basin's dimensions are 195 m by 135 m.
3. *Harbor Uses.* The turning basin of Port Hope Harbour serves as a boat mooring area for the Port Hope Yacht Club. Boat launching and fueling facilities are supplied by a marina located on the lower Ganaraska River. CAMECO, formerly Eldorado Resources Ltd., has uranium production facilities located along the western edge of the turning basin and west slip and warehousing along the eastern edge. The facility withdraws process and cooling waters from the southern end of the slip along its western bank and discharges effluents at various points along the western side of the turning basin and west slip. In Canada, issues relating to the control of radioactive substances, including uranium refinery facilities, fall under the jurisdiction of the Atomic Energy Control Board (AECB). Quality of effluent discharged by CAMECO is regulated under the plant operating license issued by the AECB. A large amount of recreational fishing takes place in the Port Hope area. The most frequently used area is the Ganaraska River which is outside the AOC. Although conditions are not as good for fishing within the harbor, various species of fish are present, particularly during the spring and fall runs of Rainbow Trout and Brown Trout in the adjacent Ganaraska River.

Environmental Conditions

As previously noted, RAPs are developed in three stages. The environmental conditions and problems in an AOC are examined in the Stage 1 RAP document. The Port Hope Harbour Stage 1 RAP document was submitted to the IJC in 1990 for review and comment. In reviewing the Stage 1 RAP document, the IJC noted that the Port Hope Harbour RAP had incorporated an ecosystem approach and had a commendable public involvement program. The IJC concurred with the Commission's Great Lakes Water Quality Board's assessment that the Port Hope Harbour RAP had produced an excellent Stage 1 document (Slyfield 1991).

The environmental condition of greatest concern for the Port Hope Harbour RAP is sediment quality. There is significant radioactive sediment contamination found in Port Hope Harbour. The area of contamination is confined to the turning basin and a portion of the west slip. Turning basin sediments contained the highest mean activity for radionuclides, followed closely by sediments collected from the west slip. Concentrations are considered normal in sediments taken from the area where Port Hope Harbour and the Ganaraska River meet and from the nearshore zone of Lake Ontario.

Contaminated sediments comprise a fairly uniform layer (about 2.5 m thick) of organic silts and fine sands in the turning basin and west slip, beneath about 3 m of water, to a native till or limestone bedrock (Fenco MacLaren Inc. 1994). The sediment is contaminated with uranium, radium, thorium, radioactive lead, and heavy metals (iron, lead, zinc, copper). PCBs are also slightly above criteria for open water disposal of dredged sediments. Due to the radionuclide content, sediments in the turning basin and west slip have been classified as low-level radioactive waste. The total volume of contaminated sediment in the turning basin and west slip is estimated at 90 000 m^3.

Studies have been conducted to determine if contaminants in the sediments were also being found in local fish. Several species of fish caught in the turning basin were analyzed for radionuclide concentrations. The Brown Bullhead Catfish and the Yellow Perch (both bottom feeding species) were found to have levels of uranium and thorium series radionuclides that were slightly higher than levels found in the same species elsewhere in Lake Ontario. All other species had levels that were similar to those found in other parts of the lake. However, all species, including catfish and perch, had radionuclide levels that are considered acceptable for human consumption (Port Hope Harbour RAP 1989).

The Stage 1 document revealed no impairment of water uses such as fishing, swimming, and use of the water for drinking purposes in areas close to the harbor. The major use impairment associated with the contaminated sediments in the harbor concerns the use of the turning basin as a boat mooring facility. Contamination of turning basin sediments have caused a cessation of maintenance dredging. Any sediments resulting from dredging of the turning basin must be stored in a low-level radioactive waste management facility licensed by the Atomic Energy Control Board (AECB), and no such facility is currently available. If dredging is not resumed, continued sedimentation will eventually make the turning basin inoperative as a boat mooring facility.

The environmental concern associated with Port Hope Harbour can be best viewed as a contained historical sediment contamination problem. Concentrations of radionuclides and heavy metals in the turning basin and west slip sediments closely resemble residues generated by the radium and uranium refining operations conducted by the former Eldorado Resources Ltd. (Cameco) at this site during the 1930s and 1940s. Between 1933 and 1948, refinery wastes were stockpiled at a number of locations throughout the Town of Port Hope, including adjacent to the turning basin. Migration of contaminants from these stockpiles is thought to be the principal source of contaminants found in the turning basin sediments.

Remediation Strategy

Currently, the Port Hope Harbour RAP is focusing on the Stage 2 phase of the RAP process—the development of remedial options, implementation schedules, and surveillance monitoring program. In developing a management strategy for the Stage 2 phase, the Port Hope Harbour RAP took into consideration:

(a) the context of the environmental restoration problem
(b) the interrelatedness of other remediation programs being undertaken in Port Hope, and
(c) the overlapping agency responsibilities.

The radionuclide contamination of the harbor sediments is part of a larger radiological contamination problem in the Town of Port Hope. During the 1930s and 1940s, radium and uranium refinery wastes were stockpiled at a number of locations throughout the town. The problem of the historical LLRW in the Town of Port Hope was recognized in the mid-1970s. Remedial work involving removal of this historical LLRW to Chalk River Laboratories (CRL) was undertaken between 1976 and 1979. As a result of the limited storage capacity at CRL, some LLRW was left in Port Hope for cleanup at a later date. These historical LLRW deposits were left at less accessible locations in the town. The remaining historical LLRW areas within the Town of Port Hope requiring remedial work can be divided into three main categories:

1. Major On-Land Areas (ravines, municipal land fill, etc.)
2. Small-Scale Sites (public roadways)
3. Harbor (turning basin and west slip).

In 1982, the federal government assumed responsibility for all "historic" low-level radioactive wastes, which includes the contaminated sediment in the harbor. Remediation plans are being developed to deal with the historic LLRW (both contaminated harbor sediments and contaminated land based sites in the town of Port Hope), as well as towards the establishment of a new disposal facility for these wastes.

The focus of the Port Hope Harbour RAP is the remediation of the harbor. Other agencies/groups involved in the broader remediation activities for the Port Hope area are summarized in Table 1.

The complex interrelationships and interdependence among all these remediation activities was recognized early in the RAP process. To ensure that these activities were complementary and reinforcing, there was a need for cooperative action and alliances between these various government agencies/groups and the local community. The first objective of the Stage 2 RAP was to focus on creating cooperative working relationships. The intent is to cooperate rather than compete in the development of remedial options.

An informal cooperative network of groups and agencies has been established. The RAP team has encouraged the various groups, stakeholders, and agencies to participate in devising and implementing the remediation strategy for Port Hope Harbour. The first task was to determine which agency/group/stakeholder was best positioned to deliver a service or engage in a particular activity. Accordingly, each of these groups and agencies has its own role/responsibility for various segments of the harbor remediation plan. Table 2 shows the management strategy set forth in the RAP.

Although these distinct roles have been identified for the development of remediation options, the RAP recognized the need for on-going involvement and exchange of information and ideas between the various agencies and groups. Establishing and sustaining effective communication and cooperation becomes essential to maintaining a cooperative and integrated management approach.

TABLE 1—*Port Hope remediation activities.*

Agency	Remediation Activities
Low-Level Radioactive Waste Management Office (LLRWMO)	A function of this federally established office is to provide technical and economic assessments and recommendations to the federal government regarding the management of historical low-level radioactive wastes in Port Hope. The long-term objectives of the LLRWMO for historic wastes in Port Hope are: (a) removal or isolation of contaminated materials and soil to allow unrestricted access to current areas of contamination, (b) minimum future restrictions on land use, or requirements for development, (c) implementation of remedial work, particularly restoration of contaminated areas, in a way that maximizes the benefit to the community.
Siting Process Task Force	A federal Siting Process Task Force on Low-Level Radioactive Waste Disposal was established in December 1986 by the federal government (Ministry of State—Forestry and Mines). The central focus of the Task Force's efforts was the development of a process for siting a disposal facility in Ontario for the existing, ongoing and historic wastes in the Port Hope area, and where advantageous, for the disposal of other existing and ongoing low-level wastes located in the province. The main objective of the process proposed by the task force is the identification of one or more voluntary host communities, each with a suitable site.
Siting Task Force	Acting on the recommendations of the Siting Process Task Force, the government of Canada established another task force in September 1988. The Siting Task Force (STF) is an independent group of private sector citizens appointed by the federal Minister of Energy—Mines and Resources (recently renamed Ministry of Natural Resources). The mandate of the STF is to find a site for a LLRW disposal facility for Ontario. The STF is implementing a five-phase siting process that encourages the voluntary/collaborative participation of communities in the search for a LLRW facility site. In addition to working with communities volunteering to accept a LLRW facility (a.k.a. host communities), the STF is actively working with communities that have historical LLRW which would be relocated to the new LLRW facility (a.k.a. source communities).
Siting Task Force Secretariat	To support the STF in the implementation of the siting process, a secretariat has been established. The secretariat provides administrative and technical support to both the host and source communities involved in the LLRW siting process.
Port Hope Community Liaison Group	Acting on the recommendations of the Siting Process Task Force, the STF established local community liaison groups (CLGs) in all volunteer communities and in all source communities. A Source CLG was established in Port Hope to undertake the following (Siting Task Force, undated): (a) initiate activities to involve community residents in the process; (b) assess the amount and nature of LLRW and to identify preferred management options; (c) assist the STF in developing guidelines for the consolidation or cleanup of LLRW and rehabilitation of waste sites; (d) in conjunction with the STF, select technical experts to provide advice and conduct environmental assessments on these activities, and; (e) develop an action plan for cleanup and remediation

TABLE 1—*Continued*

Agency	Remediation Activities
Atomic Energy Control Board	The power to regulate the Canadian nuclear industry is contained in the Atomic Energy Control Act, and this power is exercised by means of the Atomic Energy Control Regulations. The Atomic Energy Control Board (AECB) was set up by the Act in 1946 to provide the control and supervision of the development, application, and use of atomic energy. The AECB must approve any remedial work undertaken in Port Hope, such as the remediation of Port Hope Harbour. The AECB would issue a license that would define standards that must be met. Once a license is issued, the AECB would carry out compliance inspections to ensure that its requirements are met.

TABLE 2—*RAP management strategy.*

Agency/Group	Roles/Responsibilities
Port Hope Community Liaison Group (CLG)	• Develop an action plan for LLRW in Port Hope area. This would include the review of remedial options and the assessment of potential impacts and mitigation measures associated with the harbor sediments.
Port Hope Community	• Identify beneficial use goals for harbor. • Evaluate and endorse action plan developed by the Port Hope CLG.
Port Hope Harbour RAP	• Monitor and assess remedial options developed by the Port Hope CLG for the harbor. • Incorporate remedial options that meet IJC/RAP objectives into the Stage 2 document.
Atomic Energy Control Board	• Review, approve, and license remediation activities in Port Hope.
Low Level Radioactive Waste Management Office	• Implement action plan for Port Hope's historical LLRW, as approved by the federal government.

Conclusion

The RAP program has been identified by the IJC as one of the success stories of the GLWQA (International Joint Commission 1994). The RAP program is viewed both nationally and internationally as a model for local environmental planning and action (Government of Canada and Government of the Province of Ontario 1990).

The challenge for the Port Hope RAP and other remediation projects is to establish a management framework that takes into consideration the unique context of the environmental restoration problem and provides for distinct and complementary roles for various stakeholders. To meet this challenge in Port Hope, strategic partnerships were established that recognized the need to integrate and coordinate the efforts of all agencies, stakeholders, and the community. The objective was to develop an environmentally sound remediation plan through a efficient and effective management framework.

In remediation projects, the lack of a clear division of roles and responsibilities can create irritants as well as gaps and duplication of activities. The Port Hope Harbour RAP remediation management approach has resulted in a higher degree of coordination amongst agencies and stakeholders. It has been found that by clarifying roles (and responsibilities) in the remediation of this AOC, overlap and duplication of activities has been minimized. Accordingly, this has resulted in a more efficient and effective use of resources.

The Port Hope Harbour RAP has recognized the need to develop a management framework, based upon cooperation and an effective and efficient definition of role and responsibilities, that will lead to environmental restoration. Success is more common when agencies and stakeholders join forces to accomplish well defined mutually beneficial tasks.

References

Canada and the United States, 1987, *Protocol to the Great Lakes Water Quality Agreement*, Windsor, Ontario, Canada.

Ecologistics Ltd., 1992, *Port Hope Harbour Remedial Action Plan—Socio-Economic Profile of Port Hope*, prepared for Environment Canada, Burlington, Ontario.

Fenco MacLaren Inc., 1994, *Site Specific Action Plan Feasibility—RAP 3, Town of Port Hope*, Draft 2, prepared for the Technical Working Group Remedial Action Plans, March.

Government of Canada, 1993, "Canadian Great Lakes Program—Overview," *Third Report of Canada under the 1987 Protocol to the 1978 Great Lakes Water Quality Agreement*, October.

Government of Canada and Government of the Province of Ontario, 1990, "Issues Overview," *Second Report of Canada under the 1987 Protocol to the 1978 Great Lakes Water Quality Agreement*, December.

Great Lakes Science Advisory Board, 1993, *1993 Report on the International Joint Commission*, International Joint Commission, Windsor, Ontario, Canada.

International Joint Commission, 1988, *Revised Great Lakes Water Quality Agreement of 1978*.

International Joint Commission, 1990, *Fifth Biennial Report on Great Lakes Water Quality—Part II*, International Joint Commission, Windsor, Ontario, Canada.

International Joint Commission, 1991, *Summary of the Remedial Action Forum*, Traverse City, MI, 27–28 September.

International Joint Commission, 1994, *Seventh Biennial Report on Great Lakes Water Quality*, International Joint Commission, Windsor, Ontario, Canada.

Port Hope Harbour RAP, 1989, *Stage 1—Environmental Conditions and Problem Definition*.

Siting Task Force, undated, *Fact Sheet #2—The Participants*, STF Secretariat, Ottawa.

Slyfield, P., 1991, *Correspondence to Right Honourable Joe Clark (Secretary of State for External Affairs) from Philip Slyfield*, Secretary, Canadian Section, International Joint Commission, dated 13 February 1991.

Rosa Galvez-Cloutier,[1] *Raymond N. Yong,*[2] *Jeanette Chan,*[2] *and Edward Bahout*[2]

Critical Analysis of Sediment Quality Criteria and Hazard/Risk Assessment Tools

REFERENCE: Galvez-Cloutier, R., Yong, R. N., Chan, J., and Bahout, E., **"Critical Analysis of Sediment Quality Criteria and Hazard/Risk Assessment Tools,"** *Dredging, Remediation, and Containment of Contaminated Sediments, ASTM STP 1293,* K. R. Demars, G. N. Richardson, R. N. Yong, and R. C. Chaney, Eds., American Society for Testing and Materials, Philadelphia, 1995, pp. 306–318.

ABSTRACT: The Canadian Federal Environmental Assessment Review Office is funding the Geotechnical Research Centre to facilitate their participation in the analysis of the Environmental Impact Statement during the public hearings of the Lachine Canal decontamination project. A summary and critical analysis of recent research and practical experience in the interpretation and application of sediment quality criteria and hazard/risk assessment tools was prepared for use in a comparative examination for the Lachine project.

This present survey shows that various approaches can be taken when establishing both sediment quality and remediation criteria, in light of an increasing preference for flexible guidelines that can be adjusted according to specific site conditions and that will marry more than one approach.

Depending on the complexity of the decontamination scenario, there exist different levels of application in the analysis of the hazard and risk. Simple and complex models are available, but their actual application in ongoing sediment decontamination projects is minimal due to lack of understanding of the basics of contaminant exposure assessment and exposure routes.

KEYWORDS: quality criteria, contaminated sediments, risk assessment

Sediments are an important and integral component of aquatic ecosystems because they provide a substrate and habitat for a wide variety of organisms as well as species important in the food chain, such as worms, amphipodes, oligochaetes, chironomids, bivalves, and insects (Adams et al. 1992). Sediments are a resource that needs to be protected and preserved. Regulating sediment quality criteria is a course to accomplish this goal.

At the same time, an important step in any remediation attempt is the assessment of the sediments' toxicity and the potential threat to human health and the environment, before and after remediation. There is the need to accurately answer the question: How clean is clean?

From the review of the current literature and approaches, it is evident that no single approach can, by itself, be deemed adequate especially when it comes to making decisions regarding toxicity assessment and which of the various remediation techniques to use. The lack of standard protocols and field data (i.e., data obtained using innovative protocols), and the fact that results are highly site-specific, makes it almost impossible to compare any

---

[1] Professor, Department of Civil Engineering, Laval University, Adrien-Pouliot Bldg., Quebec, Canada G1K 7P4.

[2] Director, and research assistants, respectively, Geotechnical Research Centre-McGill University, Montreal, P.Q. Canada H3A 2K6.

separate two cases. Therefore, it is not surprising that part of the actual decisions are taken based solely on economical and political considerations.

For instance, the volume and the degree of contamination have a direct effect on the total cost of the project, when dredging/treatment/disposal is the adopted remedial technique. Commonly, the amount of sediment to be treated is determined based on the quality criteria. Depending on the approach used, the final numbers will result in a larger or smaller volume. Various groups prefer to describe these differences as being either underprotective or overprotective. Nevertheless, with today's economy and social concerns, we cannot afford to be in either of these extremes.

Contaminants-Sediment Constituents Interactions

Metals, synthetic organic compounds, nutrients, and many other contaminants are found in sediments all over the world. Most of these contaminants are of anthropogenic origin which, through industrial effluent discharges, point contaminant spills, and diffuse agricultural sources, have found their way to the sediments. Once in the sediments, contaminants become entrapped in various compartments of the ecosystem. Depending on the biogeochemical interaction occurring in each compartment, the fraction of a particular contaminant can be considered as a sink or source for toxicity and, in both cases, mobility and transport are likely to occur. Figure 1 schematizes some of the mechanisms that are responsible for the fate of contaminants in the river system. Figure 2 shows the detailed local geochemical distribution of heavy metals in sediments.

A toxic material is a substance that has an adverse effect on the health of living organisms. The level of toxicity of a substance is related to the amount that causes an adverse effect and to the type of effect. However, some metals, which are essential or at least beneficial to

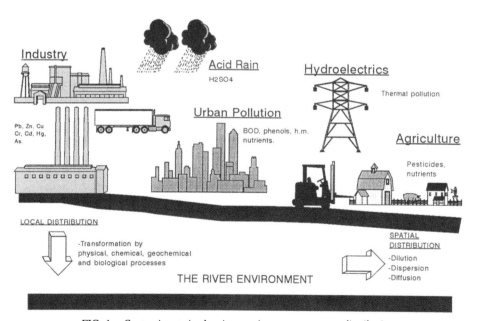

FIG. 1—*Contaminants in the river environment: sources, distribution.*

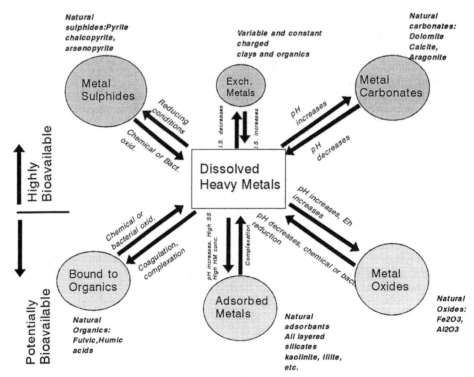

FIG. 2—*Geochemical cycle of heavy metals in sediments.*

human health, can become toxic under certain circumstances (e.g., when present in excess). Also, they can have either synergistic or antagonist effects on the biological properties of other elements. For instance, some of the health effects of heavy metals are presented in Table 1.

As one can gather from the review of the concepts discussed in the previous paragraphs and represented in Figs. 1 and 2 and Table 1, a global understanding can only be achieved if a multidisciplinary team takes in charge the interrelation of the concepts. This is not achieved yet and remains the main cause of controversy when developing sediment quality criteria and risk assessment tools.

Criteria for Characterizing Sediment Pollution

The definition of contaminated sediment has been somewhat arbitrary. Several attempts have been made to establish criteria for distinguishing natural from contaminated sediment. As mentioned by Thomas et al. (1985) and Dermont and Munawar (1992) these criteria historically included:

1. The Jensen criteria set by the U.S. Environmental Protection Agency (USEPA) in 1971. The criteria consider a sediment to be contaminated if one or more of its properties exceeds a permissible limit. These criteria have been criticized (i.e., Chapman 1989) because in some instances bulk concentrations have been found as poor indicators of the potential for biological effects.

TABLE 1—*Health effects due to the presence of heavy metals.*

Organs Affected	Heavy Metal	Health Effects	
		Deficiency	Excess
Central Nervous System	Hg Pb		Brain damage; reduced neuropsychological functioning
Peripheral Nervous System	Hg Pb		Abnormal movement and reflexes; peripheral neurological effects
Digestive System	Cu Zn	Reduced growth; reduced appetite and growth	Cirrhosis; nausea
Blood System	Pb Cd Cu Cr	Anaemia; artherosclerosis	Inhibits biosynthesis; slight anemia
Respiratory Tract	Cd As Hg Se		Emphysema; emphysema and fibrosis; bronchial effects; respiratory inflammation
Skeleton	Cd		Osteomalacia
Reproductive System	Hg As		Spontaneous abortion
Urinary System	Cd Hg As Cr		Tubular, glomerular damage; proteinuria; tubular nephrosis
Teratogenesis	Hg		Deformed brain and body

2. The elutriate test adopted by the USEPA in the early 1970s. This test was devised by the U.S. Army Corps of Engineers and the USEPA.

3. The bioassay test adopted by the USEPA in 1977. Liquid-phase, suspended solids, and solid-phase bioassay measurements are used to determine the effect of a contaminant or to reflect the cumulative influence of all contaminants present, including any possible interactions of contaminants (e.g., synergistic or antagonistic effects) on biological organisms of concern (MEC 1985; USEPA 1991).

4. The limits proposed by Vigneault (1978) for the disposal of dredged material in open sea was the first set of criteria proposed in Canada. After Vigneault, other canadian regulatory bodies proposed limits that have been applied to sediments; these criteria are summarized in Table 2.

5. It was not until very recently that Canada has produced a set of standards for the classification of contaminated sediments (MEC 1993). These standards are called Interim Criteria and are for special application to the St. Lawrence River sediments. This guide needs to be used in conjunction with the Methodology Guideline for the Characterization of Contaminated sediments (MEC 1993) which is still under revision.

In general, the rationale behind the selection of the limiting (threshold) levels has been based on reviews of past approaches used by agencies involved with the management of sediments (contaminated or not) (MEC 1992; MENVIQ 1992). Very commonly, the background approach was used for the determination of the first level, whereas second and third levels are sometimes related to fixed percentage rate of mortality according to bioassays results.

TABLE 2—Canadian sediment quality criteria.

Parameter, mg/kg	Vigneault, 1978 Sediments		MENVIQ, 1988b Cont. Soil				MEO, 1990 Contaminated Sed.			MEC, 1992 Sediments			CCME, 1992 Treat. Soil		
	A'	B'	A	B	C		NEL	MEL	TEL	SSE	SEM	SEN	AG	R/P	C/I
Total PCBs	0.05	0.1	0.1	1	10		0.02	0.07	—	0.02	0.2	1	0.5	5	50
Total PAH	—	—	1	20	200		2.5	2.5	6	1.1	—	—	0.9	26	260
Total MAH	—	—	0.1	22.5	215		—	—	—	—	—	—	0.85	22.5	225
Total O&G	1000	2000	100	1000	5000		—	1500	—	—	—	—	—	—	—
Phenols	—	—	0.1	2.5	25		—	—	—	—	—	—	0.15	1.5	15
TOC, wt%	1.5	5	—	—	—		1	1	10	—	—	—	—	—	—
Cyanides	—	—	5	50	500		0.1	0.1	0.25	—	—	—	5	50	500
Total As	3	6	10	30	50		4	6	33	3*	7*	17*	20	30	50
Total Cd	5	8	1.5	5	20		1	1	10	0.2*	0.9*	3*	3	5	20
Total Cr	70	90	75	200	800		31	31	111	55*	55*	100*	750	250	800
Total Cu	30	60	50	100	500		25	25	114	28*	28*	86*	150	100	500
Total Pb	20	60	50	200	600		23	31	250	23*	42*	170*	375	500	1000
Total Hg	0.3	1	0.2	2	10		0.1	0.2	2	0.05	0.2*	1*	0.8	2	10
Total Zn	80	175	100	500	1500		65	120	820	100*	150*	540*	600	500	1500
Total Ni	—	—	—	—	—		31	31	90	35*	35*	61*	150	100	500
TKN	—	—	—	—	—		—	550	4800	—	—	—	—	—	—
Total P	—	—	—	—	—		—	600	2000	—	—	—	—	—	—

*Extractable.

A' = Acceptable for ocean disposal.
B' = Nonacceptable.

A = Background.
B = Need for study.
C = Need for remediation.

NEL = No effect level.
MEL = Minimal effect level.
TEL = Toxic effect level.

SSE = No effect.
SEM = Minimal effect.
SEN = Toxic effect.

AG = Agricultural.
R/P = Res./Parkland.
C/I = Commercial/Ind.

Since contaminants in sediments may be present in bioavailable and nonbioavailable forms, sediment bioassessment and further correlation with physiochemical data are required to measure the real toxicity of the contaminants.

Environmental Regulations that Relate to Contaminated Sediments

In the handbook, *Remediation of Contaminated Sediments* (USEPA 1991), the USEPA enumerates the various environmental regulations that relate to contaminated sediments:

- The Comprehensive Environmental Response, Compensation, and Liability Act (CERCLA)
- The Clean Water Act (CWA) which authorizes the development of criteria that may apply to dredging and dredged material disposal, assessment, source control, and remediation
- The Resource Conservation and Recovery Act (RCRA)
- The Marine Protection, Research, and Sanctuaries Act (MPRSA)
- The Toxic Substance Control Act (TSCA)
- The Federal Insecticide, Fungicide, and Rodenticide Act (FIFRA)
- The Clean Air Act (CAA)
- The National Environmental Policy Act (NEPA) which requires an Environmental Impact Statement (EIS) for many federally funded projects, in order to incorporate environmental considerations into decision-making at the federal level. The preparation of EIS provides an opportunity to explore the options available for dredging and disposal of contaminated dredged material
- The Rivers and Harbors Act (RHA)
- The Endangered Species Act of 1973
- The Great Lakes Water Quality Agreements (GLWQA) between Canada and the United States
- The State Environmental Statues that Relate to Contaminated Sediments.

In Canada, the Environment Protection Act protects *citizen's health* against pollution in general, whether or not the threat arises from soil, water, or air. Similarly to the United States, each province may have its own provincial regulatory protocols. The bills and guidelines that carry federal and provincial jurisdiction during dredging and/or decontamination projects of sediments in Canada are:

- Décret sur les lignes directrices visant le processus fédéral d'évaluation et d'examen en matière d'environnement (PFEEE) (Bill on the Policy lines concerning the federal evaluation process for environmental examination)
- Canadian Law for the protection of the environment
- Canadian Law for the conservation and protection of migratory birds
- Articles 36, 37, 38, and 39 of the fisheries law— Protection of fish and its habitats (sections regarding the discharge of toxic substances)
- The Canadian law for the protection of navigation waters.

Sediment Toxicity Assessment

Sediment toxicity assessment is complex in that many site-specific parameters need to be considered. Most of the parameters are not related to the pore water chemistry but to the solids biogeochemistry. For instance, the adsorption/ion exchange processes as well as the kinetics, the contaminants bioavailability, the sediment mineralogy, the sediment depositional patterns, the erosion and compaction, the bioturbation, and temporal and spatial variations are all critical factors affecting the final toxicity.

Regarding the problem of assessing sediment quality, Adams et al. (1992) address two issues: (1) How can we best protect and manage our sediment resources? and (2) What is the significance of sediment contamination and under what circumstances do contaminated sites need to be remediated?

Scientifically based answers to these questions are still being sought and there is much debate on several aspects of available methodologies, including their validity and reliability. This resulted in the nonstandardization of approaches used in regulations, assessment, and remediation criteria for contaminated sediments. For example, the state of Washington has implemented State sediment quality standards based on the Apparent Effects Threshold (AET) approach, whereas the USEPA office of Water and Regulations encourages the Equilibrium Partitioning (EqP) approach. In Canada, Ontario uses the criteria based on bioassay impacts, while Quebec bases its criteria on chemical analysis.

In an effort to develop a U.S. national strategy to address the issue of sediment toxicity assessment, the USEPA is currently reviewing Sediment Assessment Methods, developing Sediment Quality Criteria (SQC) and Toxicity Identification Evaluation (TIE), and discussing the need for a consistent tiered testing approach to sediment quality assessment (USEPA 1991).

In order to assess the quality of chemically contaminated sediments, the USEPA has developed a Sediment Classification Methods Compendium (USEPA 1989) that describes various methods as well as their advantages, limitations, and existing applications. These sediment quality assessment methods are summarized in Table 3 and are broken down into three basic types: numeric, descriptive, or a combination of both. The compendium, however, provides no guidance on how the different methods can be used as part of a decision-making framework.

To establish its SQC, the EPA has selected the equilibrium partitioning (EqP) approach which relies on established water quality criteria to assess sediment toxicity. The EqP approach assumes that sediment toxicity is correlated to the concentration of the contaminants in the interstitial water and not to total sediment concentration, and that contaminants partitioned between the sediment adsorptive sites and the interstitial water are in equilibrium. Thus, the interstitial concentration for a given contaminant can be calculated when the total sediment concentration, the concentration of the sorbents (i.e., organic carbon) and the partitioning coefficient of that contaminant are known. Then, to assess sediment toxicity, the calculated interstitial concentration is compared to established water quality criteria. However, from selective sequential extraction (SSE) procedures (applied to heavy and trace metals), it has been shown that sediments contain various sorbent materials that retain metals to varying degrees (Yong et al. 1993).

TIE methods, which have been developed by the National Effluent Toxicity Assessment Centre, are used to identify specific causes of acute toxicity in aqueous sediment fractions, e.g., in pore water and elutriates. There is the hope that the tiered sediment assessment strategy will help provide a mechanism for achieving cleaner sediments and water. Burton and Scott (1992) suggest that reconnaissance surveys (i.e., Microtox) or tiered testing approaches may produce more efficient assessments of sediment toxicity than multiple toxicity assays.

Adams et al. (1992) have used a conceptual framework to develop an approach for assessing the significance of chemicals sorbed to sediments. They recommended that sediment assessment begin with a TIE to derive Sediment Assessment Values (SAVs). The SAVs can be obtained from the EqP approach for non-ionic organics, the cation exchange capacity,

TABLE 3—*Sediment quality assessment methods (USEPA 1991).*

Method	Numeric	Descriptive	Combination	Concept
Bulk Sediment Toxicity	●			Test organisms are exposed to sediments that contain unknown quantities of potentially toxic chemicals. At the end of a specific time period the response of the test organisms is examined in relation to a specified biological end point.
Spiked Sediment Toxicity	●			Dose-response relationships are established by exposing test organisms to sediments that have been spiked with known amounts of chemicals or mixtures of chemicals.
Interstitial Water Toxicity	●			Toxicity of interstitial water is quantified and identification evaluation procedures are applied to identify chemical components responsible for sediment toxicity. The procedures are implemented in three phases: (1) characterization of interstitial water toxicity, (2) identification of the suspected toxicant, and (3) confirmation of toxicant identification.
Equilibrium Partitioning	●			A sediment quality value for a given contaminant is determined by calculating the sediment concentration of the contaminant that would correspond to an interstitial water concentration equivalent.
Tissue Residue	●			Safe sediment concentrations of specific chemicals are established by determining the sediment chemical concentration that will result in acceptable tissue residues. Methods to derive unacceptable tissue residues are based on chronic water quality criteria and bioconcentration factors, chronic dose response experiments or field correlations, and human health risk levels from the consumption of fresh fish or seafood.
Sediment Quality Triad	●	●	●	Sediment chemical contamination, sediment toxicity, and benthic infauna community structure are measured on the same sediment. Correspondence between sediment chemistry, toxicity, and biological effects is used to determine sediment concentrations that discriminate conditions of minimal, uncertain, and major biological effects.
Apparent Effects Threshold	●		●	An AET is the sediment concentrations of a contaminant above whose statistically significant biological effects (e.g., amphibod mortality in bioassay, depressions in the abundance of benthic infauna) would always be expected.
Fresh Water Benthic Community Structure		●		Environmental degradation is measured by evaluating alterations in freshwater benthic community structure.
Marine Water Benthic Community Structure		●		Environmental degradation is measured by evaluating alterations in marine benthic community structure.

and critical micelle concentration for ionic organics, or the Acid Volatile Sulfide (AVS) normalization for metals. The Tier I SAVs would then be used as screening-level concentrations to be compared against sediment chemical concentrations. If the SAV is exceeded by the sediment concentration, additional assessment of the sediment would be appropriate (Tier II). Finally, if necessary, one can move from Tier II, which is referred to as an investigative tier, to Tier III for final decision.

Risk, Hazard, and Exposure Assessment

The general conceptual framework for risk assessment has been developed with a focus on health risk. Most of the work has been developed for contaminated air and soil studies (Ricci 1985). However, there are only few examples applied to contaminated sediment projects.

Risk, hazard, and exposure assessment protocols and frameworks must include all critical steps from the input of contaminants in the ecosystem to their final impact on human health. In order to accurately assess the public health risk, Swider and Kastenberg (1993) suggested the accountability of the following issues:

- characterization of the pollution sources
- quantification of the transport and fate of the pollutants in the multimedia environment, including any biological, chemical, and physical transformations
- assessment of any possible human exposure pathway with its associated dose-uptake
- comparing the dose-uptake via each pathway to the health effect (dose/response) for the population at risk
- quantification of the uncertainties associated with each of the above-mentioned steps
- other aspects related to the review of administrative law and its effect on science policy, as well as on cost-benefit analysis

On the second issue, Onishi (1985) summarized important pathways, or mechanisms that affect the chemical migration and fate in surface water and sediments as well as ground water, for the modeling of transport and pathways of pollutants emitted into the ecosystem. Some of the complex interactions among various transporting media were illustrated in Fig. 1. The transport and fate mechanisms start with the chemical source characterization which includes: (1) the chemical source quantification and (2) the release scenario determination. For example, the contribution from point and nonpoint sources may include:

- dry and wet deposition from atmosphere
- runoff and soil erosion from land surface
- seepage from and into ground water
- direct discharge

Regarding the actual chemical transport and fate mechanisms in surface waters the following issues are of concern:

- Transport: water and sediment movements
- Intermedia Transfer: adsorption-desorption, precipitation-dissolution, and volatilization
- Degradation and Decay: radionuclide decay, hydrolysis, oxidation, photolysis, and biological activities
- Transformation: yield of degradation product, yield of chemical reaction product, and yield of daughter product

These mechanisms can be incorporated into mathematical simulation models that can then be used to predict the transport and fate of chemicals released to the environment. Specifically, simulation models can be used as: (1) a scientific decision-making tool to regulate toxic chemicals, (2) a research tool to evaluate the relative significance of various transport and fate mechanisms (3) a tool to evaluate the effectiveness of chemical management practices, (4) a tool to optimize the facility operations by balancing risks and benefits, and (5) a means for identifying current gaps and future needs in chemical risk assessment. In selecting a model, the analyst must choose an appropriate scale length, a time frame, and a mathematical model(s) that reflects the real system.

In general, available mathematical models address the movement problem of water, chemicals, and other substances (e.g., sediments) by solving continuity and momentum equations. The law of mass conservation is solved either deterministically or nondeterministically. Deterministic solutions can involve analytical or numerical techniques (finite difference), or finite element method. Nondeterministic solutions use the stochastic approach by using stochastic models or direct simulation models.

A recent list of available computer hydrogeological and hydrogeochemical models is found in an EPA review (USEPA 1993). The compendium discusses the applications, special features/options, and the basis of each computer model. It also presents numerous intermodel comparisons.

Other related aspects include the quantification of risk as a conditional probability (quantal dose-response functions), the classification of dose-response functions (e.g., low-dose linear, low-dose supralinear, etc.), and the divergence of dose-response functions at low doses (e.g., the threshold dose hypothesis, the low-dose linear responses, and the pharmacokinetic considerations which include the relationship between the exposure dose and the amount metabolized, i.e., target tissue dose).

In trying to overcome the estimation of human risk from animal data (i.e., animal-to-human extrapolation), several methods are under continuous development. Some of these methods attempt to combine data from various studies involving the same or different type of species. For instance, the quantification of carcinogenic risk to humans is calculated from exposures to carcinogens by estimating animal risks from low doses (based upon high-dose animal data). These estimated animal risks are then converted to human risks.

The EPA has adopted a list of risk factors where the risk is calculated based on the ratio of the quantity of a particular contaminant that is ingested, inhaled, and/or absorbed to its reference dose, at which no adverse effect has been observed. This ratio is referred to as the hazard index. In general, a hazard index >1.0 is considered excessive. However, the risk factor, as quantified by this method, does not give round numbers for the various exposed groups (i.e., children, adults, male, or female). New indexes are then calculated from the decontamination criteria targets and then evaluated in function of risk reduction.

The main constraint associated with the methods used to quantify the estimates of health effects lies in correlating the animal dose-response functions to human beings (i.e., the interspecies comparison).

Despite all the recent advances, most measurements of technological risk are fraught with controversy as to the definition of risk and the uncertainties associated with the quantifying methods by which risks are expressed (e.g., social or economical issues).

To the best of the authors' knowledge most of the criteria for aggregating risks or for the use of cumulative distribution functions are based on the expertise of the professional performing the analysis.

Since an uncertainty factor is associated with each and every step of any risk/hazard evaluation, its quantification procedure may under- or over-estimate the final risk.

Most of the currently available risk assessment models are specifically designed to solve toxic air emission problems. Thus, they are not well equipped to address the problem of contaminated sediments. The Wastewater Technology Centre in Canada has developed the AERIS model for the management purposes of contaminated soils. However, this model has not been widely used. Hence, no information regarding its performance in real conditions is yet available.

To conclude, most risk assessment frameworks and models are inherently fraught with uncertainties. In most studies, risk is evaluated with simple models which generally use the EPA hazard index (USEPA 1989a, 1989b).

References

Adams, W. J., Kimerle, R. A., and Barnett, J. W. Jr., 1992, "Sediment Quality and Aquatic Life Assessment," *Environmental Science and Technology*, Vol. 26, No. 10, pp. 1864–1875.
Burton, G. A. Jr. and Scott, K. J., 1992, "Sediment Toxicity Evaluations: Their Niche in the Ecological Assessment," *Environmental Science and Technology*, Vol. 26, No. 11, pp. 2068–2075.
Centre Saint-Laurent, 1993, "Guide pour l'évaluation et le choix des technologies de traitement des sédiments contaminés" (Guidelines for the evaluation on the choice of treatment technologies for contaminated sediments), Document rédigé par Jean-René Michaud, Direction du développement technologique, No. de catalogue: En40-450/1993F.
Chapman, M. P., 1989, "Current Approaches to Developing Sediment Quality Criteria," *Environmental Toxicology and Chemistry*, Vol. 28, pp. 589–593.
Dermont, R. and Munawar, M., 1992, "A Simple and Sensitive Assay for Evaluation of Sediment Toxicity Using Lumbriculus Variegatus (Muller)," *Hydrobiologia*, Vol. 235/236, pp. 407–414.
Ministry Environment Canada (MEC), 1992, "Review and Recommendations for Canadian Interim Environmental Quality Criteria for Contaminated Sites," Inland Water Directorate, Water Quality Branch, Scientific Series No. 197.
Ministry of Environment Canada (MEC), 1985, "Problems Associated with Dredging Operations on the St. Lawrence: Situation, Methods and Priority Areas for Research" *Report EPS 4/MA/1*, August.
Ministry of Environment Ontario (MEO), 1990, "Provincial Sediment Quality Guidelines (draft project)," p. 21.
Ministère de l'Environnement du Québec (MENVIQ), 1988, "Critères indicatifs de la contamination des sols et de l'eau souterraine, dans la politique de réhabilitation des terrains contaminés" (Quality Criteria for contaminated soil and ground water. Policy for the restoration of contaminated soils).
Onishi, Y., 1985, "Chemical Transport and Fate in Risk Assessment," in *Principles of Health Risk Assessment*, P. F. Ricci, Ed., Prentice-Hall Inc., NJ, pp. 117–156.
Principles of Health Risk Assessment, 1985, P. F. Ricci, Ed., Prentice-Hall Inc., NJ.
Swider, J. and Kastenberg, W. E., 1993, "Propagation of Uncertainty in Air Toxics Risk Assessment," in *Proceedings of the Joint CSCE-ASCE National Conference on Environmental Engineering*, Queen Elizabeth Hotel, Montreal, Quebec, Canada, 12–14 July.
Thomas, D. J., Martin, L. C., and Ruffell, J. P., 1985, "Environmental Assessment of Dredging Technology," prepared by Arctic Laboratories Limited, ESL Environmental Science Limited and EBA Engineering Consultants Ltd. for the Environmental Protection Service, Yellowknife, N.W.T. (unpublished manuscript), 95 pp. + appendices.
USEPA, 1993, *Compilation of Ground-water Models*, P. K. M. van der Heijde, and O. Elnawawy, U.S. Environmental Protection Agency, OK.
USEPA, 1991, *Remediation of Contaminated Sediments*, EPA/625/6-91/028, U.S. Environmental Protection Agency, Cincinnati, OH, April.
USEPA, 1989, *Sediment Classification Methods Compendium*, Final Draft Report, U.S. Environmental Protection Agency, June.
USEPA, 1989a, *Risk Assessment Guidance for Superfund, Vol. 1. Human Health Evaluation Manual (Part A0)*, EPA/540/1-89-002, U.S. Environmental Protection Agency.
USEPA, 1989b, *Exposure Factors Handbook*, EPA/600/8-89/043, U.S. Environmental Protection Agency.
USEPA, 1988, *Guidance for Conducting Remedial Investigations and Feasibility Studies under CERCLA*, Office of Emergency and Remedial Responses, U.S. Environmental Protection Agency, Washington, DC, EPA 540 G-890040.

Vigneault, Y., 1978, "Répercussions écologiques du dragage dans le Saint-Laurent" (Ecological impact of dredging in the St. Lawrence River), Congrès conjoint: AQTE-FACE; 8.1-8.32.

Yong, R. N., Galvez-Cloutier, R., and Phadungchewit, Y., 1993, "Selective Sequential Extraction Analysis of Heavy Metal Retention in Soil," *Canadian Geotechnical Journal*, Vol. 30, No. 5, pp. 834–847.

Discussion

Dennis W. Lawson[1] *(written discussion)*—The hallmark of the superb collection of papers in Session III on Management Strategies is its central theme; namely, that the various technical alternatives for handling contaminated sediments should be embodied within a democratic decision-making process that weighs the perspectives of both professional specialists and the general lay public. This one somewhat zealous and more technical presentation did not seem to take up the larger cause.

The presentation was indeed critical and, while refreshing and stimulating, its narrower focus was also mildly disturbing, As such it warrants what is intended as hopefully some broadening, stabilizing, and more balanced commentary. Perhaps the published paper will better set the public consultation scene and diminish the uncompromising technical overtone of the presentation.

The talk was interesting from a conceptual perspective, but seemed to fail to reach a meaningful conclusion. There also appeared to be minor technological imperfections (perhaps simplifications and consolidations) which detracted from accepting its overall philosophical message.

The authors' apparent objections to virtually all technical solutions for contaminated sediments would seem to lead to a single "fix" that is morally and ethically "clean"; namely, returning the contaminants to their place of origin (e.g., taking natural radionuclide contaminants back to the former subsurface location of their parent uranium ore body). While this grand conclusion has been recognized by several philosophers, cannot be denied, and should be pursued, there are other solutions on a lower theological plane that are worthy of pursuit by responsible technocrats, concerned citizens, and the harmonized tiers of our elected governments (e.g., various other site-specific forms of assuredly safe underground or subaqueous disposal after any necessary pretreatment).

Perceptions are reality for everyone who comes to formulate and hold them, and all such realities need to be addressed by rational discussion within a free society. Otherwise, we will unnecessarily constrain our range of reasonable alternatives and not be led to an agreement on the best solution, where "best" embraces all philosophical, theological, technological, economic, and social considerations. In order to properly commence the debate, professional analyses are required that embody all of these subjects. The authors' paper should serve to moderate the debate and move it in the right direction (i.e., towards prevention, adequate cleanup, and any necessary long-term satisfactory containment of contaminated sediments).

Sound technological alternatives can, to some extent, address moral dilemmas and assist the public in forming or altering their opinions. Environmental impact assessment processes and subsequent regulatory licensing actions are such that the finally selected alternative should have come to pass a multitude of tests of blended professional, public, and govern-

---

[1] Environment Canada, Prairie and Northern Region, Regina, Saskatchewan, Canada S4P 4K1.

ment scrutiny and acceptance. The wheels of this scrutiny grind long and fine, but "war" remains too important to be left to "generals."

The authors' paper could provide some part of many packages for future public hearings over the fate of contaminated sediments, but the consensus or (more likely) compromise that marks all final solutions will likely have to accommodate a multitude of different, if not conflicting, perspectives. While we should continue the search, there is as yet no single tract of "holy ground" that is universally accepted.

Howard Zar[1]

Regulatory Strategies for Remediation of Contaminated Sediments

REFERENCE: Zar, H., "**Regulatory Strategies for Remediation of Contaminated Sediments**," *Dredging, Remediation, and Containment of Contaminated Sediments, ASTM STP 1293,* K. R. Demars, G. N. Richardson, R. N. Yong, and R. C. Chaney, Eds., American Society for Testing and Materials, Philadelphia, 1995, pp. 319–328.

ABSTRACT: A number of federal and state laws may be used to obtain remediation of contaminated sediments in the United States. Until recently, the most prominent approaches at the federal level were the use of Superfund authorities for sites on the National Priority List and navigational dredging activity by the Corps of Engineers. However, with the increasing concern about contaminated sediments, regional offices of the U.S. Environmental Protection Agency (EPA) and state agencies have begun to use a greater variety of regulatory approaches, both individually and in combination. These efforts have been particularly evident in the Great Lakes and are now being extended nationwide, as embodied in the EPA's Contaminated Sediment Management Strategy. This paper will describe some of the regulatory approaches being applied, case examples in the Great Lakes area, and the expected directions of these efforts, as embodied in the national strategy.

KEYWORDS: sediments, sediment contamination, sediment remediation, environmental regulation, sediment strategy, sediment criteria, dredged material, Great Lakes

Background

Previous papers at this conference described sites at which contaminated sediments have been found and for which efforts at management and remediation are being made. These efforts are indicative of an aggressive and growing interest in the problem. In the United States and elsewhere, discharges from industries and municipalities that once dominated discussions of environmental impact are viewed to a large extent as on their way to abatement, while the historical sediment contamination caused by such discharges often remains as an important barrier to environmental improvement. As is widely known, sediment contamination can be the source of a wide variety of impacts, including contaminated fish and wildlife, toxicity to benthic organisms, and threats to human health.

As recognition of sediment problems has grown, the efforts of governments and private entities to understand and clean up contaminated sediments has also been increasing on both a national and a worldwide basis. In the United States, individual offices of the U.S. Environmental Protection Agency (EPA) and state agencies began responding more strongly to the problem in the 1970s. The responses initially took place in older industrialized areas around the country, such as those in the Great Lakes Basin, where the earliest problems with contaminated sediments were identified. In 1990, as part of the accelerating efforts in the area of sediments, EPA also began developing a contaminated sediment management strategy.

---

[1] Senior technical advisor, U.S. Environmental Protection Agency, Region 5, Water Division, WS-16J, Chicago, IL 60604.

This paper will discuss the elements of EPA's national strategy, indicate some of the regulatory approaches available for sediment management and remediation, and provide a number of specific examples of application of these approaches by the Region 5 Office of EPA, most of them involving contaminated sediment problems within the Great Lakes Basin.

EPA Contaminated Sediment Management Strategy

EPA's *Contaminated Sediment Management Strategy* (U.S. EPA 1994b) (Strategy), is a comprehensive, multimedia document dealing with all of the contaminated sediment programs that EPA manages. The document does not propose regulations but rather describes how existing statutory and regulatory authorities will be used by EPA to deal with contaminated sediment problems. In view of the variety of laws, regulations, and programs that apply to sediments, one of the document's purposes is to streamline decision-making, within and among EPA programs. A central purpose of publishing the document is to solicit comment from interested parties.

A *Federal Register* notice and the strategy itself were published on 30 August 1994. Copies of the full document are available from U.S. Environmental Protection Agency, National Center for Environmental Publications and Information, 11029 Kenwood Road, Building 5, Cincinnati, OH, 513-891-6561. Included in the strategy are the specific elements of EPA's intended approaches to assess sediment contamination. These include:

Assessment—efforts to develop test methods, develop sediment criteria, inventory sites and sources, and to improve monitoring.
Prevention and source control—efforts to keep existing and future sources from making existing sediment deposits worse or creating new zones of contamination.
Remediation and enforcement activity—efforts to clean up zones of contamination that are causing adverse effects.
Sediment dredging and dredged material management—environmentally sound dredging and disposal activities.
Research and demonstration—efforts to support the above objectives.
Outreach—efforts to assure adequate communication and coordination with other state and federal agencies, industry, and the public.

The strategy discusses various contaminated sediment programs and activities intended by EPA, including a number of activities that have begun or were expanded in the several years that the strategy has been under development. Specifics are given for the manner in which individual agency programs will implement various aspects of the strategy. The following discussion deals principally with those elements of the strategy aimed at remediation of contaminated sediments. For a more detailed description of these elements and a more complete set of references, refer to the Strategy document itself.

Assessment

The assessment portion of the strategy includes development of sediment criteria and also the development of standard sediment testing methods, including sediment toxicity tests. The importance of these elements to remediation is in the tools that they provide to assist in prioritizing sites for action and determining sediment cleanup objectives. Five proposed sediment criteria were published in the *Federal Register* for comment on 18 January 1994 (U.S. EPA 1994a).

The assessment portion also includes the efforts being made to inventory contaminated sediment sites and sources. An effort is underway at the national level in response to the requirements of the Water Resources Development Act of 1992. EPA Region 5 has completed a partial inventory, including sites in the Lake Superior and Lake Michigan Basins, which is available through EPA's Nonpoint Source Bulletin Board (U.S. EPA Region 5 1993a). Inventories provide an important opportunity to prioritize sites and sources for possible remedial action.

Field studies to assess sediment contamination are being undertaken by EPA and the states for a variety of purposes. Many of these studies are intended to support remediation efforts at specific sites. EPA also is considering a nationwide sediment monitoring program and intends to assure that sediment databases developed by the agency at the regional and national levels are compatible.

Prevention and Source Control

The category of prevention and source control refers to efforts made to reduce contaminant sources in order to keep them from contributing to existing deposits of contamination or forming new deposits. The strategy places great importance on these efforts for the long-run resolution of the sediment contamination problem. Prevention and source control efforts are important for the management aspects of sediment remediation as well. Remediation activity will not generally be justified if sources remain which recontaminate sediments at the site, although short-term remediation measures may be necessary to deal with short-term impacts. In practice, efforts to improve a site must consider both source control and remediation, in phases, individually or in combination, to assure that adverse impacts are ultimately abated.

Remediation and Enforcement

EPA may take actions directed at remediation of contaminated sediments through multiple statutes, either applied individually or in concert. Applicable authorities include the Comprehensive Emergency Response, Compensation, and Liability Act (CERCLA) known as Superfund, the Resource Conservation and Recovery Act (RCRA), the Clean Water Act (CWA), the Rivers and Harbors Act (RHA), the Toxic Substances and Control Act (TSCA), and the Oil Pollution Act of 1990 (OPA). It is recognized in the strategy that sediment contamination will not necessarily justify a remediation effort. If a combination of pollution prevention and source controls, coupled with natural recovery, will allow the sediments to clean up in an acceptable period of time, a remediation effort may not be necessary. However, EPA will seek to remediate contaminated sediments that are the source of unacceptable adverse impacts.

EPA can use these authorities to: (1) compel parties to clean up the sites that they have contaminated, (2) recover costs from responsible parties for EPA-performed cleanups, and (3) coordinate with natural resource trustees to seek restitution from responsible parties for natural resource damages. The agency's ability to obtain sediment remediation within a reasonable time frame may be enhanced through the coordinated use of contractor listing authority, debarment and suspension, state or local laws and regulations, other federal laws and regulations, and EPA's criminal enforcement authority.

To date, EPA has successfully used only section 309(b) of CWA, RCRA Corrective Action Authority, and section 106 of CERCLA in conjunction with contractor listing authority to require sediment cleanups. Settlements of CWA unauthorized discharger enforcement cases have incorporated sediment cleanup as part of the injunctive relief. EPA will use all of its authorities, individually or in combination, to require sediment remediation by responsible parties where justified. Assessment efforts and inventories, when available, will assist in the targeting of actions for remediation.

The strategy will contain a summary of the legal elements and application of the noted authorities. Some highlights are as follows:

CERCLA or Superfund—provides one of the most comprehensive authorities available to EPA to obtain sediment cleanup, reimbursement of EPA cleanup costs, and compensation to natural resource trustees for damages to natural resources affected by contaminated sediments. Once EPA determines that there is a release or substantial threat of release to the environment, EPA may undertake response action necessary to protect public health and the environment, and, if necessary, to compel the potentially responsible parties (PRPs) to undertake the cleanup. Liability is strict, meaning responsible parties are liable without fault, and "joint and several," meaning that they are collectively responsible for the entire cost of the cleanup.

Most CERCLA sediment cleanups thus far have been long term (remedial) actions associated with complete site cleanup. With the development of the Superfund Accelerated Cleanup Model (SACM), recent emphasis in Region 5 has been on short-term (removal) actions to control immediate threats. The implementation of SACM in 1994 is bridging the remedial and removal approaches, in order to achieve more timely and efficient cleanups.

RCRA—Subtitle C of RCRA provides EPA with the authority to assess whether releases from a hazardous waste treatment, storage, or disposal facility have contaminated sediments and to require "corrective action," which could include sediment remediation. RCRA corrective action provisions address releases of hazardous waste or constituents to all environmental media, including sediment, and they are implemented through either administrative orders or RCRA permits. The RCRA corrective action process is initiated by requiring the facility owner/operator to conduct extensive investigations on-site as well as off-site. If solid waste management units at a RCRA facility are then shown to be the source of contamination, sediment remediation can be required. RCRA corrective action authorities for sediments are expected to get more use in the future.

CWA—Section 309 of CWA authorizes EPA to take civil action for discharges in violation of permit limits and seek appropriate relief, including environmental remediation. If environmental harm is demonstrated, EPA can seek sediment remediation as part of the injunctive provisions of the administrative or judicial order. Enforcement actions for permit effluent violations can also encourage sediment cleanups in lieu of civil penalties. The facility may be willing to clean up the sediments even if the sediment contamination is the result of permitted discharges as an offset to civil penalties or to limit possible liability under other statutes. Region 5 has been leveraging penalties to require sediment cleanups.

TSCA—TSCA does not explicitly require cleanup of regulated substances other than PCBs. PCB spills that occurred after the effective date of the TSCA regulations (18 April 1978) are subject to the TSCA disposal rules. PCB spills and discharges that occurred before the effective date of TSCA may also be subject to TSCA disposal rules. An Agency position is currently under development to determine how such authority will be applied.

RHA—The Rivers and Harbors Act includes provisions that may be used to address sediment contamination. The injunctive relief available under the Act includes the ability to order the removal of obstructions to navigation and the removal of refuse.

OPA—Under the Oil Pollution Act of 1990, EPA may require responsible parties to clean up contaminated sediments resulting from oil spills and discharges. There are important limitations, based on when the spill occurred.

Dredged Material Management

The Marine Protection, Research, and Sanctuaries Act (MPRSA) and Section 404 of the CWA establish a framework to regulate dredging and disposal of dredged material in the oceans and in inland waters, respectively. Although these authorities are not set up for the purpose of sediment remediation, they do offer a number of important sediment cleanup opportunities. Navigation channels are frequently located in polluted waterways so that normal dredging and disposal of contaminated sediments can effectively reduce environmental impact. Also, as sediments near navigation channels are often contaminated as well, cooperative ventures among EPA, the states, the U.S. Army Corps of Engineers (USACE), and port authorities offer the opportunity to combine navigational dredging and enforcement authorities to obtain additional remediation with important economies of scale. A further consideration is that contaminated sediments can increase the costs of navigational dredging and disposal and attract public opposition on environmental grounds. In such circumstances, cooperative efforts among agencies have the possibility of combining navigational and environmental objectives in a way that satisfies such concerns.

Research and Demonstration

Research and demonstration continues to evaluate technologies that promise cost-effective approaches to contaminated sediment dredging and disposal. The most active programs are those in EPA's Office of Research and Development, the Superfund SITE program, and the Assessment and Remediation of Contaminated Sediments (ARCS) program of the Great Lakes National Program Office in Chicago.

Outreach

Participation and support from the public and industry is necessary to make remediation efforts a success. EPA will make regular reports to the public on sediment management activities and involve the public in decisions on individual sites and remedial actions.

Sediment Remediation in Region 5 and the Great Lakes System

The contaminated sediment problem has special importance in the Great Lakes system and in the Region 5 Office of EPA, which covers the states of Illinois, Indiana, Michigan, Minnesota, Ohio, and Wisconsin, encompassing the U.S. portions of most of the Great Lakes, as well as portions of the Mississippi and Ohio River basins. Of the Areas of Concern (AOCs) identified by the International Joint Commission on the Great Lakes, virtually all have significant sediment problems. In many of the AOCs, sediments have become the principal issue of focus. Considerable efforts are being made to clean up contaminated sediments, using available regulatory tools. Some highlights of Region 5 remediation efforts are given below with details on specific sites provided in Table 1. These efforts provide concrete evidence of the directions the EPA Strategy is heading in dealing with contaminated sediments.

Regulatory approaches to sediment cleanup once were very limited. When the PCB problem at Waukegan Harbor (IL) was discovered in 1976, the Clean Water Act (CWA) was successful in stopping the active discharge of PCBs but turned out to be ineffective in dealing with the sediment contamination problem. The Waukegan site was one of the first to make the National Priorities List after the passage of CERCLA in 1980 but it was only after

TABLE 1—*Examples of remediation actions now underway in Region 5.*

Indiana Harbor Canal and Grand Calumet River	• Corps Navigation Project; EPA and Corps working together; ECI bankruptcy settlement lodged Oct. 93 may provide a disposal site. • LTV Steel; May 1992 CWA consent decree; characterization and cleanup of an intake flume. • Inland Steel; March 1993 multimedia consent decree (CWA, RCRA, SDWA); includes sediment remediation. • USX Gary Works; October 1990 CWA consent decree; characterization of 13 miles (21 km), remediation of 5 miles (8 km). • Gary Sanitary District; October 1992 CWA/TSCA consent decree; characterization and remediation of 4 miles (6.4 km). • Hammond Sanitary District; August 1993 compliant under CWA and Rivers and Harbors Act. Relief sought includes sediment remediation.
Other Lake Michigan Basin	• Sheboygan Harbor; Wisconsin NPL site (Tecumseh Products); CERCLA orders resulted in partial cleanup, storage, and studies. • Waukegan Harbor; Illinois (OMC) NPL Site; 1988 CERCLA consent decree resulted in dredging, treatment, and disposal. Sediment remediation completed in 1993. • Manistique River/Harbor, Michigan (Manistique Paper, Edison Sault Elec.); CERCLA removal action placed temporary plastic cover on a hot spot. Decision on a final response action for the full site is now being made by EPA.
Lake Erie Basin	• River Raisin, Michigan (Ford Monroe); the PRPs have agreed to complete remediation of the site under CERCLA authority in 1995. • Ashtabula River/Harbor and NPL site in Fields Brook; CERCLA remedial action at Fields Brook in design stage. The Ashtabula River is in investigation phase while a private public partnership approach is being established, in combination with navigational dredging. • Black River, Ohio (USX Lorain); a CWA consent decree resulted in dredging in November 1990.

CERCLA was amended in 1986 that USEPA was able to compel a sediment cleanup, completed just last year.

As concerns about contaminated sediments grew, regulators looked for further approaches to obtain cleanup. An opportunity presented itself in 1985. Enforcement penalties assessed at the USX steel mill in Lorain, OH for CWA and Clean Air Act (CAA) violations were turned into a consent decree requiring dredging of sediments contaminated with heavy metals and PAHs from the Black River, which drains into Lake Erie. This is the kind of supplemental environmental project or environmentally beneficent expenditure conducted in lieu of civil penalties that is discussed in more detail in EPA's Strategy.

Northwest Indiana

Based on the success in the Black River, USEPA Region 5 began looking for other opportunities to apply CWA authorities and also to use the other available regulatory authorities discussed in EPA's Strategy. The Indiana Harbor Canal and Grand Calumet River in Indiana is an area that is severely contaminated with heavy metals, PCBs, PAHs, volatiles, oil, and grease. As discussed in another paper in the symposium, these sediments are among the worst found anywhere. The sediments are highly toxic, have pore spaces filled with as much oil as water, have caused fish contamination, and are losing contaminants at a significant rate to Lake Michigan. Due to these impacts and others, the area is the subject of a cooperative geographically targeted "Northwest Indiana Environmental Initiative" being conducted by the EPA and the Indiana Department of Environmental Management. A number of enforcement actions have been taken in the course of this initiative which are resulting in the necessary studies and remedial actions as indicated in Fig. 1.

Currently, USX Gary is required by a consent decree under the CWA to remediate five miles of the Grand Calumet River's East Branch. A characterization study of the entire East Branch also required by the consent decree has been largely completed. Under judicial consent decrees, the Gary Sanitary District is required to remediate four miles of the East Branch and LTV Steel is required to remediate an intake flume to Indiana Harbor. Inland

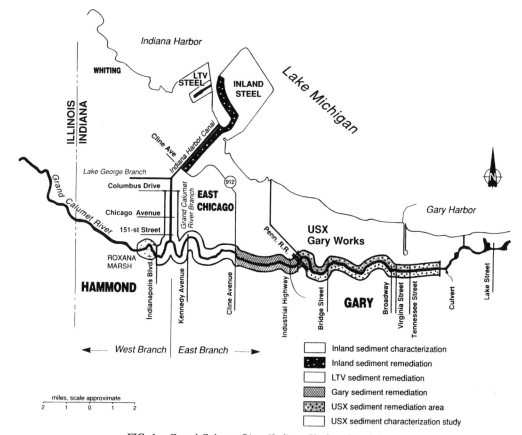

FIG. 1—*Grand Calumet River/Indiana Harbor Canal Area.*

Steel, under a multimedia consent decree (CWA, RCRA, and Safe Drinking Water Act), is required to remediate sections of the Indiana Harbor Canal. The City of Hammond is the subject of a 1993 complaint under CWA and RHA related to the West Branch of the Grand Calumet River.

Although the initial enforcement actions in this area were taken principally under the CWA, later actions have begun to make use of multiple authorities in concert, called multimedia actions. As noted, the 1993 consent decree involving Inland Steel results from a multimedia complaint. The use of the strongest authority available at any site and multiple authorities is now a part of the USEPA approach to sediment remediation. A USEPA/State Great Lakes Enforcement Strategy covering these approaches was issued in September 1993 (U.S. EPA Region 5 1993b). Using multiple authorities where environmental damage is known will increase the possibility of obtaining the desired environmental remediation.

Corps of Engineers Coordination

The enforcement activities noted above for Northwest Indiana can resolve only part of the sediment contamination problem. Much of the contamination of the Indiana Harbor Canal is in an area that has been intended for navigational dredging by the U.S. Army Corps of Engineers (USACE) for some years. The sediments in this area are close to Lake Michigan and subject to effects of boat movements, making them a high priority for remedial action. Unfortunately, concerns about contamination levels have made it difficult for USACE to find suitable disposal locations.

EPA and USACE began working on the problem jointly and have made considerable progress. A bankruptcy settlement lodged in October 1993 may provide a suitable disposal site. Issuance of a Draft Environmental Impact Statement, developed by USACE with assistance from USEPA, is expected shortly. USEPA and USACE are working together increasingly in other areas where dredging of contaminated sediments is desired for both navigational and environmental reasons.

Cooperative efforts at Indiana Harbor Canal illustrate the opportunities available for obtaining cleanups through dredged material management in cooperation with USACE and also with states, port authorities, and other parties. As an indication of the national commitment to progress on improving dredged material management, the issue is the subject of an options paper drafted by the Federal Interagency Working Group on the Dredging Process convened by the Secretary of Transportation in the fall of 1993 (U.S. Department of Transportation 1994). The objectives of the group are to provide greater certainty and predictability in the dredging and dredged material process and to facilitate long-term management strategies.

CERCLA

Enforcement is only one of the available regulatory tools. As discussed above, Superfund legislation (CERCLA) provides comprehensive authority for obtaining cleanup, reimbursement for cleanup, and compensation for natural resources trustees. The implementation of the Superfund Accelerated Cleanup Model (SACM) this year is bridging the remedial and removal approaches, in order to achieve more timely and efficient cleanups.

The Waukegan Harbor site and actions currently in progress at Sheboygan Harbor, WI (PCBs) and Fields Brook, Ashtabula, OH (PCBs, HCBs, PAHs) are examples of remedial actions being undertaken in Region 5. The Manistique River, MI (PCBs), a SACM site, and River Raisin, MI (PCBs) are examples of removal actions, now in progress. The Some details on these sites are also in Table 1.

RCRA

RCRA authority was involved in obtaining the multimedia consent decree for contaminated sediment cleanup at Inland Steel in East Chicago, IL. There are also several RCRA corrective action sites within Region 5 where 3004(u) and (v) and 3008(h) authorities are being used to require characterization of contaminated sediments, among them the U.S.S. Lead plant in East Chicago. If these studies show that contamination emanated from the facility and poses an environmental risk, EPA will require remediation of contaminated sediments. It is anticipated that corrective action authority for sediment cleanup will be used further in both a permits and enforcement mode.

TSCA

To date, TSCA involvement in Region 5 sediment remediation activities has primarily been assistance to the CERCLA and CWA programs concerning PCB cleanup, storage, and disposal issues. TSCA authority has also been used on small-scale PCB sediment cleanups as part of broader settlements of TSCA Civil Administrative Actions. Future use of TSCA authority for large-scale sediment remediation projects will be through this same settlement process or by referral to the Department of Justice.

Cooperative Approach/Public Private Partnership

While the Enforcement and Superfund approaches noted above have been effective in sediment cleanup, they can involve considerable resources and time expenditures by agencies and responsible parties. A number of entities, including the State of Wisconsin, have advocated working closely with responsible parties in a nonadversarial mode. The objective is to turn the potential costs of adversary proceedings into real cleanup and to avoid the delays involved in contentious lawsuits.

As a recent example, the Citizens Advisory Group for the Ashtabula Remedial Action Plan, working with state and federal agencies and potentially responsible parties, has begun to forge a Public Private Partnership to combine different statutory approaches in a more cost-effective solution for the Ashtabula River, with partial focus on a multiuser combined disposal facility.

Site Prioritization

Sites in Region 5 with contaminated sediments have become the subject of regulatory attention by EPA and the states thus far because a combination of obvious environmental impact and the availability of clear-cut regulatory tools made action cost-effective. As the agencies make progress in attacking this first group of sites, it is necessary to identify which sites come next. The experience with regulatory authorities that has been gained and the greater availability of information about sites, including the Region 5 and National Sediment Inventories, is increasing our ability to prioritize further sites and to deal with them effectively.

Conclusion

Sediment contamination can be a daunting problem from both a technical and regulatory standpoint. Agencies at various levels, working together with industries and the public, have made progress in developing improved regulatory and technical approaches to clean up the most contaminated sites and to identify sites that require the most rapid action. No single regulatory or technical approach will work in all situations. However, with the variety of authorities, additional resources, and new approaches that are being applied on all fronts, as reflected in EPA's forthcoming Contaminated Sediment Management Strategy, we appear to be on the way to reducing the problem of contaminated sediments in Region 5, the Great Lakes Basin, and across the United States.

Interested parties are welcome to contact the author at U.S. Environmental Protection Agency, Region 5, WS-16J, 77 West Jackson St., Chicago, IL 60604, telephone 312-886-1491, fax 312-886-7804, internet zar.howard@epamail.epa.gov.

References

U.S. Department of Transportation, Maritime Administration, 1994, "The Interagency Working Group on the Dredging Process—Options Paper," May.

U.S. Environmental Protection Agency, 1994a, "Notice of Availability and Request for Comment on Sediment Quality Criteria," *Federal Register*, 18 January 1994.

U.S. Environmental Protection Agency, 1994b, *EPA's Contaminated Sediment Management Strategy*, EPA 823-R-94-001, August.

U.S. Environmental Protection Agency Region 5, 1993a, *Region 5 Sediment Inventory*, available on the EPA Non-Point Source Bulletin Board, telephone 301-589-0205.

U.S. Environmental Protection Agency Region 5, 1993b, *Great Lakes Enforcement Strategy*, 15 September 1993.

Author Index

A
Averett, Daniel E., 128

B
Bachus, Robert C., 239
Bahout, Edward, 306
Boscardin, Marco D., 220
Boyle, William C., 155

C
Chaney, R. C., 195
Chan, Jeanette, 306
Collingwood, Bruce I., 220

D
Davison, William, 170
Demars, K. R., 195
Deschênes, Jean-Hughes, 252

F
Fritschel, Paul R., 155
Fukue, Masaharu, 58, 136

G
Galvez-Cloutier, Rosa, 306
Garbaciak, Stephen, Jr., 145
Grime, Geoffrey W., 170

H
Herbich, John B., 77
Hollenweger, Reto, 271
Hollifield, Matthew B., 155

J
Jacobson, Brian D., 239

K
Kato, Yoshihisa, 58

M
Martinenghi, Lisa, 271

Massièra, Michel, 252
Miller, Jan, 289
Miller, Jan A., 145
Murdock, Richard F., 220

N
Nakamura, Takaaki, 58

O
Ohtsubo, Masami, 50

P
Palermo, Michael R., 289
Park, Jae K., 155
Pelletier, Jean-Pierre, 112
Petrovski, David M., 40, 195

R
Rad, Nader S., 239
Reible, D. D., 227
Richardson, G. N., 195

T
Thibodeaux, L. J., 227
Tournier, Jean-Pierre, 252

V
Valsaraj, K. T., 227

W
Weston, Sandra M. C., 297

Y
Yamasaki, Shoichi, 58, 136
Yasui, Hiroshi, 136
Yong, Raymond N., 13, 306
Yoshinaga, Kiyoto, 182

Z
Zar, Howard, 319
Zhang, Hao, 170

Subject Index

A

Adsorption
  toxic pollutants, 13
  zinc, marine clays, 50
Air emissions, contaminated dredged materials, modeling, 227
Anchored sheet pile wall, 220
Areas of Concern
  Canadian Great Lakes, 112
  Great Lakes, 1
  Port Hope Harbour, 297
  potential, U. S., 1
Assessment and Remediation of Contaminated Sediments Program, 145

B

Background concentrations, heavy metals, 58
Bathymetry, sediment characterization, 40
Bay sediments, heavy metal concentrations, 58
Biochemical oxygen demand, 136
Biodegradation, 145
Bioremediation, polychlorinated biphenyls, 155
Biostimulant, 155

C

Cement-bentonite cutoff wall, 252
Clamshell bucket, 112
Coal tar, seepage abatement, surface water bodies, 220
Compatibility, soil-bentonite slurry cutoff walls, 239
Compression, unconfined, 252
Compressive strength, unconfined, 136
Confined disposal facility, contaminant pathway control, 195
Containment, 1
  cement-bentonite cutoff wall, 252
  soil-bentonite slurry cutoff walls, 239
Contaminants
  dredged materials, air emission modeling, 227
  hazard/risk assessment, 306
  hydraulic isolation, 271
  management strategies, 289
  pathway control, 195
  releases during dredging, 128
  remediation, 297, 319
  removal by dredging, 77
  soil-bentonite slurry cutoff walls, 239
  toxic, 13
Contaminated Sediment Removal Program, 112
Cutoff wall
  cement-bentonite, 252
  soil-bentonite, 239, 271

D

Dechlorination, *in-situ* anaerobic, 155
Dehalogenation, reductive, 155
Design mix methodology, 239, 252
Diaphragm walls, 271
Diffusion, 271
Diffusive equilibration in thin-film, 170
Diffusive gradient in thin-film, 170
Dredged materials
  contaminated, 128, 289
    air emission modeling, 227
    pathway control, 195
    remediation, 319
Dredging, 289
  contaminant releases during, 128
  cutterless suction, 182
  equipment, 77
  field studies, 77
  mercury-contaminated sludge, 182
  sediment characterization, 40
  solidification technique, 136
  technologies, 1

E

Environmental media, 40
Environmental regulation, remediation, 319
Environmental restoration, 297
Equipment, dredging of contaminated sediments, 77

F

Fate, toxic pollutants, 13
Federal Navigation Project, 40
Fox River, Wisconsin, anaerobic dechlorination of PCBs, 155
Functional groups, 13

G

Great Lakes
  Areas of Concern, 1
  Canada, 112
  regulatory approaches, 319

H

Hazard/risk assessment, contaminated sediments, 306
Heavy metals
  adsorption, marine clays, 50
  in bay sediments, 58
  fate, 13
  Port Hope Harbour, 297
  removal efficiency, 145
Hydraulic conductivity, 252
Hydraulic isolation, contaminated sites, 271

I

Indiana Harbor, 40
Interlocks, sealed, coal tar seepage abatement, 220
Ion-exchange resin, 170
Iron oxide, marine clays, 50
Isolation, 1

J

James Bay Development Project, 252
Japan
  heavy metal concentration, bay sediments, 58
  Minamata Bay, mercury contamination, 182

L

Laboratory testing
  cement-bentonite cutoff wall, 252
  soil-bentonite slurry cutoff walls, 239, 271
Lachine Canal decontamination project, 306

M

Management strategies, 1
  contaminated sediments, 289
Marine clays
  pyrite oxidation, 50
  zinc adsorption, 50

Mercury-contaminated sludge, dredging, 182
Minamata Bay, mercury-contaminated sludge treatment, 182
Minamata Disease, 182
Monitoring program, dredging, 128

N

Navigation projects, 40
Nutrients, dechlorination stimulation, 155

O

Oil products, soil-bentonite slurry cutoff walls, 239
Organic chemicals, fate, 13
Oxidation, pyrite, in marine clays, 50
Oxygen, dissolved, 136

P

Partitioning, 13
Pathways, contaminant control, 195
Permeability, 271
Persistence, toxic pollutants, 13
pH
  change, pyrite oxidation in marine clays, 50
  leachate from solidification, 136
Pilot-scale demonstrations, treatment technologies, 145
Plastic barrier, cement-bentonite cutoff wall, 252
Pneuma pump, 112
Pollution
  criteria, 1
  degree of, 58
Polyacrylamide gel, metals measurement, 170
Polychlorinated biphenyls
  in-situ anaerobic dechlorination, 155
  Port Hope Harbour, 297
  removal efficiency, 145
Polynuclear aromatic hydrocarbons, removal efficiency, 145
Pore water, trace metals, in-situ measurement procedures, 170
Port Hope Harbour Remedial Action Plan, 297
Pyrite, oxidation, in marine clays, 50

SUBJECT INDEX

Q

Quality criteria, 306

R

Radioactive waste, low-level, remediation, 297
Regulatory determinations, 40
Regulatory strategies, remediation, 319
Remediation, 1, 289
  contaminated sediments, 319
  Port Hope Harbour Remedial Action Plan, 297
Removal technologies
  clamshell bucket, 112
  pneuma pump, 112
Resource Conservation and Recovery Act, 40
Revetment, 182
Risk assessment, 306

S

Sediment characterization, bathymetry, 40
Sediment disposal, 182
Sediment resuspension, 128
Sediment washing, 145
Seepage, coal tar, abatement, surface water bodies, 220
Selectivity, 13
Sheet piling, coal tar seepage abatement, 220
Site encapsulation, 271
Slurry cutoff wall
  cement-bentonite, 252
  soil-bentonite, 239, 271
Soil-bentonite slurry cutoff walls, 239, 271

Solidification technique, dredging, 136
Solvent extraction, 145
Storage capacity, of sediments, 13
Sulfuric acid, produced by pyrite oxidation, 50
Surface water bodies, coal tar seepage abatement, 220
Surfactants, dechlorination stimulation, 155
Suspended sediment, 128
Suspended solid, 136

T

Thermal desorption, 145
Toxic pollutants, fate, 13
Trace metals, *in-situ* measurement procedures, 170
Treatment technologies, 1
  field demonstrations, 145
Triaxial compression, 252

V

Volatile organic compounds, air emissions from contaminated dredged materials, 227

W

Water quality, contaminant release during dredging, 128
Water Quality Act of 1987, 145
Water Resources Development Act, 1
Wisconsin, *in-situ* anaerobic dechlorination of PCBs, 155

Z

Zinc, adsorption, marine clays, 50